Layered Double Hydroxides (LDHs)

Layered Double Hydroxides (LDHs)

Editors

Roberto Pizzoferrato
Maria Richetta

MDPI • Basel • Beijing • Wuhan • Barcelona • Belgrade • Manchester • Tokyo • Cluj • Tianjin

Editors
Roberto Pizzoferrato
University of Rome Tor Vergata
Italy

Maria Richetta
University of Rome Tor Vergata
Italy

Editorial Office
MDPI
St. Alban-Anlage 66
4052 Basel, Switzerland

This is a reprint of articles from the Special Issue published online in the open access journal *Crystals* (ISSN 2073-4352) (available at: https://www.mdpi.com/journal/crystals/special_issues/LDHs).

For citation purposes, cite each article independently as indicated on the article page online and as indicated below:

LastName, A.A.; LastName, B.B.; LastName, C.C. Article Title. *Journal Name* **Year**, *Volume Number*, Page Range.

ISBN 978-3-0365-0476-6 (Hbk)
ISBN 978-3-0365-0477-3 (PDF)

Cover image courtesy of Roberto Pizzoferrato.

© 2021 by the authors. Articles in this book are Open Access and distributed under the Creative Commons Attribution (CC BY) license, which allows users to download, copy and build upon published articles, as long as the author and publisher are properly credited, which ensures maximum dissemination and a wider impact of our publications.

The book as a whole is distributed by MDPI under the terms and conditions of the Creative Commons license CC BY-NC-ND.

Contents

About the Editors vii

Preface to "Layered Double Hydroxides (LDHs)" ix

Roberto Pizzoferrato and Maria Richetta
Layered Double Hydroxides (LDHs)
Reprinted from: *Crystals* **2020**, *10*, 1121, doi:10.3390/cryst10121121 1

Brenda Antoinette Barnard and Frederick Johannes Willem Jacobus Labuschagné
Exploring the Wet Mechanochemical Synthesis of Mg-Al, Ca-Al, Zn-Al and Cu-Al Layered Double Hydroxides from Oxides, Hydroxides and Basic Carbonates
Reprinted from: *Crystals* **2020**, *10*, 954, doi:10.3390/cryst10100954 5

Alberto Tampieri, Matea Lilic, Magda Constantí and Francesc Medina
Microwave-Assisted Aldol Condensation of Furfural and Acetone over Mg–Al Hydrotalcite-Based Catalysts
Reprinted from: *Crystals* **2020**, *10*, 833, doi:10.3390/cryst10090833 21

Bianca R. Gevers and Frederick J.W.J. Labuschagné
Green Synthesis of Hydrocalumite (CaAl-OH-LDH) from $Ca(OH)_2$ and $Al(OH)_3$ and the Parameters That Influence Its Formation and Speciation
Reprinted from: *Crystals* **2020**, *10*, 672, doi:10.3390/cryst10080672 35

Ligita Valeikiene, Marina Roshchina, Inga Grigoraviciute-Puroniene, Vladimir Prozorovich, Aleksej Zarkov, Andrei Ivanets and Aivaras Kareiva
On the Reconstruction Peculiarities of Sol–Gel Derived $Mg_{2-x}M_x/Al_1$ (M = Ca, Sr, Ba) Layered Double Hydroxides
Reprinted from: *Crystals* **2020**, *10*, 470, doi:10.3390/cryst10060470 61

Anna Maria Cardinale, Cristina Carbone, Sirio Consani, Marco Fortunato and Nadia Parodi
Layered Double Hydroxides for Remediation of Industrial Wastewater from a Galvanic Plant
Reprinted from: *Crystals* **2020**, *10*, 443, doi:10.3390/cryst10060443 81

Octavian D. Pavel, Ariana Şerban, Rodica Zăvoianu, Elena Bacalum and Ruxandra Bîrjega
Curcumin Incorporation into Zn_3Al Layered Double Hydroxides—Preparation, Characterization and Curcumin Release
Reprinted from: *Crystals* **2020**, *10*, 244, doi:10.3390/cryst10040244 91

Maleshoane Mohapi, Jeremia Shale Sefadi, Mokgaotsa Jonas Mochane, Sifiso Innocent Magagula and Kgomotso Lebelo
Effect of LDHs and Other Clays on Polymer Composite in Adsorptive Removal of Contaminants: A Review
Reprinted from: *Crystals* **2020**, *10*, 957, doi:10.3390/cryst10110957 109

Mokgaotsa Jonas Mochane, Sifiso Innocent Magagula, Jeremia Shale Sefadi, Emmanuel Rotimi Sadiku and Teboho Clement Mokhena
Morphology, Thermal Stability, and Flammability Properties of Polymer-Layered Double Hydroxide (LDH) Nanocomposites: A Review
Reprinted from: *Crystals* **2020**, *10*, 612, doi:10.3390/cryst10070612 149

About the Editors

Roberto Pizzoferrato is Associate Professor at the Department of Industrial Engineering of University of Rome Tor Vergata. His research focuses on the synthesis and characterization of pristine and functionalized nanomaterials, especially Layered Double Hydroxides and carbon-based nanoparticles. He also worked on the optical properties of innovative materials, optical sensors for the detection of heavy metals, nonlinear optical materials and hybrid organic/inorganic materials for optical emitters. He has authored more than 130 publications. He is a member of the international laboratory "Laboratory Ionomer Materials for Energy (LIME)" established between the University of Rome "Tor Vergata", the Aix Marseille Université and the CNRS, and a member of the editorial board of *Sensors*.

Maria Richetta is Assistant Professor at the Department of Industrial Engineering of the University of Rome Tor Vergata, Italy. She graduated with honours in physics from the University of Rome "La Sapienza" and obtained a Ph.D. title in Thermophysical Properties of Materials at the University of L'Aquila.

Since the beginning of her Ph.D., she has been carrying out experimental research activity mainly related to the following topics: biomedical materials, characterization of biological tissues, membranes materials for nuclear fusion reactors, mechanical properties of materials (mechanical spectroscopy, instrumented nanoindentation), X-ray spectroscopy and interferometry, anticorrosion coatings, and LDHs, as demonstrated by more than 200 publications.

For about ten years, she has been concentrating part of her interest on Layered Double Hydroxides (LDH) nanomaterials, regarding both their preparation and growth, and their characterization in terms of morphological study and investigation of the growth mechanism. She also dealt with the anticorrosive properties of the ZnAl-LDH deposited on different metal substrates (anodized and non-anodized 2024T3 steel), as well as the intercalation of drugs within the LDH structures for "in situ" release.

She has been a member of the Scientific Committee of Thermec, International Conference on Processing & Manufacturing of Advanced Materials since 2016, of the Ph.D. Board in "Industrial Engineering" of the University of Rome Tor Vergata since 2012, and of the international laboratory "Laboratory Ionomer Materials for Energy (LIME)", established between the University of Rome "Tor Vergata", the Aix Marseille Université and the CNRS, for four years.

Preface to "Layered Double Hydroxides (LDHs)"

Their exceptional characteristics and uniqueness make Layered Double Hydroxides (LDHs) and their derivatives promising two-dimensional layered materials that are suitable for several existing and future applications in diverse fields. To name a few, we can mention environmental monitoring and preservation, biotechnology, pharmaceutical and chemical processes, the design and realization of new functional polymers and new magnetic materials with a high-saturation magnetic field, and new composite materials for sustainable concrete infrastructure.

The number of research and review articles in high-impact journals highlights the growing attention paid to these materials by not only academic, but also applied-science researchers. This is due to the peculiar properties of LDHs, such as their ease of synthesis even on a large scale, their chemical and thermal stability, their uniform distribution of metal cations, and their ability to intercalate anionic species within the interlayer space and possibly release them, together with their high biocompatibility.

This growing interest has led to a parallel growth in the number of publications, some of which appear in this Special Issue of *Crystals*, presenting both research and review papers. In particular, deeper attention has been paid to innovative synthesis techniques, focusing on those with a low environmental impact, to applications for renewable energy sources, and to interlayer anions' exchange capability for drug release. The ability of Layered Double Hydroxides to form hybrid inorganic/organic nanomaterials is stressed in the review articles.

We are confident that reading the articles published in this Special Issue, written by accomplished researchers working for years in the field, will be a source of inspiration to any scientist who studies LDHs within any discipline.

Roberto Pizzoferrato, Maria Richetta
Editors

Editorial

Layered Double Hydroxides (LDHs)

Roberto Pizzoferrato * and Maria Richetta *

Department of Industrial Engineering, Università degli Studi di Roma Tor Vergata, 00133 Rome, Italy
* Correspondence: pizzoferrato@uniroma2.it (R.P.); richetta@uniroma2.it (M.R.)

Received: 4 December 2020; Accepted: 7 December 2020; Published: 9 December 2020

Hydrotalcite, the first natural mineral belonging to the family of layered materials, was discovered in Sweden by Hochstetter in 1842, but it was not until 1930 that the first study on its synthesis, solubility, stability, and structure was carried out by Feitknecht.

Since then, the family of layered materials has become wider and wider, and also been variously named over time. Layered Double Hydroxides (LDHs), Layered Hydroxicarbonates, Hidrotalcite-like Materials, Anionic Clays, etc., are just some of the many examples, although none of them are sufficiently exhaustive and reflect the current situation. Regardless of the denomination, these materials are not as naturally occurring as cationic clays are, nonetheless they are easy to prepare and are low-cost.

What made and makes these materials extremely interesting is the fact that the nature of the layer cations can be varied within a wide selection, and the nature of the interlayer anion can be chosen, almost at will, between organic and inorganic anions, polymetalates, simple anionic coordination compounds, etc. Like the cationic clays, they can be pillared and, even more importantly, the interlayer anions can be easily exchanged. This property increases the possible applications and opens new routes to the synthesis of derivatives. Furthermore, unlike cationic clays, they are able to recover the lamellar structure after undergoing thermal decomposition. This property can also be used as a synthesis technique.

Since the pioneering work of Feitknecht, LDHs have been synthesized by direct and indirect methods, such as coprecipitation, hydrothermal growth, sol–gel synthesis, soft chemistry, electrochemical synthesis, anion exchange, and those in which LDHs are used as precursors.

The opportunities offered by these properties are extremely ample, and it is precisely for this reason that the applications of LDHs are constantly growing. The main areas of interest range from renewable energy production to water purification and remediation, including functionalized materials for piezoelectric nanogenerators and gas sensing. Great attention is paid to biomedical applications and to the synthesis of hybrid smart nanocomposites, which involve expanding sectors such as drug-delivery, food packaging and safety.

Within this Special Issue, eight articles are collected, and are divided between synthesis techniques [1–3], applications [4–6], and review works [7,8].

The first work, related to the synthesis of Mg-Al, Ca-Al, Zn-Al and Cu-Al LDHs, is the one carried out by Barnard and Labuschagne [1]. The authors, in order to propose a green synthesis technique, implement the wet mechanochemical method through the use of a Netzsch LME 1 horizontal bead mill, designed "ad hoc" for wet grinding applications. In this way they are able to eliminate the production of salt-rich effluent and the control of pH. Furthermore, an aging phase allows a better conversion of raw materials into LDH structures, as well as a morphological improvement of the structures. Another notable point is that the selected mill can be easily scaled up for the production of large quantities of LDH products.

In addition, in the work proposed by Gevers and Labuschagné [2], the authors take care to adopt an environmentally friendly synthesis. In particular, they present the results obtained through the hydrothermal synthesis of hydrocalumite (HC) and $Al(OH)_3$ in water, examining the parameters that impact the formation process of CaAl-OH-LDH, i.e., reaction temperature and time, molar ratio, mixing ratio, water/solid ratio, and morphology/crystallinity of reactants. They show that the use of oxides and hydroxides as starting materials allows one to reduce the production of polluting waste streams, while permitting one to obtain HC formation in each experiment conducted. Furthermore, the carbonate content present in CaAl-OH-LDH essentially comes from calcite or $Al(OH)_3$, such as surface adsorbed carbonate species, rather than from air. Regarding the significant parameters, the authors show how the use of a low water/solid ratio, an increase in time and temperature of reaction, the use of amorphous and large surface $Al(OH)_3$, as well as a stoichiometric ratio calcium/aluminium, favor the formation of katoite, and the purity of HC.

In Reference [3], Valeikiene et al. present a study on the reconstruction peculiarities of $Mg_{2-x}M_x/Al_1$ (M = Ca, Sr, Ba) LDH made by means of the indirect sol–gel synthesis route. In particular, the results of two different sol–gel synthesis procedures are presented. First, the mixed metal oxides (MMO) are obtained by directly heating the precursor Mg(M)–Al–O to (650, 800, 950) °C. All the samples obtained, once immersed in water at 50 °C, were reconstructed to $Mg_{2-x}M_x/Al_1$ LDH. However, the spinel phases remained as impurities and a small quantity of carbonates formed. During the second phase, the reconstructed LDHs were heated to the same temperatures as before. The composition, morphology, and surface properties of these MMOs were then compared with the analogues obtained by the first method. The results showed that the Ca and Sr substituted MMOs contain multiple side phases. The most interesting result, however, lies in the "memory effect" exhibited by the microstructures of MMOs reconstructed from sol–gel-derived LDH, i.e., the microstructural properties of the MMOs were found to be practically identical to those of LDH and, moreover, independent of the annealing temperature.

The other five papers explore the wide field of the present and potential applications of LDHs, and provide some interesting examples of how the many peculiar properties of these materials can be exploited in very different sectors.

In Reference [4], Tampieri et al. address the search for renewable energy sources by investigating the properties of LDHs as catalysts for the synthesis of biofuels. The authors report a microwave-assisted batch process for the neat aldol condensation of furfural and acetone over Mg:Al hydrotalcites (HTs) and derivatives. Differently from previous studies, they prepared HTs in the laboratory and carried out calcination and rehydration to produce mixed metal oxides (MMOs) and meixnerite-like (MX) LDHs, respectively. This allowed them to study how the catalyst activity and selectivity varied over the different derivatives. In addition, an exhaustive analysis of the influence of other reaction parameters was performed. MX resulted in by far the most active catalyst, followed by MMO and HT. Interestingly, HT was generally reported as inactive. In comparison with conventional heating, microwave-assisted condensation is more selective and faster, and also works well at temperatures below 100 °C, even though it requires a longer reaction time.

The capability to exchange interlayer anions is another remarkable characteristic of LDHs, and can be exploited to either adsorb or release anionic species from or into the environment. In Reference [5], Cardinale et al. explore the adsorption properties of $MgAl-CO_3$ and $NiAl-NO_3$ LHDs for the removal of some heavy metals from real wastewater, supplied by a galvanic treatment company. The authors found a certain degree of selectivity of these LDHs, in that Cr(VI) is more efficiently removed by the NiAl LDH through an exchange with the interlayer nitrate. On the other hand, Fe(III) and Cu(II) are removed in higher amounts by the MgAl LDH, probably through a substitution with Mg. However, other mechanisms, such as sorption on the OH^- functional groups, surface complexation, and/or precipitation on the surface of LDH, could not be completely excluded. In the first case, the ionic concentration of Cr(VI) is lowered to a value close to the legal limit, while the concentrations of Fe(III) and Cu(II) are reduced well below the legal limit.

In Reference [6], Pavel et al. report the incorporation of Curcumin (CR) in the Zn_3Al-LDH matrix in order to investigate the release of anionic species for application in drug delivery. The promising antioxidant activity of this natural polyphenol is hindered by its poor solubility in water at neutral pH, which could be overcome by the incorporation of suitable nanocarriers, such as

LDHs. Specifically, the authors used Zn$_3$Al-LDH, in place of the commonly studied Mg$_x$Al-LDH, to increase the antioxidant activity due to the antiseptic properties of Zn. By performing incorporations in both a pristine and a reconstructed matrix, with the addition of CR either as an aqueous alkaline solution (Aq) or as an ethanolic solution (Et), the authors investigated the conditions for the lowest degradation and highest release of CR. They found that reconstruction with a CR-ethanolic solution, which does not restore the layered LDH structure, is the preferable method to obtain CR-loaded Zn$_3$Al solids from LDH precursors.

Finally, two review papers conclude this Special Issue by reporting on some of the many different applications that derive from the capability of LDHs to combine with organic polymers and form hybrid organic/inorganic nanocomposites materials. In Reference [7], Mohapi et al. review and compare the role of LDHs and natural nanoclays in forming polymer-based materials for water purification systems. Specific attention is paid to the preparation methods and the corresponding influence of external parameters in the adsorption process. A solution blending technique and in situ polymerization strategies seem to provide a better dispersion of clay layers in the polymer matrix compared to the melt blending technique. However, melt blending is considered more industrially viable as well as eco-friendly, and shows high economic potential. In Reference [8], Mochane et al. review the utilization of LDHs as nanofillers in polymer-based matrices to improve mechanical and thermal stability, flame retardancy and gas barrier characteristics. While these properties are key factors in a wide field of use, such as in food packaging and safety, other applications, including energy, water purification, gas sensing, biomedical and piezoelectric nanogenerators, are also reviewed. The synergy between polymers and LDHs with peculiar characteristics is especially discussed. It is also pointed out that there are few studies investigating the thermal conductivity of LDHs in combination with other well-known conductive fillers, such as expanded graphite, carbon nanotubes, carbon black, and carbon fibers, which could widen the applications of LDHs nanocomposites.

In summary, we believe that this Special Issue highlights some of the recent lines of a topic as broad as a peculiar type of layered nanomaterial with a large range of possible compositions, many different methods of synthesis and functionalization, several interesting physicochemical properties, and ample opportunities for present and potential application. The present articles show that remarkable progress has been and is still being made on all these aspects, to allow the considering of LDHs as one of the most interesting and versatile inorganic materials.

We would like to thank all authors who have contributed for having submitted manuscripts of such excellent quality. We also wish to thank the large number of reviewers and the editorial staff at *Crystals*, especially the Section Managing Editor, for the fast and professional handling of the manuscripts and for the help provided throughout.

Funding: This research received no external funding.

Conflicts of Interest: The authors declare no conflict of interest.

References

1. Barnard, A.A.; Labuschagné, F.J.W.J. Exploring the Wet Mechanochemical Synthesis of Mg-Al, Ca-Al, Zn-Al and Cu-Al Layered Double Hydroxides from Oxides, Hydroxides and Basic Carbonates. *Crystals* **2020**, *10*, 954. [CrossRef]
2. Gevers, B.R.; Labuschagné, F.J.W.J. Green Synthesis of Hydrocalumite (CaAl-OH-LDH) from Ca(OH)$_2$ and Al(OH)$_3$ and the Parameters That Influence Its Formation and Speciation. *Crystals* **2020**, *10*, 672. [CrossRef]
3. Valeikiene, L.; Roshchina, M.; Grigoraviciute-Puroniene, I.; Prozorovich, V.; Zarkov, A.; Kareiva, A. On the Reconstruction Peculiarities of Sol–Gel Derived Mg$_{2-x}$M$_x$/Al$_1$ (M = Ca, Sr, Ba) Layered Double Hydroxides. *Crystals* **2020**, *10*, 470. [CrossRef]
4. Tampieri, A.; Lilic, M.; Costantini, M.; Medina, F. Microwave-Assisted Aldol Condensation of Furfural and Acetone over Mg–Al Hydrotalcite-Based Catalysts. *Crystals* **2020**, *10*, 833. [CrossRef]
5. Cardinale, A.M.; Carbone, C.; Consani, S.; Fortunato, M.; Parodi, N. Layered Double Hydroxides for Remediation of Industrial Wastewater from a Galvanic Plant. *Crystals* **2020**, *10*, 443. [CrossRef]

6. Pavel, O.D.; Şerban, A.; Zăvoianu, R.; Bacalum, E.; Bîrjega, R. Curcumin Incorporation into Zn_3Al Layered Double Hydroxides—Preparation, Characterization and Curcumin Release. *Crystals* **2020**, *10*, 244. [CrossRef]
7. Mohapi, M.; Shale Sefadi, J.; Mochane, M.J.; Magagula, S.I.; Lebelo, K. Effect of LDHs and Other Clays on Polymer Composite in Adsorptive Removal of Contaminants: A Review. *Crystals* **2020**, *10*, 957. [CrossRef]
8. Mochane, M.J.; Magagula, S.J.; Shale Sefadi, J.; Rotimi Sadiku, E.; Mokhena, T.C. Morphology, Thermal Stability, and Flammability Properties of Polymer-Layered Double Hydroxide (LDH) Nanocomposites: A Review. *Crystals* **2020**, *10*, 612. [CrossRef]

Publisher's Note: MDPI stays neutral with regard to jurisdictional claims in published maps and institutional affiliations.

© 2020 by the authors. Licensee MDPI, Basel, Switzerland. This article is an open access article distributed under the terms and conditions of the Creative Commons Attribution (CC BY) license (http://creativecommons.org/licenses/by/4.0/).

Article

Exploring the Wet Mechanochemical Synthesis of Mg-Al, Ca-Al, Zn-Al and Cu-Al Layered Double Hydroxides from Oxides, Hydroxides and Basic Carbonates

Brenda Antoinette Barnard * and Frederick Johannes Willem Jacobus Labuschagné

Department of Chemical Engineering, University of Pretoria, Lynnwood Rd, Hatfield, Pretoria 0002, South Africa; johan.labuschagne@up.ac.za
* Correspondence: u14037948@tuks.co.za

Received: 21 September 2020; Accepted: 16 October 2020; Published: 20 October 2020

Abstract: The synthesis of Mg-Al, Ca-Al, Zn-Al and Cu-Al layered double hydroxides (LDHs) was investigated with a one-step wet mechanochemical route. The research aims to expand on the mechanochemical synthesis of LDH using a mill designed for wet grinding application. A 10% slurry of solids was added to a Netzsch LME 1 horizontal bead mill and milled for 1 h at 2000 rpm. Milling conditions were selected according to machine limitations and as an initial exploratory starting point. Precursor materials selected consisted of a mixture of oxides, hydroxides and basic carbonates. Samples obtained were divided such that half was filtered and dried at 60 °C for 12 h. The remaining half of the samples were further subjected to ageing at 80 °C for 24 h as a possible second step to the synthesis procedure. Synthesis conditions, such as selected precursor materials and the M^{II}:M^{III} ratio, were adapted from existing mechanochemical methods. LDH synthesis prior to ageing was successful with precursor materials observably present within each sample. No Cu-Al LDH was clearly identifiable. Ageing of samples resulted in an increase in the conversion of raw materials to LDH product. The research offers a promising 'green' method for LDH synthesis without the production of environmentally harmful salt effluent. The synthesis technique warrants further exploration with potential for future commercial up-scaling.

Keywords: layered double hydroxide; mechanochemistry; bead mill; green chemistry; synthesis; wet grinding

1. Introduction

Layered double hydroxides (LDHs) are clay-like minerals commonly referred to as anionic clays with a wide range of physical and chemical properties. They are represented by the general formula $[M^{II}_{1-x}M^{III}_{x}(OH)_2][X^{q-}_{x/q} \cdot H_2O]$ in which M^{II} and M^{III} represent the selected divalent and trivalent metal elements and $[X^{q-}_{x/q} \cdot H_2O]$ denotes the interlayer composition. LDHs often find application in pharmaceuticals, as polymer additives, as additives in cosmetics, and in catalysis. This is due to having variable layer charge density, reactive interlayer space, ion exchange capabilities, a wide range of chemical compositions and rheological properties [1]. LDH materials can be synthesised using various different techniques of which the most common are co-precipitation, reconstruction, hydrothermal methods and urea decomposition-homogenous precipitation. The primary principle associated with these methods include the precipitation of various types of metal ions which makes large scale production difficult. Challenges associated with these methods include differing precipitation rates of metal ions, need for inert environments, production of environmentally harmful waste and high production costs [2]. Novel, 'green' synthesis techniques are therefore often sought after. Recently the use of mechanochemistry as an alternative synthesis procedure has gained wide-spread attention. Mechanochemistry is considered a versatile method of synthesis with the promise of producing LDH materials with unique elemental combinations [3,4]. The most common types of mechanochemical synthesis techniques include single-step or one-pot grinding [5,6], mechano-hydrothermal synthesis [7–10] and two step grinding. Two-step grinding can consist of an initial grinding step followed by an additional treatment step or a second grinding step [11–13]. Grinding of raw materials can be conducted wet, dry or as a paste. Various techniques and combinations involving the wet or dry milling of raw materials have been attempted and found to be successful [2]. Studies have shown that the type of grinding technique can largely affect the success of LDH synthesis, with some techniques not producing sufficient mechanical energy for the synthesis to occur readily [11]. Research has indicated that a large amount of mechanochemical methods explored typically involve the use of ball mills, mixer mills or a mortar and pestle as the primary grinding technique [2]. The final properties of LDH are further influenced by the selected method of grinding [14]. It is therefore of interest to expand on the effect of milling techniques on the synthesis of LDH materials. The success associated with the formation of an LDH phase for single step grinding procedures are further influenced by the selected starting materials [2]. The use of metallic salts of chlorides or nitrates allows for LDH synthesis but introduces a washing step that could produce an undesirable waste solution [5,6]. The use of hydroxides and oxides eliminates the production of waste solution promoting 'green' synthesis of LDH materials, however, has proven to be challenging [2]. The addition of water to existing grinding techniques, such that wet grinding occurs, is considered unsuitable for solid state chemistry as it may reduce the degree of amorphitization and prevent active site formation [15]. Dry grinding is therefore typically conducted as an initial mechanochemical step when synthesising LDHs. The absence of water allows for sufficient active site formation and amorphitisation. Dry grinding of the precursor materials is regularly used in conjunction with a second synthesis step. A variation of secondary synthesis steps have been explored. LDH materials have successfully been synthesised with the dry grinding of raw materials and agitating the milled material in a solution containing the desired anion for intercalation [16–19]. Similarly, LDH synthesis methods have involved dry grinding followed by washing or thermal treatment of the sample [2,20]. Unique methods have also involved a combination of the initial dry grinding step with that of a wet grinding step [15,21] or methods involving ultrasonic irradiation [22–24]. Limited research has been conducted on single-step or one-pot wet grinding and low conversion rates obtained warrant the need for further research [2,25]. Incomplete conversion or no LDH formation have been attributed to the quantity of water present with insufficient mechanochemical activation of the precursor materials occurring [15]. The study therefore aims to expand on the one-step wet mechanochemical synthesis of layered double hydroxides, from oxides, hydroxides and basic carbonates, by making use of a Netzsch LME 1 horizontal bead mill. The selected mill is designed specifically for wet grinding application and allows for the continuous, semi batch or batch synthesis of LDH materials. The process could be easily up scaled to produce large volumes of consistent and commercially viable LDH product. Precursor materials and M^{II}:M^{III} ratios were adapted from mechanochemical techniques in which LDH synthesis was successful [15,17,18,21]. The performance of the selected mill and synthesis conditions could therefore be investigated. Samples obtained were further subjected to ageing at 80 °C to determine the effects of including a thermal step to the selected mechanochemical method.

2. Materials and Methods

2.1. Milling Operation

Selected raw materials were wet batch milled with the use of a Netzsch LME 1 horizontal bead mill, under air atmosphere. The milling chamber (1.225 L) was loaded to a capacity of 60% (by volume) with 2 mm diameter yttrium stabilised zirconia beads. Cooling water was allowed to circulate through the outer jacket of the milling chamber at a constant inlet temperature of 30 °C and flow rate of

525 L·h^{-1}. A water slurry consisting of 10% solids (reactants) was added to the milling chamber and milled for 1 h at 2000 rpm. Samples obtained were divided such that half was filtered and dried at 60 °C for 12 h. The remaining half of the sample was subjected to ageing at 80 °C for 24 h. The principle of the mill is similar to that of agitator bead mills in which the grinding media is accelerated with the use of an agitator shaft. The energy supplied to the media is then transferred to the solids via collisions and de-acceleration. The vessel is placed in a horizontal position to allow for even grinding activity and activation. Figure 1 depicts a technical schematic of the Netzsch LME 1 horizontal bead mill. The product inlet and outlet to the grinding chamber were sealed to allow for batch milling. Raw materials and grinding media were added to the vessel through the 'bead filling connection'. At the end of every experimental run, the 'tank floor' or front cap of the milling chamber was removed and the sample and beads collected. The grinding media and mill were washed in preparation for the next experimental run. The pump set-up provided by Netzsch was not used.

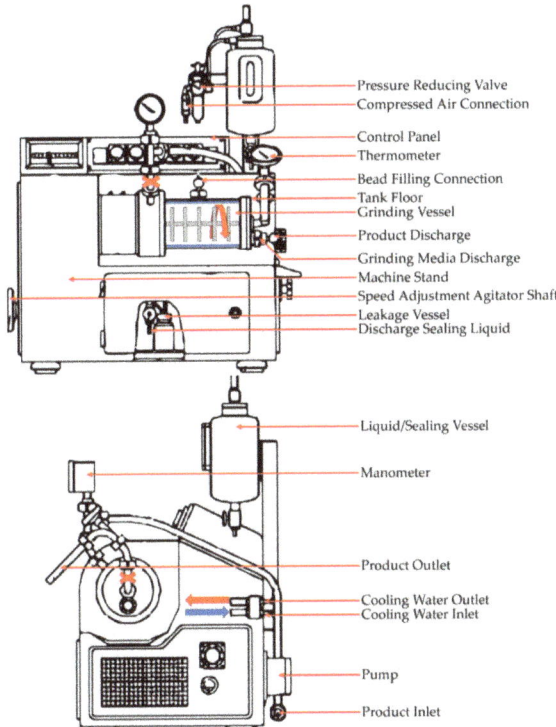

Figure 1. Technical Schematic of Netzsch LME 1 Horizontal Bead Mill as modified from Netzsch.

2.2. Ageing Process

The ageing step was conducted by making use of a bench-top Lasec digital hotplate stirrer. Samples obtained from the milling chamber were divided such that half was immediately filtered and dried and the other half subjected to an ageing step. Samples were placed in a glass beaker and agitated, at 400 rpm, for 24 h. Sample temperature was elevated and kept constant at 80 °C. A thin plastic film was placed over the beaker to prevent excessive moisture loss. Experiments were performed without the use of an inert gas under air atmosphere. All samples were filtered and dried at 60 °C for 12 h.

2.3. Mg-Al LDH

Commercial grade MgO (86%, Chamotte Holdings, JHB, GP, ZA) was initially calcined at 800 °C for 1 h to eliminate carbonate and hydroxide contaminants. This was then milled with Al(OH)$_3$ (Hindalco, Belgaum, India) making use of a 2:1 (28.63 g MgO, 23.83 g Al(OH)$_3$) (S1) and 3:1 (33.79 g MgO, 18.75 g Al(OH)$_3$) (S2), M^{II}:M^{III} metal ratio [15,26]. The selected MgO contained SiO$_2$ as an impurity and was prevalent in all relevant samples collected.

2.4. Ca-Al LDH

Commercial grade Ca(OH)$_2$ (LimeCo. Minerals, JHB, GP, ZA) was first calcined at a temperature of 900 °C for 1 h to remove any hydroxide and carbonate impurities, to form CaO. This was then reacted with 100 mL water for 15 min to form Ca(OH)$_2$. This step eliminated the possibility of vapor formation within the milling chamber due to the extremely exothermic CaO hydration reaction. The Ca(OH)$_2$ and Al(OH)$_3$ (Hindalco, Belgaum, India) were milled with and without the addition of a carbonate source, CaCO$_3$ (Kulubrite 45, Idwala Carbonates, Port Edward, KZN, ZA). The selected metal starting ratios were Ca:Al:CaCO$_3$ of 2:1:0 (35.80 g CaO, 16.60 g Al(OH)$_3$)(S4) and 3:2:1 (25.93 g CaO, 15.91 g Al(OH)$_3$, 10.58 g CaCO$_3$)(S3) [21].

2.5. Zn-Al LDH

The synthesis of Zn-Al LDH was conducted with Zn$_5$(CO$_3$)$_2$(OH)$_6$ (Sigma-Aldridge, St. Louis, MO, USA). This was milled at a 1:1 (Zn:Al) metal ratio with Al(OH)$_3$ (Hindalco, Belgaum, India). The sample was further referred to as S5 (30.69 g Zn$_5$(CO$_3$)$_2$(OH)$_6$, 21.77 g Al(OH)$_3$) [18].

2.6. Cu-Al LDH

Commercial grade Cu$_2$(OH)$_2$CO$_3$ (Adchem, MELB, AU) and Al(OH)$_3$ (Hindalco, Belgaum, India) were milled making use of a 2:1 (Cu:Al) ratio with the aim of synthesising Cu$_2$Al(OH)$_5$CO$_3 \cdot$XH$_2$O (38.97 g Cu$_2$(OH)$_2$CO$_3$, 13.75 g Al(OH)$_3$) (S6) [17].

2.7. Material Characterisation

2.7.1. Particle Size Analysis (PSA)

Samples collected were analysed wet and fully dispersed, before the filtration and drying steps, with the use of a Mastersizer 3000 (Malvern Panalytical, Malvern, UK) using a Hydro LV liquid unit.

2.7.2. Scanning Electron Microscopy (SEM)

SEM imaging was used to observe the morphology of the prepared samples. A Zeiss Gemini 1 cross beam 540 FEG SEM (Oberkochen, Germany). Powdered samples were placed secured onto an aluminium sample holder and graphite coated 5 times with a Polaron Equipment E5400 SEM auto-coating sputter system (Quorum, East Sussex, UK).

2.7.3. X-ray Diffraction Analysis (XRD)

Reaction products of powdered samples were identified using a PANalytical X'Pert Pro powder diffractometer in θ-θ configuration fitted with an X'Celerator detector and variable divergence- and fixed receiving slits (Malvern Panalytical, Malvern, UK). The system made use of Fe filtered Co-Kα (λ = 1.789Å) source. Samples were prepared using the standardised PANalytical backloading system, providing a random distribution of particles. Samples were scanned from 5° to 90° with a step size of 0.008°. Sample mineralogy was determined using the ICSD database in correlation with X'Pert Highscore plus software.

2.7.4. Fourier Transform Infrared Spectroscopy (FT-IR)

FT-IR spectra for the samples were obtained using a PerkinElmer 100 Spectrophotometer (Massachusetts, USA) over a range of 550–4000 cm^{-1} and represent an average of 32 scans, at a resolution of 2 cm^{-1}.

2.7.5. X-ray Fluorescence (XRF)

XRF was used for elemental analysis of the samples. Samples were dried at 100 °C and roasted at 1000 °C to determine mass loss on ignition. In addition, 1 g of the sample was mixed with 6 g Lithumtetraborate flux and fused at 1050 °C to form a stable fused glass bead. Analysis was conducted using a Thermo Fisher ARL Perform 'X Sequential instrument (Massachusetts, USA). Samples were characterised using UNIQUANT software.

3. Results and Discussion

3.1. Particle Size Analysis

The particle size of the raw material mixtures as well as that of the sample obtained is depicted in Tables 1 and 2, respectively. It was noted that overall particle size reduction occurred for most samples, with the exception of S3, with an increase in the grinding time as expected. This could possibly be attributed to the formation and agglomeration of Ca-Al LDH present within the sample. Raw material mixtures exhibited large D_{90} measurements that could be attributed to immediate reaction with water, as well as agglomeration.

Table 1. Particle size analysis of raw material mixtures prior to milling, relevant to each sample.

Sample	D_{10} [µm]	D_{50} [µm]	D_{90} [µm]
S1	2.35	7.79	17.6
S2	1.66	7.22	18.8
S3	1.98	7.84	23.6
S4	2.29	9.17	686
S5	1.71	4.13	10.5
S6	1.34	5.55	15.6

Table 2. Particle size analysis of each sample after 1 h of wet milling in a Netzsch LME 1 horizontal bead mill.

Sample	D_{10} [µm]	D_{50} [µm]	D_{90} [µm]
S1	0.962	3.39	6.21
S2	0.693	2.33	4.94
S3	0.594	1.70	36.9
S4	0.661	2.71	86.0
S5	0.77	2.51	4.83
S6	0.764	2.43	4.93

3.2. X-ray Fluorescence

The elemental composition and metal ratios of the samples were obtained via XRF analysis and listed in Table 3. All samples were found to have a small amount of zirconium, yttrium, and iron contamination from the milling media and the milling chamber. Samples S1 and S2 contained SiO_2 introduced by the selected commercial grade MgO reagent. XRF analysis was conducted to ensure that the correct metal ratios were applied to the raw materials added to the system and are therefore not an indication of the composition of the LDH phases present within each sample. They are an indication of the metal ratios within the overall sample obtained. Calculated metal ratios were observed to correlate with those adapted from literature.

Table 3. Calculated $M^{II}:M^{III}$ ratios of each sample, after wet milling for 1 h with a Netzsch LME 1 horizontal bead mill, as obtained through XRF analysis.

Sample	Expected $M^{II}:M^{III}$ Ratio	Calculated $M^{II}:M^{III}$ Ratio
S1	2.00:1.00	2.00:1.04
S2	3.00:1.00	3.00:1.04
S3	2.00:1.00	2.00:1.09
S4	2.00:1.00	2.00:0.94
S5	1.00:1.00	1.00:1.01
S6	2.00:1.00	2.00:10.00

3.3. X-ray Diffraction Analysis

Samples analysed were observed to have minor unindexed peaks present. This could be attributed to unidentified phases present or impurities due to mill degradation. Future studies should be conducted in attempt to adequately investigate all phases present within samples obtained.

Mg-Al LDH. Figures 2 and 3 show the XRD spectra obtained for samples S1 and S2. The reaction had not yet reached completion at the selected synthesis conditions prior to ageing. The sample synthesised with a 2:1 metal ratio (S1) exhibited no clear LDH peaks with no crystalline LDH phase detected within the sample. Ageing of the sample resulted in a clear LDH pattern with primary and secondary peaks located at 2θ values of 13.49° and 27.22°, respectively. This correlated with a basal spacing of 0.759 nm. Comparatively, S2 exhibited primary and secondary LDH peaks prior to ageing at 2θ values of 13.37° and 26.99°, respectively. This correlated with a basal spacing of 0.767 nm. Ageing of the sample resulted in more complete conversion of raw materials to LDH product, with clear peaks observed at 2θ values of 13.48° and 27.10°. The basal spacing was calculated to be 0.760 nm. Basal spacing has been found to be influenced by numerous factors including milling time, the amount of water present and possible carbonate contamination [15]. Lattice imperfections as a result of mechanically induced amorphitisation could further contribute to larger basal spacing values [5]. The addition of water has been known to decrease the degree of supersaturation, which could negatively impact morphology and crystallinity of the synthesised LDHs. It has further been observed that the crystallinity of LDHs can pass through a maximum, with lattice imperfections increasing with an increase in milling time [26]. The calculated basal spacing values obtained for both S1 and S2, after ageing, were found to be similar with those reported in literature [15]. Similarly, spectra for the aged samples obtained further correlated with existing literature [9,15,27].

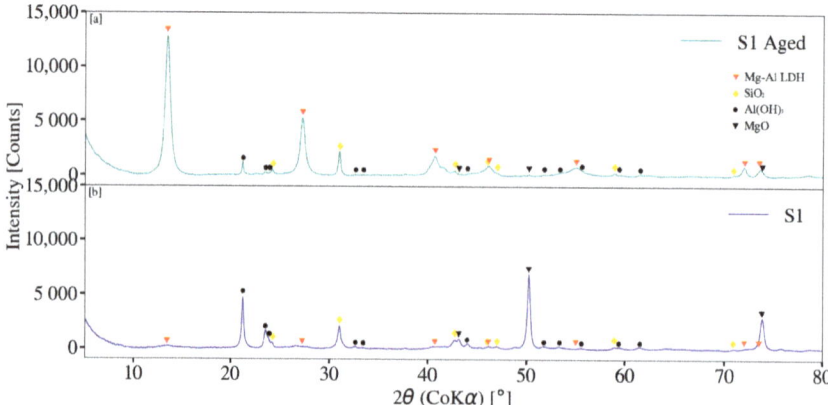

Figure 2. (a) X-ray diffraction analysis of the Mg-Al LDH sample (S1), synthesised with a 2:1 $M^{II}:M^{III}$ ratio, after ageing for 24 h at 80 °C.; (b) X-ray diffraction analysis of sample S1 prior to ageing after 1 h of wet milling at 2000 rpm.

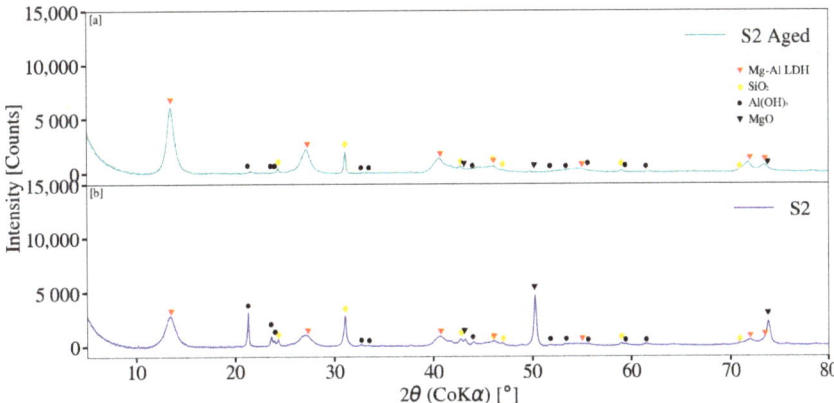

Figure 3. (a) X-ray diffraction analysis of Mg-Al LDH sample (S2), synthesised with a 3:1 $M^{II}:M^{III}$ ratio after ageing for 24 h at 80 °C; (b) X-ray diffraction analysis of sample S2 prior to ageing after 1 h of wet milling at 2000 rpm.

Ca-Al LDH. XRD reflection patterns for samples S3 and S4 are as depicted in Figures 4 and 5, respectively. The presence of LDH was observed prior to ageing for S3 with a primary peak at a 2θ of 13.50°. This correlated to a basal spacing of 0.759 nm. Ageing of the sample improved conversion of raw materials to LDH product, with a decrease in $Al(OH)_3$ and $Ca(OH)_2$ peak intensity observed. Twinning primary peaks were observed at 2θ values of 13.54° and 13.20° corresponding to basal spacing's of 0.757 nm and 0.776 nm, respectively. Comparatively, the XRD spectra for S4 depicted the presence of LDH within the sample despite the lack of a direct carbonate source. The primary peak was observed to occur at 2θ of 13.47°, with a basal spacing of 0.761 nm. The presence of $CaCO_3$ was further noted and likely due to atmospheric carbonate contamination. Ageing of the sample resulted in the formation of katoite ($Ca_3Al_2(OH)_{12}$) as well as twinning primary LDH peaks. These were observed to occur at 2θ of 13.27° and 13.58°, corresponding to basal spacing of 0.772 nm and 0.755 nm. Twinning peaks could be attributed to the presence of different LDH phases within the sample and differ through carbonate content [22,27,28]. Basal spacing values obtained for both samples, before and after ageing, suggest the presence of either or a mixture of calcium monocarboaluminate and a dehydrated polymorph of calcium hemicarboaluminate that forms upon ageing. Basal spacing reported for each of these were 0.750 nm [22,27] and 0.760–0.770 nm [22,28], respectively. Ca-Al LDH synthesised in the presence of a carbonate source formed more readily when compared to that synthesised with no $CaCO_3$. Previous studies have shown that Ca-Al LDH and katoite (tricalcium aluminate) can result when reacting $Ca(OH)_2$ and $Al(OH)_3$ in the absence of an additional phase or carbonate source, with little to no LDH formation occurring at times [11,21,29]. Studies have also shown that, upon the addition of a third phase such as $CaCO_3$ or $CaCl_2 \cdot 2H_2O$ LDH, formation occurs more readily with little to no katoite formation. This suggests a complex relationship between the formation of Ca-Al LDH and katoite. It has been suggested that the presence of pillared anions such as Cl^- and CO_3^{2-} assist in the formation of the layered structure of Ca-Al LDH. The third phase therefore stabilises the Ca-Al LDH structure allowing for formation to occur more readily [21].

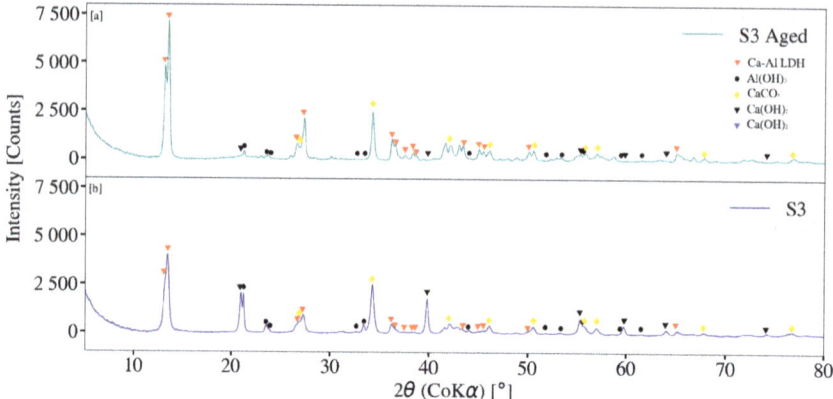

Figure 4. (a) X-ray diffraction analysis of sample Ca-Al LDH sample (S3), synthesised with a 2:1 $M^{II}:M^{III}$ ratio, in the presence of a carbonate source, after ageing for 24 h at 80 °C.; (b) X-ray diffraction analysis of sample S3 prior to ageing, after 1 h of wet milling at 2000 rpm.

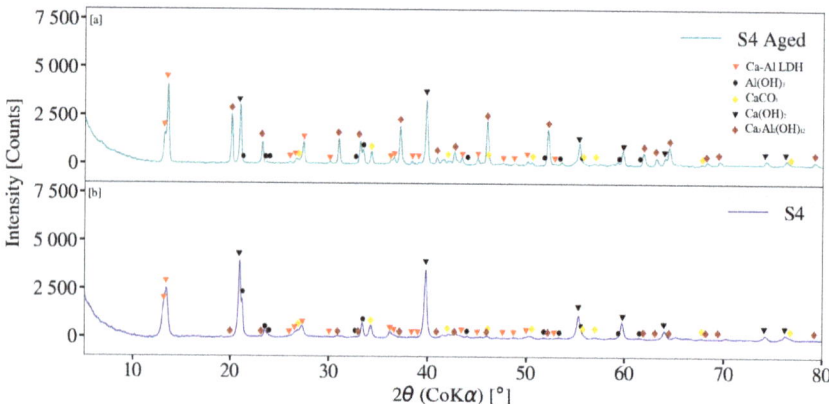

Figure 5. (a) X-ray diffraction analysis of sample Ca-Al LDH sample (S4), synthesised with a 2:1 $M^{II}:M^{III}$ ratio, without the presence of a carbonate source, after ageing for 24 h at 80 °C.; (b) X-ray diffraction analysis of sample S4 prior to ageing after 1 h of wet milling at 2000 rpm.

Zn-Al LDH. The XRD patterns of samples S5 are depicted in Figure 6. The primary LDH peak was observed to occur at a 2θ value of 13.86°, corresponding to a basal-spacing of 0.749 nm. Conversion prior to ageing was observed to be incomplete with $Zn_5(CO_3)_2(OH)_6$ and $Al(OH)_3$ peaks observed at 2θ values of 15.08° and 21.31°, respectively. Ageing of the sample resulted in the increase in LDH peak intensity with a primary peak at 2θ of 13.70 which corresponds to a basal spacing of 0.748 nm. A decrease in raw material peaks were observed with ageing; however, conversion remained incomplete for the selected synthesis conditions. Metal ratios ($M^{II}:M^{III}$) typically employed for the synthesis of Zn-Al LDH are between 2:1 and 4:1 for conventional methods such as co-precipitation [18]. It has recently been suggested that molar ratios suitable for mechanochemical synthesis range between 1:1 and 2:1, with nearly pure phase Zn-Al LDH as the result [18]. Slight differences were observed for basal spacing values obtained. These were observed to differ from those reported in literature (0.758 nm, $Zn_4CO_3(OH)_6 \cdot H_2O$ as starting material) with the Zn content influencing the basal spacing obtained [18].

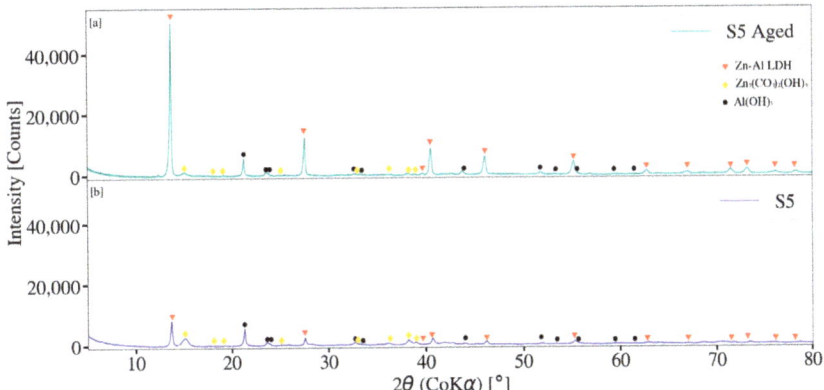

Figure 6. (a) X-ray diffraction analysis of Zn-Al LDH sample (S5), synthesised with a 1:1 $M^{II}:M^{III}$ ratio, after ageing for 24 h at 80 °C.; (b) X-ray diffraction analysis of sample S5 prior to ageing after 1 h of wet milling at 2000 rpm.

Cu-Al LDH. The XRD spectra obtained for samples S6 and $Cu_2(OH)_2CO_3$ can be seen in Figure 7. The results for S6 prior to ageing were considered to be inconclusive as no obvious LDH peaks were identified. A decrease in $Cu_2(OH)_2CO_3$ peak intensities were, however, observed to occur after 1 h of milling activity. Ageing of the sample resulted in the formation of a peak at 13.74° with a second peak present at approximately 27.76°. Identification of the LDH peaks were difficult due to prominent and overlapping $Cu_2(OH)_2CO_3$ peaks. Basal spacing associated with the observed primary peak was determined to be 0.746 nm. This was observed to be smaller than that reported in literature (0.754 nm) [17]. The formation of Cu-Al LDH was reported to be dependent on the selected rotational speed and therefore the degree of amorphitization [17].

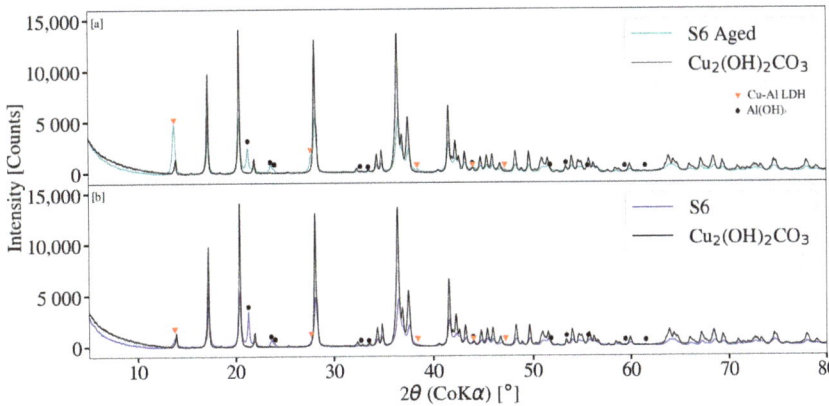

Figure 7. (a) X-ray diffraction analysis of Cu-Al LDH sample (S6), synthesised with a 2:1 $M^{II}:M^{III}$ ratio, after ageing for 24 h at 80 °C.; (b) X-ray diffraction analysis of sample S6 prior to ageing after 1 h of wet milling at 2000 rpm.

3.4. Fourier Transform Infrared Spectroscopy

The main purpose of the FT-IR data was to support the notion that LDH was present within each sample. This was due to the fact that not all LDH peaks were easily identifiable when conducting XRD analysis.

Mg-Al LDH. The FT-IR spectra for S1 and S2, before and after ageing, are depicted in Figures 8 and 9. Prior to ageing peaks were observed to occur between 3500 cm^{-1} and 3700 cm^{-1} for both samples and could be attributed to the stretching vibrations of free –OH groups [15]. Similarly, peaks located between 3250 cm^{-1} and 3600 cm^{-1} are likely due to bonded –OH within both samples [15]. Peaks located at 1367 cm^{-1} (S1) and 1365 cm^{-1} (S2) could be attributed to carbonate interactions (CO_3^{2-} $v3$ vibrations) [9,15,20]Ageing of the samples at 80 °C for 24 h, resulted in the intensification of these peaks. A broad peak, from 3250–3700 cm^{-1}, specifically 3425 cm^{-1} (S1) and 3460 cm^{-1} (S2), was observed to form upon ageing of both samples. This could be assigned to the –OH stretching vibrations that occur within the layered brucite like structure of the LDH as well as interlayer water molecules [9,15,20]. Peaks observed between 500 and 900 cm^{-1} could be attributed to M-O and MO-H (M = Mg, Al) vibrations [30]. Peaks located from 1100–900 cm^{-1} are typical of Si-O interactions from SiO_2 impurities in the MgO raw material [30]. The FT-IR spectra for both S1 and S2 after ageing coincide with spectra observed in literature [9,15,20,27].

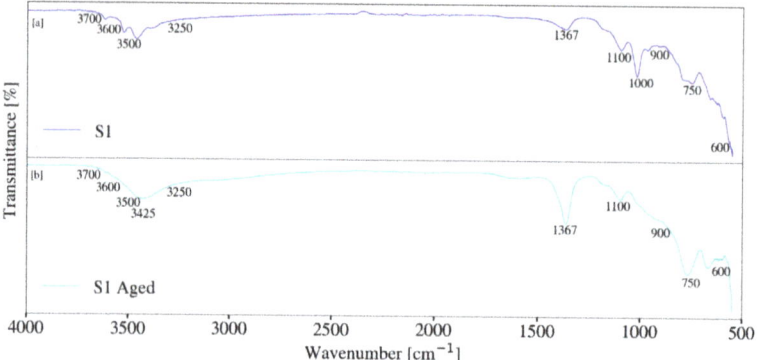

Figure 8. (a) FT-IR analysis of the Mg-Al LDH sample (S1) prior to ageing, synthesised with a 2:1 M^{II}:M^{III} ratio, after 1 h of wet milling at 2000 rpm.; (b) FT-IR analysis of sample S1 after ageing for 24 h at 80 °C.

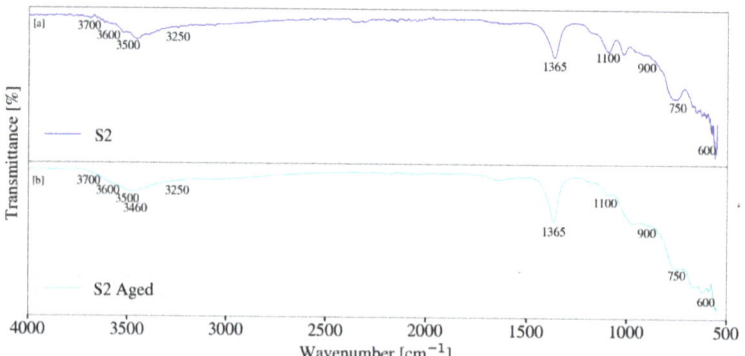

Figure 9. (a) FT-IR analysis of the Mg-Al LDH sample (S2) prior to ageing, synthesised with a 3:1 M^{II}:M^{III} ratio, after 1 h of wet milling at 2000 rpm.; (b) FT-IR analysis of sample S2 after ageing for 24 h at 80 °C.

Ca-Al LDH. FT-IR spectra for S3 and S4, before and after, ageing are depicted in Figures 10 and 11. Peaks observed between 3700–3300 cm^{-1} could be due to MO-H vibrations within each sample [5,21] as well as the vibration of –OH ($v2$) within the inorganic main layers of the LDH structure [5]. Prior to ageing, S3 depicted peaks at 1418 cm^{-1} and 876 cm^{-1}, which could be due to carbonate vibrations on the surface of the LDH structure present [5,21]. Similarly peaks at 1370 cm^{-1} could be attributed to CO_3^{2-} $v3$ vibrations within the interlayer of the LDH structure [21]. Ageing of the sample resulted in similar spectra to that obtained prior to ageing. The twinning carbonate interactions observed near 1366 cm^{-1} have been suggested further to be the result of two different environments for carbonate present, likely due to different Ca-Al LDH phases present [27]. The spectra for S4 prior to ageing were observed to be similar to that of S3. Peaks observed at 1414 cm^{-1} could be the result of carbonate within the system [5,21]. Synthesis, drying, and ageing were conducted, without the use of an inert gas, under air atmosphere. Carbonate contamination was therefore possible. Ageing of the sample resulted in the formation of twinning peaks at 1366 cm^{-1} and 1415 cm^{-1}. These could once again be attributed to interlayer and surface carbonate interactions of LDH formed within the system [5,21].

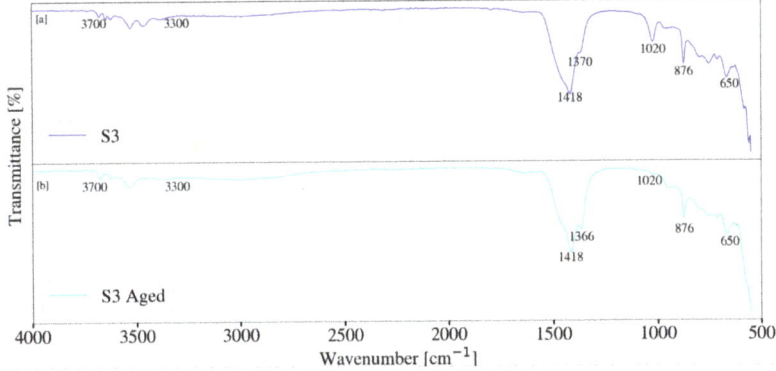

Figure 10. (a) FT-IR analysis of the Ca-Al LDH sample (S3) prior to ageing, synthesised with a 2:1 M^{II}:M^{III} ratio in the presence of a Carbonate source, after 1 h of wet milling at 2000 rpm.; (b) FT-IR analysis of sample S3 after ageing for 24 h at 80°.

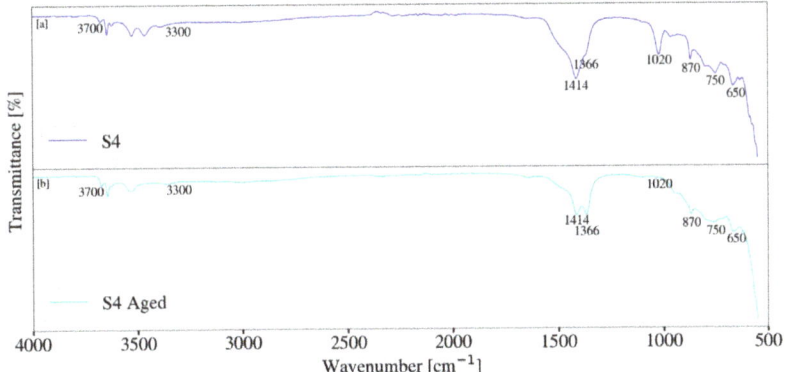

Figure 11. (a) FT-IR analysis of the Ca-Al LDH sample (S4) prior to ageing, synthesised with a 2:1 M^{II}:M^{III} ratio in the absence of a Carbonate source, after 1 h of wet milling at 2000 rpm.; (b) FT-IR analysis of sample S4 after ageing for 24 h at 80 °C.

Zn-Al LDH. The FT-IR spectra for S5, as well as that of $Zn_5(CO_3)_2(OH)_6$, is depicted in Figure 12. The peaks at and before 3468 cm^{-1} could be attributed to the stretching vibrations of –OH groups within the sample [18]. Spectra of the sample was observed to resemble that of the $Zn_5(CO_3)_2(OH)_6$ raw material before implantation of the ageing step. Ageing of the sample resulted in more complete conversion of raw materials to LDH product. Broadening of peaks between 3000 cm^{-1} and 3700 cm^{-1} were noted and are due to O-H stretching of hydroxyl groups [31]. Peak formation at approximately

1356 cm^{-1} and 1620 cm^{-1} was observed and is likely due to the asymmetric stretching vibrations of CO_3^{2-} within the interlayer of the LDH [18,30,31]. Peaks below 1000 cm^{-1} could further be attributed to M-O vibrations (M = Zn, Al) [18,31].

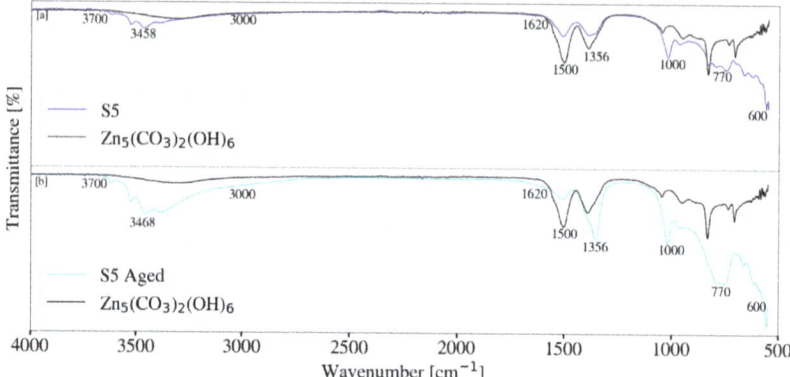

Figure 12. (a) FT-IR analysis of the Zn-Al LDH sample (S5) prior to ageing, synthesised with a 1:1 M^{II}:M^{III} ratio, after 1 h of wet milling at 2000 rpm; (b) FT-IR analysis of sample S5 after ageing for 24 h at 80 °C.

Cu-Al LDH. The FT-IR spectra for S6 and $Cu_2(OH)_2CO_3$, before and after ageing, are depicted in Figure 13. Samples obtained depicted similar spectra to that of the $Cu_2(OH)_2CO_3$. Identification of bond interactions associated specifically with LDH was therefore difficult and inconclusive. It was noted, however, that additional and broadening of peaks occurred between 3300 cm^{-1} and 3700 cm^{-1} and was consistent with Cu-Al LDH spectra reported in literature [17,32]. Additional peaks could further be attributed to bonded and free –OH groups within the sample [30]. Ageing resulted in the formation of a minor peak at 1632 cm^{-1} which could be due to the vibrations of water molecules [17].

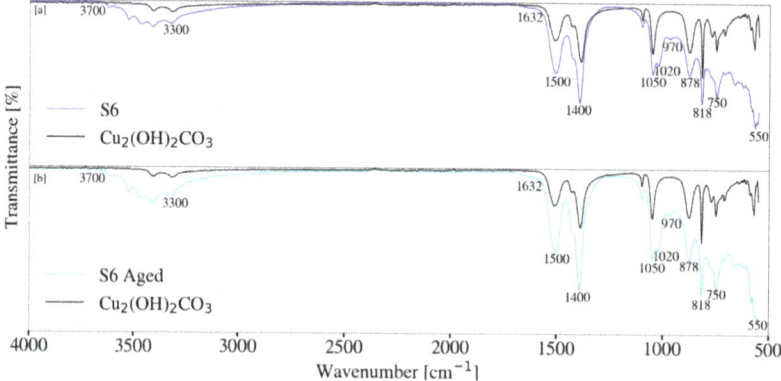

Figure 13. (a) FT-IR analysis of the Cu-Al LDH sample (S6) prior to ageing, synthesised with a 2:1 M^{II}:M^{III} ratio, after 1 h of wet milling at 2000 rpm.; (b) FT-IR analysis of sample S6 after ageing for 24 h at 80 °C.

3.5. Scanning Electron Microscopy

SEM imaging was conducted to provide insight on the morphology of the samples obtained before and after ageing and is depicted in Figure 14. Prior to ageing, both S1 and S2 depicted similar morphology with no obvious differences observed. Ageing of the samples resulted in the formation of thin platelet like structures. These were observed to correlate with SEM images associated with Mg-Al LDH reported in literature [9]. No obvious morphological differences were observed between S3 and S4 prior to ageing. Ageing of S3 resulted in the formation of thin crystalline platelets which correlate to those reported in literature for Ca-Al LDH[1]. Comparatively, S4 depicted structures associated with katoite, calcite, and Ca-Al LDH, and was observed to coincide with the XRD data obtained. No clear morphological differences were observed for S5 and S6 before and after ageing. Platelet like structures could be construed as being present.

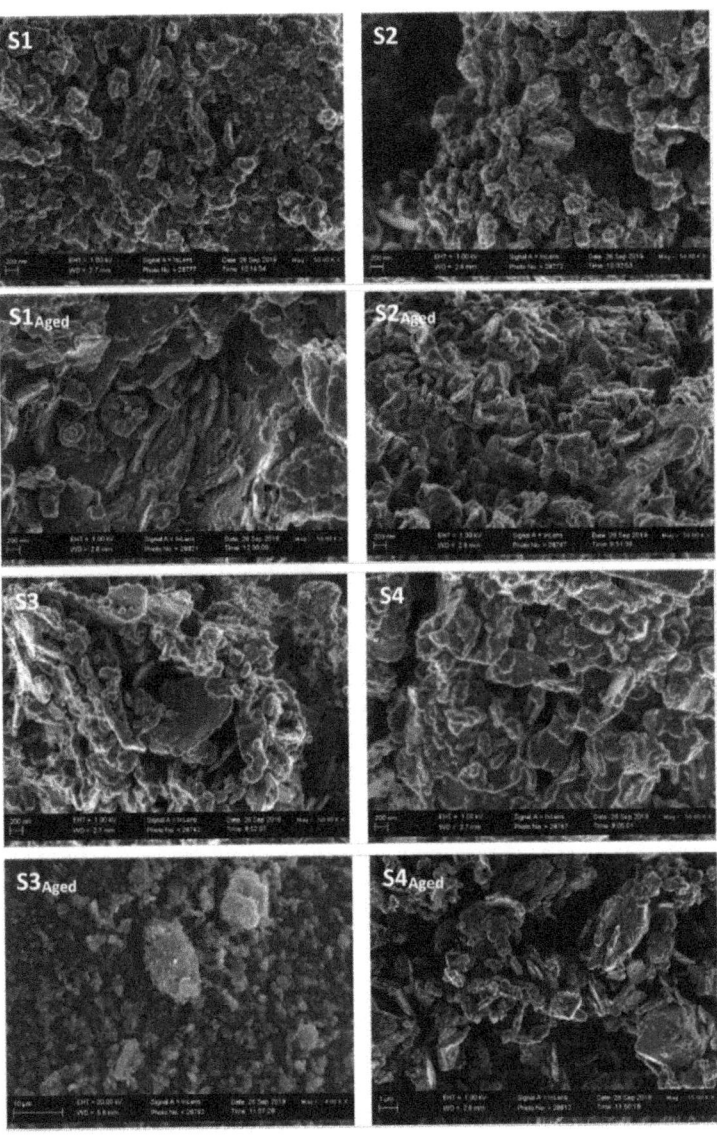

Figure 14. *Cont.*

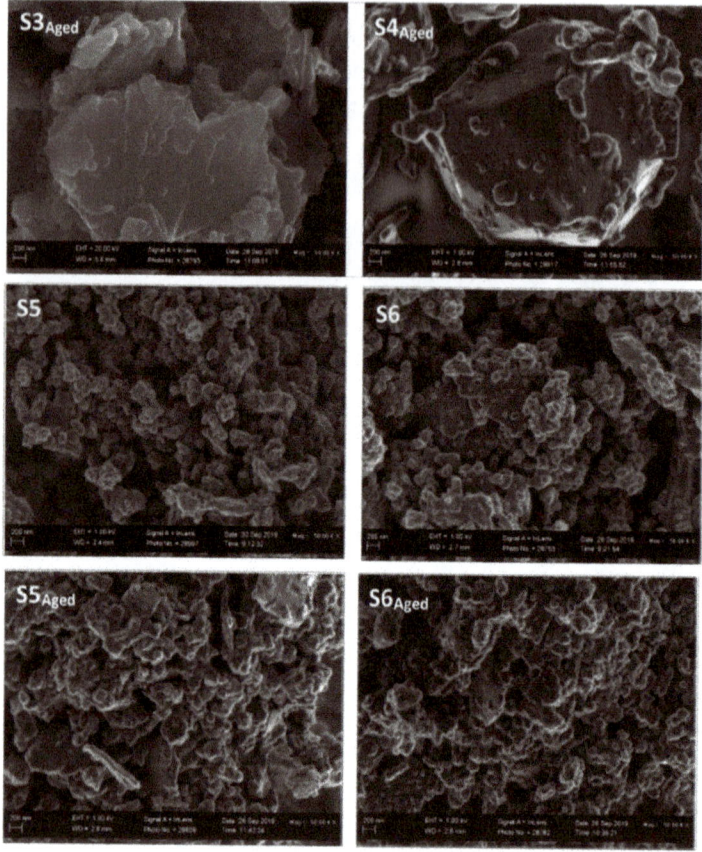

Figure 14. SEM imaging of samples before and after ageing. Imaging was conducted at a magnification of 50.00 K with a 200 nm scale and EHT of 1.00 kV for all samples. Additional imaging was conducted for S3$_{aged}$ at a scale of 10 µm, EHT of 20.00 kV and magnification of 4.00 K. Similarly S4$_{aged}$ was imaged with a scale of 1.00 µm, EHT 1.00 kV and magnification of 15.00 K.

4. Conclusions

A facile mechanochemical method for the synthesis of Mg-Al, Ca-Al, Zn-Al, and Cu-Al LDH was successfully developed. Implementation of ageing resulted in more complete conversion of raw materials to LDH product. Starting materials selected were a mixture of oxides, hydroxides, and basic carbonates, promoting 'green' synthesis of LDH materials. The synthesis method eliminates the use of pH control and avoids the production of salt rich effluent. The selected mill offers potential for commercial up-scaling with a variety of iso mills available, capable of processing large volumes of liquid reagent. Mill operation is versatile offering the potential for batch, semi-batch, or continuous synthesis of LDH materials. Traditional methods have predominantly made use of ball mills, mixer mills, or a mortar and pestle [2]. Wet milling allows for ease of transfer of reagents to downstream processes with little to no process interruption. Inclusion of the ageing step has allowed for an increase in conversion of raw materials to LDH product and improved sample morphology. This offers potential for converting the one-step wet mechanochemical route to that including a second step. Previous studies conducted have indicated that a grinding step is necessary for LDH formation to occur more readily [9,17,18]. The grinding activity allows for mechanochemical activation of raw materials, creating active sites. The selected horizontal bead mill has successfully resulted in the formation of multiple types of LDH product; however, raw material peaks were still observably present under the selected synthesis conditions. Samples S1, S2, S3, and S5 resulted in successful LDH formation with high intensity peaks observed after ageing. Samples S4 resulted in LDH formation prior to ageing; however, katoite formation occurred during the ageing step. The presence of a carbonate source was therefore necessary for LDH formation to occur and continue successfully. Cu-Al LDH did not form readily, with no clear indication of its presence prior to ageing. LDH peaks were observed to have formed upon ageing with precursor materials still notably present. The mill therefore offers potential for a successful one-step wet mechanochemical technique. Further investigation of a change in milling parameters such as retention time, jacket water temperature, bead size, solids loading and bead loading is warranted.

Author Contributions: B.A.B.: Methodology, Formal Analysis, Investigation, Data Curation, Writing—Original Draft and Final Version, Visualization, Conceptualisation. F.J.W.J.L.: Supervision, Project administration, Funding Acquisition, Writing—Review and Editing, Resources, Conceptualisation, Moral support. All authors have read and agreed to the published version of the manuscript.

Funding: This research project was funded by TeckSparks Pty Ltd. and THRIP.

Conflicts of Interest: The authors declare no conflict of interest.

References

1. Bergaya, F.; Lagaly, G. *Handbook of Clay Science*, 1st ed.; Elsevier Science & Technology Books: San Diego, CA, USA, 2013; pp. 1021–1069.
2. Qu, J.; Zhang, Q.; Li, X.; He, X.; Song, S. Mechanochemical approaches to synthesize layered double hydroxides: A review. *Appl. Clay Sci.* **2015**, *119*, 185–192. [CrossRef]
3. Ferencz, Z.; Szanados, M.; Adok-Sipiczki, M.; Kukovecz, A.; Konya, Z.; Sipos, P.; Palinko, I. Mechanochemical assisted synthesis of pristine Ca(II)Sn(IV)-layered double hydroxides and their amino acid intercalated composites. *Mater. Sci.* **2014**, *49*, 8479–8486. [CrossRef]
4. Li, Z.; Zhang, Q.; Liu, X.; Chen, M.; Wu, L.; Ai, Z. Mechanochemical synthesis of novel heterostructured Bi_2S_3/Zn-Al layered double hydroxide nano-particles as efficient visible light reactive Z-scheme photocatalysts. *Appl. Surf. Sci.* **2018**, *452*, 123–133. [CrossRef]
5. Fahami, A.; Beall, G.W.; Enayatpour, S.; Tavangarian, T.; Fahami, M. Rapid preparation of nano hexagonal-shaped hydrocalumite via one-pot mechanochemistry method. *Appl. Clay Sci.* **2017**, *136*, 90–95. [CrossRef]
6. Fahmi, A.; Duraia, E.M.; Beall, G.W.; Fahami, M. Facile synthesis and structural insight of chloride intercalated Ca-Al Layered double hydroxide nanopowders. *J. Alloys Compd.* **2017**, *727*, 970–977. [CrossRef]
7. Du, W.; Zheng, L.; Li, X.; Fu, J.; Lu, X.; Hou, Z. Plate like Ni-Mg-Al layered double hydroxides synthesised via a solvent free approach and its application in hydrogenolysis of D sorbitol. *App. Clay Sci.* **2016**, *123*, 166–172. [CrossRef]
8. Qu, J.; Li, X.; Lei, Z.; Li, Z.; Chen, M.; Zhang, Q. Mechano-Hydrothermal synthesis of tetraborate pillared Li-Al layered double hydroxide. *Am. Ceram. Soc.* **2016**, *99*, 1151–1154. [CrossRef]
9. Zhang, F.; Du, N.; Song, S.; Liu, J.; Hou, W. Mechano-hydrothermal synthesis of Mg_2Al-NO_3 layered double hydroxides. *Solid State Chem.* **2013**, *206*, 45–50. [CrossRef]
10. Zhang, F.; Hou, W. Mechano-hydrothermal preparation of Li-Al-OH layered double hydroxides. *Sol. State Sci.* **2018**, *79*, 93–98. [CrossRef]
11. Ferencz, Z.; Kukovecz, A.; Konya, A.; Sipos, P.; Palinko, I. Optimisation of the synthesis parameters of mechanochemically prepared Ca-Al-layered double hydroxide. *Appl. Clay Sci.* **2015**, *112*, 94–99. [CrossRef]
12. Kuramoto, K.; Intasa-ard, S.; Bureekaew, S.; Ogawa, M. Mechanochemical synthesis of finite particles of layered double hydroxide-acetate intercalation compound: Swelling, Thin film ion exchange. *Sol. State Chem.* **2017**, *253*, 147. [CrossRef]
13. Tongamp, W.; Zhang, Q.; Saito, F. Mechanochemical route for synthesizing nitrate form layered double hydroxide. *Powder Technol.* **2008**, *185*, 43–48. [CrossRef]
14. Pagano, C.; Marmottini, F.; Nocchetti, M.; Ramella, D.; Perioli, L. Effects of different milling techniques on the layered double hydroxides final properties. *Appl. Clay Sci.* **2018**, *151*, 124–133. [CrossRef]

15. Tongamp, W.; Zhang, Q.; Saito, F. Preparation of meixnerite (Mg–Al–OH) type layered double hydroxide by a mechanochemical route. *Mater. Sci.* **2007**, *42*, 9210–9215. [CrossRef]
16. Li, Z.; Chen, M.; Ai, Z.; Wu, L.; Zhang, Q. Mechanochemical synthesis of CdS/MgAl LDH-precursor as improved visible light driven photocatalyst for organic dye. *App. Clay Sci.* **2018**, *163*, 265–272. [CrossRef]
17. Qu, J.; He, X.; Chen, M.; Hu, H.; Zhang, Q.; Liu, X. Mechanochemical synthesis of Cu-Al and methyl orange intercalated Cu-Al layered double hydroxides. *Mater. Chem. Phys.* **2017**, *191*, 173–180. [CrossRef]
18. Qu, J.; He, X.; Chen, M.; Huang, P.; Zhang, Q.; Liu, X. A facile mechanochemical approach to synthesize Zn-Al layered double hydroxide. *Solid State Chem.* **2017**, *250*, 1–5. [CrossRef]
19. Zhong, L.; He, X.; Qu, J.; Li, X.; Lei, Z.; Zhang, Q.; Liu, X. Precursor preparation for Ca-Al layered double hydroxide to remove hexavalent chromium co-existing with calcium and magnesium chloride. *Solid State Chem.* **2017**, *245*, 200–206. [CrossRef]
20. Fahami, A.; Beall, G.W. Mechanosynthesis and characterization of Hydrotalcite like Mg-Al-SO4-LDH. *Mater. Lett.* **2016**, *165*, 192–195.
21. Qu, J.; Zhong, L.; Li, Z.; Chen, M.; Zhang, Q.; Liu, X. Effect of anion addition on the syntheses of Ca–Al layered double hydroxide via a two-step mechanochemical process. *Appl. Clay Sci.* **2016**, *124*, 267–270. [CrossRef]
22. Szabados, M.; Mészáros, R.; Erdei, S.; Kónya, Z.; Kukovecz, Á.; Sipos, P.; Pálinkó, I. Ultrasonically enhanced mechanochemical synthesis of Ca-Al-layered double hydroxides intercalated by a variety of inorganic anions. *Ultrason. Sonochem.* **2016**, *31*, 409–416. [CrossRef] [PubMed]
23. Szabados, M.; Pásztor, K.; Csendes, Z.; Muráth, S.; Kónya, Z.; Kukovecz, Á.; Carlson, S.; Sipos, P.; Pálinkó, I. Synthesis of high-quality, well-characterised CaAlFe layered triple hydroxide with the combination of dry-milling and ultrasonic irradiation in aqueous solution at elevated temperature. *Ultrason. Sonochem.* **2016**, *32*, 173–180. [CrossRef] [PubMed]
24. Szabados, M.; Varga, G.; Kónya, Z.; Kukovecz, Á.; Carlson, S.; Sipos, P.; Pálinkó, I. Ultrasonically-enhanced preparation, characterization of Ca-Fe-layered double hydroxides with various interlayer halide, azide and oxo anions (CO_3^{2-}, NO_3^-, ClO_4^-). *Ultrason. Sonochem.* **2018**, *40*, 853–860. [CrossRef]
25. Iwasaki, T.; Yoshii, H.; Nakamura, H.; Watano, S. Simple and rapid synthesis of Ni-Fe Layered double hydroxide by a new mechanochemical method. *Appl. Clay Sci.* **2012**, *58*, 120–124. [CrossRef]
26. Zhang, X.; Qi, F.; Li, S.; Wei, S.; Zhou, K. A mechanochemical approach to get stunningly uniform magnesium aluminium layered double hydroxides. *Appl. Surf. Sci.* **2012**, *259*, 245–251. [CrossRef]
27. Labuschagne, F.J.W.J.; Molefe, D.M.; Focke, W.W.; van der Westhuizen, I.; Wright, H.C.; Royeppen, M.D. Heat stabilising flexible PVC with layered double hydroxide derivatives. *Polym. Degrad.* **2015**, *113*, 49–54. [CrossRef]
28. Rosenberg, S.P.; Wilson, D.J.; Heath, C.A. Some Aspects of Calcium Chemistry in the Bayer Process. *Light Met.* **2001**, *1*, 210–216.
29. Kano, J.; Yamashita, J.; Saito, F. Effect of heat-assisted grinding of calcium hydroxide-gibbsite mixture on formation of hydrated calcium aluminate and its hydration behavior. *Powder Technol.* **1998**, *98*, 279–280. [CrossRef]
30. Socrates, G. *Infrared and Ramen Characteristic Group Frequencies: Tables and Charts*, 1st ed.; John Wiley and Sons: Hoboken, NJ, USA, 2001; pp. 287–295.
31. Qu, J.; He, X.; Li, X.; Ai, Z.; Li, Y.; Zhang, Q.; Liu, X. Precursor preparation of Zn–Al layered double hydroxide by ball milling for enhancing adsorption and photocatalytic decoloration of methyl orange. *RSC Adv.* **2017**, *7*, 31466–31474. [CrossRef]
32. Qu, J.; He, Z.; Lei, Q.; Zhang, Q.; Liu, X. Mechanochemical synthesis of dodecyl sulfate anion intercalated Cu-Al layered double hydroxide. *Solid State Sci.* **2017**, *74*, 125–130. [CrossRef]

Publisher's Note: MDPI stays neutral with regard to jurisdictional claims in published maps and institutional affiliations.

© 2020 by the authors. Licensee MDPI, Basel, Switzerland. This article is an open access article distributed under the terms and conditions of the Creative Commons Attribution (CC BY) license (http://creativecommons.org/licenses/by/4.0/).

Article

Microwave-Assisted Aldol Condensation of Furfural and Acetone over Mg–Al Hydrotalcite-Based Catalysts

Alberto Tampieri, Matea Lilic, Magda Constantí and Francesc Medina *

DEQ, ETSEQ, Universitat Rovira i Virgili, Avinguda dels Països Catalans 26, 43007 Tarragona, Spain; alberto.tampieri@urv.cat (A.T.); lilicmatea@gmail.com (M.L.); magdalena.constanti@urv.cat (M.C.)
* Correspondence: francesc.medina@urv.cat

Received: 4 August 2020; Accepted: 16 September 2020; Published: 18 September 2020

Abstract: The depletion of fossil fuel resources has prompted the scientific community to find renewable alternatives for the production of energy and chemicals. The products of the aldol condensation between bio-based furfural and acetone have been individuated as promising intermediates for the preparation of biofuels and polymeric materials. We developed a protocol for the microwave-assisted condensation of these two compounds over hydrotalcite-based materials. Mg:Al 2:1 hydrotalcite was prepared by co-precipitation; the obtained solid was calcined to afford the corresponding mixed metal oxide, which was then rehydrated to obtain a meixnerite-type material. The prepared solids were characterized by PXRD, ICP-AES, TGA-DSC and N_2 physisorption, and tested as catalysts in the aldol condensation of acetone and furfural in a microwave reactor. The performance of the catalysts was assessed and compared; the meixnerite catalyst proved to be the most active, followed by the mixed metal oxide and the as-synthesized hydrotalcite, which has often been reported to be inactive. In all cases, the reaction is quite fast and selective, which makes our protocol useful for rapidly converting furfural and acetone into their condensation products.

Keywords: microwave-assisted organic synthesis; biofuel production; rehydrated hydrotalcite; layered double hydroxide; heterogeneous basic catalysis; green chemistry

1. Introduction

Non-renewable resources are slowly, but inexorably, depleting [1]. This has prompted researchers to find alternatives to fossil fuel-derived matter, which industry traditionally uses to produce energy and chemicals [2]. Although it is theoretically possible to switch fossil fuels for renewable sources, such as solar, wind, and hydrothermal energy [3], for the post-oil production of chemicals, alternative raw materials are required. Carbon is present on Earth in many forms. It is mostly stored in the lithosphere as carbonate minerals, whereas the part which actively takes part in the carbon cycle is distributed between the soil, atmosphere, and water as carbon dioxide, carbonates, and their intermediates, and as organic carbon in the fossil pool and biomass [4]. Hence, it comes as no surprise that numerous biomass conversion strategies have been developed in recent decades [5]. The extensive use of this strategy may also be effective at reducing the emission of greenhouse gases and tackling the climate change problem [6].

Besides this, renewable energy resources may not always be suitable substitutes for fossil fuels in specific contexts. This may be the case for transportation fuels, for which biomass could instead provide valuable alternatives, namely biomass-derived fuels (or biofuels) [7]. Mixtures must meet some specifications (e.g., viscosity, boiling point, etc.) if they are to serve as diesel or jet fuels [8]; since these fuels are made up of liquid hydrocarbons, the mixtures must contain alkenes with definite carbon chain ranges to meet the aforementioned specifications [9]. The oligomerization of bioethanol [8] and the Fischer–Tropsch reaction of biomass-derived syngas [10] are some of the C–C bond forming strategies for producing diesel and jet fuels from biomass. Another option is to use bio-based aldehydes as coupling partners in aldol condensations with enolizable ketones such as acetone [11]; the enones obtained after cross-condensation have chain lengths that are suitable for liquid alkane transportation fuels [9]. The total hydrodeoxygenation (HDO) of these products affords the corresponding linear hydrocarbons [12]. The enones themselves, and the derivatives obtained by partial HDO, may also be used as novel organic dyes, monomers, and cross-linking agents [13].

A suitable electrophilic partner for this transformation is furfural, a furanic aldehyde obtained by the dehydration of hemicellulose-based pentoses, such as xylose [14]. Furfural is considered to be one of the most important bio-derived chemical platforms, and can be converted into a wide variety of products with the most disparate applications [15]. The reaction of furfural and acetone under basic conditions affords the aldol C8-OH (Scheme 1), the dehydration of which leads to the formation of the enone C8. C13-OH is obtained after a second aldol reaction on the other side of the ketone, and C13 is ultimately formed by the dehydration of the corresponding aldol. In industry, aldol condensations are traditionally carried out in alkaline aqueous solution [16], which presents such disadvantages as reaction corrosion and the difficult separation of the products. For this reason, various reports on the cross-condensation of furfural and acetone have focused on heterogeneous catalysis [17–21].

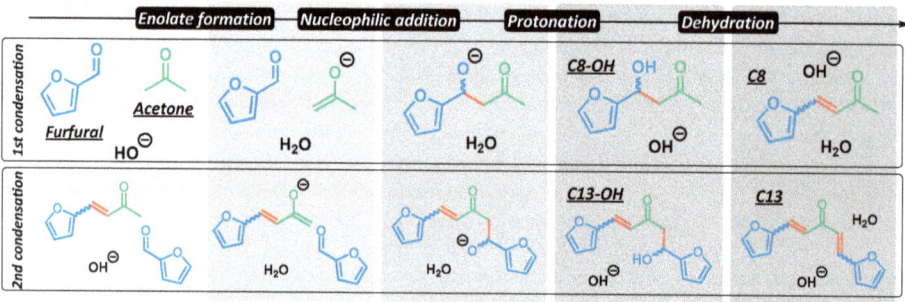

Scheme 1. Base-catalyzed aldol condensation of furfural and acetone.

We developed a microwave-assisted batch process for the neat aldol condensation of furfural and acetone over hydrotalcites and derivatives. Mg:Al hydrotalcites (HTs) are anionic clays which belong to the category of layered double hydroxides with the general formula $\left[Mg_{1-x}^{2+}Al_x^{3+}(HO^-)_2\right]^{x+}\left[CO_3^{2-}\right]_{x/2}^{x-}\cdot m(H_2O)$ [22]. HTs are prepared by co-precipitation of Mg and Al metal salts, and have been used as adsorbents, ion-exchangers, and heterogeneous basic catalysts. In the present study, we prepared 2:1 Mg:Al HTs with different degrees of modification (as-synthesized, calcined, and rehydrated). These solids have different types of basicity [23] and proved to be valuable catalysts for aldol condensation, especially the mixed metal oxides (MMOs) obtained after calcination of HTs [18,24] and the meixnerite-like (MX) layered double hydroxides obtained after rehydration of MMOs [25–27]. The solids prepared were characterized and tested as catalysts in aldol condensation reactions performed in a microwave (MW) reactor. Microwave irradiation has become a popular heating method in organic synthesis [28]. Moreover, as most of the solid catalysts are good MW absorbers, the use of microwave irradiation in heterogeneous catalytic processes is becoming increasingly popular [29]. Neat reaction of furfural and acetone with conventional heating over catalysts similar to ours has already been reported [27], although the hydrotalcites used were purchased, not prepared in the laboratory, and had different properties than the conventional hydrotalcites reported in the literature. We used MW irradiation as a heating method, and used hydrotalcites synthesized in our laboratories.

2. Materials and Methods

All starting materials were obtained from commercial sources. The starting materials for the preparation of the Mg:Al HT were $Mg(NO_3)_2 \cdot 6H_2O$ (extra pure, Acros Organics, Fair Lawn, NJ, USA), $Al(NO_3)_3 \cdot 9H_2O$ (≥98%, Fisher Chemical, Hampton, VA, USA), NaOH (Pellets, 98.9%, Fisher Chemical), and Na_2CO_3 (Anhydrous, ≥99.5%, Fisher Chemical). Deionized water was used for this preparation (18.2 MΩ·cm at 25 °C, Simplicity UV water purification system, Merck Millipore, Burlington, NJ, USA). Furfural (99%, Sigma Aldrich, St. Louis, MO, USA) and acetone (99.5+%, Acros Organic) were used in the aldol condensation. Pure N_2 (≥99.9992%, 200 bar at 15 °C, Carburos Metálicos, Cornellà de Llobregat, Spain) was used to increase the chamber pressure of the microwave reactor during the condensation experiments. Toluene (≥99.3%, Honeywell, Charlotte, NC, USA) was used as an internal standard for the GC-FID analysis of samples obtained after the microwave-assisted process.

2.1. Catalyst Preparation

Hydrotalcite (HT, 2:1 Mg:Al) was prepared by the co-precipitation of Mg and Al nitrates in deionized water. NaOH and Na_2CO_3 were added dropwise, and the pH was kept constant at 10. The resulting white slurry was left to age overnight under magnetic stirring and without heating: this aging is more advantageous than aging under heating from the point of view of energy consumption; it also produces more amorphous solids that are generally better for catalytic purposes. The slurry was then filtered, and the cake obtained was washed with abundant deionized water. The solid obtained was dried at 100 °C for 24 h, ground, and stored in a vial. Calcination at 450 °C for 4 h in a muffle furnace (HD—150 muffle furnace, Hobersal, Caldes de Montbui, Spain) transformed HT into the corresponding mixed metal oxide (MMO), which was used for the catalysis as it was, or rehydrated. MMO was rehydrated in the liquid phase by immersing it in deionized water (2 mg MMO/mL), which had previously been bubbled with pure Ar. The mixture was left stirring at 500 rpm for 24 h at room temperature in a sealed vessel, and finally centrifuged, filtered, and dried under vacuum.

2.2. Catalyst Characterization

The prepared catalysts were characterized by PXRD, ICP-AES, TGA, and N_2 physisorption. PXRD measurements were performed using a Siemens D5000 diffractometer (Bragg–Brentano parafocusing geometry and vertical θ-θ goniometer) fitted with a curved graphite diffracted-beam monochromator, incident, and diffracted-beam Soller slits, a 0.06° receiving slit, and a scintillation counter as a detector. The angular 2θ diffraction range was between 5 and 70°. The data were collected with an angular step of 0.05° at 3 s per step and sample rotation. A low background Si (510) wafer was used as a sample holder. CuKα radiation was obtained from a copper X-ray tube operating at 40 kV and 30 mA.

ICP-AES analyses were performed on samples obtained by dissolving the solids in concentrated HNO_3, and digesting the resulting solution in an Ethos Easy (Milestone, Sorisole, Italy) MW digester. These samples were then analyzed by a 160 charge coupled device (CCD) (Arcos, Spectro, Kleve, Germany) spectrometer to obtain the content of Na, Mg, and Al.

TGA-DSC was performed using a SENSYS evo TG-DSC (S60/58129, Setaram Instrumentation, Caluire-et-Cuire, France). All analyses started with a conditioning step (5 min at 30 °C), after which the chamber was heated to 800 °C at a constant heating rate of 10 °C/min. This temperature was kept for 5 min, and then the chamber was left to cool for 25 min. The analyses were conducted under Ar at a flow of 30 standard cubic centimeters.

N_2 physisorption was performed on solids outgassed at 120 °C for 5 h under vacuum (6 mTorr) in the instrument pre-chamber (FloVac Degasser, Quantachrome Instruments, Boynton Beach, FL, USA) to remove the adsorbed species from the samples. The N_2 physisorption analysis was performed at 77 K using a Quadrasorb *SI* Models 4.0 with QuadraWin Software (Quantachrome Instruments, v. 5.0 + newer).

2.3. Furfural Distillation

Furfural spontaneously undergoes degradation when in contact with air [30]. For this reason, furfural was periodically purified, by means of distillation, to remove the degradation products from the starting material. A simple distillation was carried out by heating furfural in a round-bottom flask with a heating mantle (Fibroman-N, JP Selecta, Abrera, Spain). The vapors were channeled into a condenser, and the purified furfural was collected in another round-bottom flask. The flask was sealed, its content was bubbled with argon, and it was stored in a refrigerator at 4 °C.

2.4. Microwave-Assisted Aldol Condensation

The aldol condensations were performed in an MW reactor (single reaction chamber microwave synthesis system, Milestone SynthWAVE, controlled by a Terminal 660 control panel with easyCONTROL software, Figure 1a). The reaction mixtures consisted of 0.1 g of furfural (86 uL, 1.04 mmol) in 5 mL of acetone, and the specified amounts of catalyst. The mixtures were placed in oven-dried glass reaction tubes together with magnetic stir bars; caps were put on to prevent the reaction mixtures from coming out of the tubes (Figure 1b). These tubes were placed in a rack, which was immersed in a heating bath. In all cases, this bath consisted of 250 mL of deionized water. The stirring was operated by a rotating magnetic stirrer, and the stirring rate was set to 50%. The whole PTFE reaction vessel was pressurized with 10 bars of N_2. The maximum MW irradiation power was set to 300 W. The power was modulated by the system so that the real bath temperature matched the corresponding temperature in the program. Indeed, the temperature probe measures the temperature of the bath, and all temperatures mentioned in the rest of this article refer to the bath ones. The system did not measure the temperature of the reaction mixtures. The general structure of the temperature programs (Figure S1) was the following: in the first step, the temperature was increased to the desired temperature in 10 min; in the second step, this temperature was maintained for the rest of the reaction time; in the third step, the irradiation was stopped, and the bath temperature was allowed to decrease for 20 min to below 56 °C, the boiling point of acetone. Once the third step was over, the pressure was released, the reaction vessel was opened, and the reaction tubes were collected.

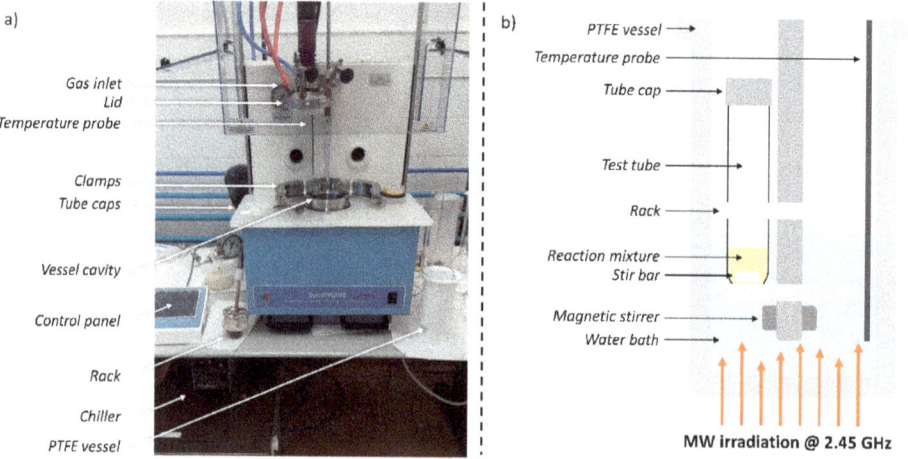

Figure 1. (a) Photograph of the microwave (MW) reactor, and (b) scheme of the reaction vessel during operation.

2.5. Analysis of the Reaction Mixtures

Samples were prepared from the final reaction mixtures by adding 100 µL of an internal standard (toluene, which is not formed in the reaction and is not reactive towards any of the reaction components), and then performing 1:5 dilutions with acetone to reduce the concentration of the reaction species and achieve better peak shapes and resolution in the GC analysis. The obtained samples were analyzed with a GC-2010 (Shimadzu, Kyoto, Japan) using a TRB-5 column (TR-120232, Teknokroma, Sant Cugat del Vallès, Spain; length: 30 m; film thickness: 25 µm; inner diameter: 0.25 mm). The product concentration was estimated from the calibration curve obtained for furfural, using the effective carbon number as a correction factor [31]. The conversion X of furfural was calculated as follows:

$$X_{FUR} = \frac{n^0_{FUR} - n_{FUR}}{n^0_{FUR}} \cdot 100 \tag{1}$$

The yields, Y, of the various products and intermediates were calculated as follows:

$$Y_{C8} = \frac{n_{C8}}{n^0_{FUR}} \cdot 100 \quad Y_{C8-OH} = \frac{n_{C8-OH}}{n^0_{FUR}} \cdot 100 \quad Y_{C13} = \frac{2 \cdot n_{C13}}{n^0_{FUR}} \cdot 100 \tag{2}$$

The C8:C13 selectivity was defined as the ratio between the yields of the mono- and the double-condensation products. With our protocol, the standard deviations of both conversions and yields are below 5%. In addition, GC-MS was used to aid peak assignments. The samples were analyzed with a GCMS-QP-2010 (Shimadzu) using a Zebron ZB-5ms column (Phenomenex, Torrance, CA, USA). The results of these analyses can be found in the Supplementary Materials (Figure S2).

3. Results and Discussion

3.1. Catalyst Characterization

The catalysts were analyzed by PXRD in order to characterize their crystalline structure (Figure 2).

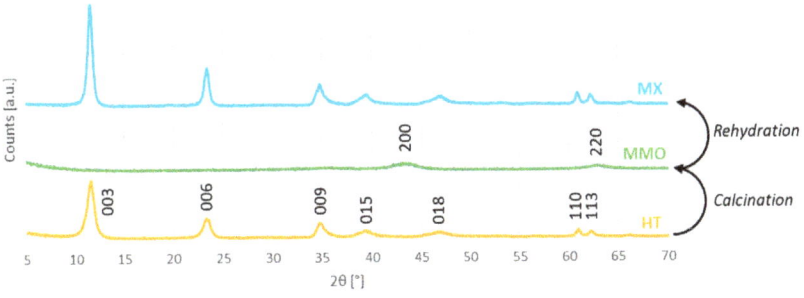

Figure 2. X-ray diffractograms of the prepared catalysts.

The original, as-synthesized HT catalyst has the typical hydrotalcite structure, that is, a layered double hydroxide structure of Mg–Al mixed hydroxides organized in brucite-like layers containing water and carbonate anions in the interlayer [32]: sharp (003), (006), (009), (110), and (113) diffraction lines, and two broad (015) and (018) ones. The crystalline structure of hydrotalcites have a high morphological anisotropy because of the layered structure. Hence, it is common to report the crystallite size for every <hkl> direction (Table 1).

Table 1. Results of the ICP-AES analysis of the prepared catalysts.

HT			MMO	MX		
2θ [°]	<hkl>	Size [nm]	Size [nm]	2θ [°]	<hkl>	Size [nm]
11.75	003	9.16		11.77	003	17.71
23.46	006	7.95		23.50	006	14.06
34.96	009	8.33		34.93	009	9.42
39.40	015	4.11	2.48	39.43	015	4.90
46.84	018	3.04		46.90	018	3.63
60.90	110	23.43		60.84	110	33.18
62.21	113	13.64		62.17	113	20.23
Lattice parameters [Å]						
$a = 3.05$ $c = 22.83$			$a = 4.19$	$a = 3.05$ $c = 22.74$		

The calculated lattice parameter c corresponds to the basal spacing of the layered double hydroxide; dividing the crystallite size in the 003 direction by c gives the number of layers in one crystallite, corresponding to about four in this case. The calcination of HT causes the collapse of the layered double hydroxide structure. As a consequence, the MMO X-ray diffractogram does not include any of the aforementioned reflections, but instead presents two broad (200) and (220) diffraction lines, corresponding to the MgO cubic phase (periclase) [33]. Immersion of MMO in aqueous media enables the double hydroxide structure layer to be reconstructed, thanks to the memory properties of these oxides which, for this reason, are also called layered double oxides. XRD analysis of MX does indeed confirm that the rehydration has induced the reconstruction of the original layered double hydroxide structure. As the rehydration was carried out in an inert atmosphere, the calcination–rehydration cycle has the overall effect of replacing the initial carbonate anions with hydroxyls, which enhance the Brønsted basicity of the catalyst. Even so, this anion exchange does not seem to have a notable effect on the diffractograms. Nevertheless, the lattice parameter c slightly changes as a result of the different interlayer composition. Moreover, the layered double hydroxide obtained after calcination and rehydration appears to have a bigger crystallite size in all directions, indicating agglomeration of the layers: the number of layers per crystallite is now about eight, approximately double the value obtained with HT (Table 1).

Next, the catalysts were analyzed by ICP-AES to obtain the atomic composition of the prepared materials. The information of interest was the final Mg: Al ratio in the solids and the Na content, which may be due to incomplete HT washing after the co-precipitation. In particular, this parameter is known to have a remarkable influence on the catalytic performance of the solids [34]. Table 2 summarizes the results of the ICP-AES analyses.

Table 2. Results of the ICP-AES analysis of the prepared catalysts.

Entry	Catalyst	Mg [w/w%]	Al [w/w%]	Na [w/w%]	Mg/Al [at/at]
1	HT	21.0	11.2	n.d	2.07
2	MMO	34.7	18.6	0.06	2.07
3	MX	20.6	11.1	n.d.	2.06

The atomic ratio of Mg and Al matches the theoretical one (2) in all cases. The Na content of the original HT sample is below the detection limit (Table 2, entry 1). The Na, Mg, and Al concentrations increase with calcination as a result of the loss of water, hydroxyls, and carbonates, although their percentage is still very low (Table 2, entry 2). On rehydration, the introduction of water and hydroxyls in the solid pushes the Na content back to below the detection limits (Table 2, entry 3). It can therefore be argued that Na has little or no effect on the aldol condensation results, which will be discussed in the sections below, since the sodium content in HT is below the level of that which is considered to have an impact on the catalytic activity (0.04%).

TGA-DSC was used to study the non-metal composition of the catalysts, and the degradation behavior of the solids. The results of the analysis are reported in Figure 3.

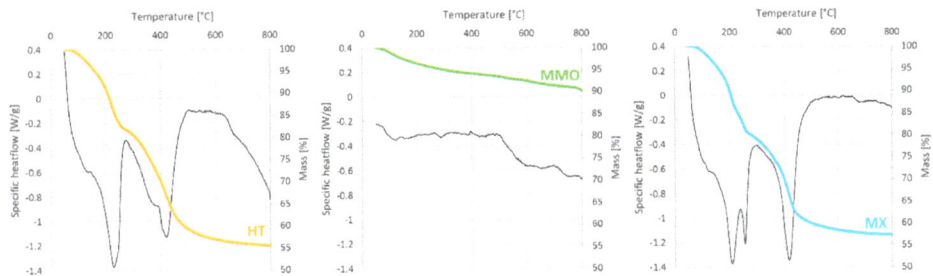

Figure 3. TGA (colored lines) and DSC (black lines, exo up) curves of the prepared catalysts.

The thermogravimetric curve obtained for HT fits the previously reported ones well [25]. The curve points to two distinct mass loss events: one of these is at a DSC negative peak at about 200 °C, which has been attributed to the loss of physisorbed and inter-layer water; the other is at about 400 °C, and should correspond to decarboxylation and dehydroxylation events. The MMO was analyzed as a freshly calcined material, so little or no mass loss was expected. As anticipated, there was no distinct mass loss, and the sample mass drifted slowly to 90% of its original value. As for MX, the profile very much resembled HT's. Yet there were some differences: for example, the different mass loss of water, and the different shape of the second mass loss step, which is probably due to the absence (or at least a reduced quantity) of carbonates in the structure. These observations are in agreement with previously reported data [25].

N_2 physisorption was used to obtain the nitrogen adsorption–desorption isotherms, Brunauer–Emmett–Teller (BET) surface area, and pore volume of the catalysts. The results obtained are summarized in Figure 4 and Table 3.

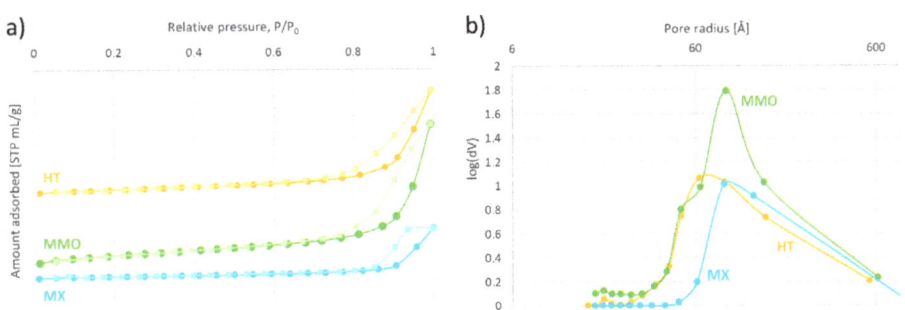

Figure 4. (a) Isotherms of N_2 adsorption–desorption, and (b) Barrett–Joyner–Halenda (BJH) pore size distributions of the prepared catalysts.

Table 3. Results of the N_2 physisorption analysis of the prepared catalysts.

Entry	Catalyst	S_{BET} [m^2/g]	V_{pore} [mL/g]	r_{pore} [Å]
1	HT	120.605	0.719	62.321
2	MMO	256.880	0.951	87.679
3	MX	45.739	0.353	85.812

Generally, the results of N_2 physisorption analysis of HT and related materials very much depend on the procedure followed for the preparation and the sample degassing [25,34–36]. The shape of the HT isotherm resembles the shape of type III (Figure 4a), which commonly corresponds to macroporous materials (even though HT has pores in the mesopore range (Table 3, entry 1)) with weak adsorbent–adsorbate interactions [37]. The same considerations apply to the isotherms of MMO and MX (Table 3, entries 2 and 3, respectively). As for adsorption hysteresis, the loops of HT and MMO seem to be type H3, which is associated with clays made of non-rigid aggregates of plate-like particles [37]. On the other hand, the hysteresis loop of MX appears to be more pronounced and seems to correspond to type H2b, which indicates pore blocking. The BET area of the prepared HT is quite high (Table 3, entry 1), and calcination of the hydrotalcite is supposed to increase this value [25,34,36], which is indeed the case for our MMO (Table 3, entry 2). The surface area of MX depends on the preparation conditions [36], and in our case, it was lower than the area of both its parent catalysts (Table 3, entry 3). Moreover, the Barrett–Joyner–Halenda (BJH) results indicated that while HT and MMO are overall quite porous, the porosity of MX is considerably lower.

3.2. Evaluation of the Catalytic Performance of HT at Different Concentrations

The as-synthesized HT was tested as a catalyst in the microwave-assisted aldol condensation of acetone and furfural. In all the GC-FID chromatograms obtained after analysis, only three products were present in detectable amounts: the mono-aldol product, C8-OH, the mono-condensation product, C8, and the double-condensation product, C13. The only other product detected was diacetone alcohol, the product of the self-aldolization of acetone, which was produced in small amounts. Since, in our system, acetone was present in large excess, the formation of diacetone alcohol did not compromise the selectivity of the reaction. However, the presence of this side product in the final reaction mixtures means that a purification may be required at the end of a potential industrial process.

The concentration of furfural is the first parameter that was investigated. We expected that the different acetone:furfural ratio would have an effect on the C8:C13 selectivity. The chosen reaction temperature was 100 °C. This temperature was reached in about 10 min by irradiating the water bath, and it was then kept constant for 20 min. The central value of concentration was 0.1 g of furfural in 5 mL of acetone (0.02 g/mL, 0.21 M), with a furfural:HT weight ratio of 2:1. Two other concentrations were explored, namely 0.11 and 0.42 M, corresponding to half and double the central value, respectively. The volume of acetone and the furfural:HT ratio was kept constant for all experiments. Figure 5 reports the results of the concentration optimization.

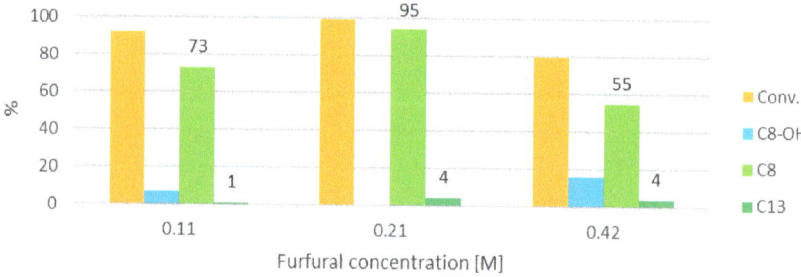

Figure 5. Influence of furfural concentration (at 100 °C, 30 min, furfural:HT 2:1).

The best concentration in terms of activity appears to be 0.21 M. The reaction is slightly slower at lower concentration, perhaps as a result of a slower dehydration, whereas the lower conversion obtained at 0.42 M may be the result of a lower irradiation power per mass of catalyst, which may influence the local temperature of the solid, so fewer or less intense hotspots are formed in the catalyst. Hence, the central concentration value was chosen for further optimization. As expected, the C8:C13 selectivity seems to decrease when the concentration is increased from 54 to 22 to 14. This means that although it is possible to obtain almost complete selectivity to C8 with a somewhat lower concentration, it is not feasible in an HT-catalyzed neat reaction to shift the selectivity to C13, because that would require an excess of furfural and, therefore, to use a solvent.

3.3. Influence of the Catalyst Quantity

Next, the effect of decreasing the furfural:HT ratio on the reaction outcome was studied (Figure 6).

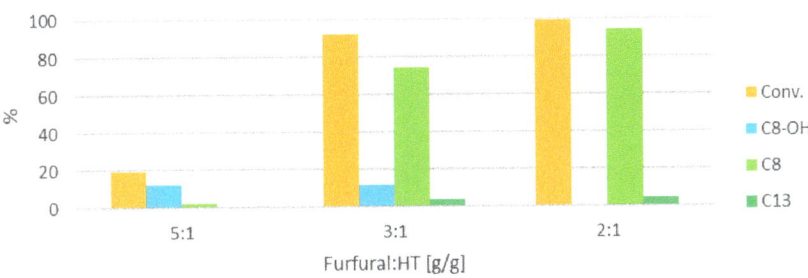

Figure 6. Influence of the furfural:HT ratio (at 100 °C, 30 min, concentration 0.21 M).

The reaction rate decreases dramatically when the furfural:HT ratio is increased to 5:1, and only traces of the condensation products are detected. However, with an intermediate ratio of 3:1, the activity is still fairly high, although a consistent part of the converted furfural is still present as the intermediate C8-OH.

3.4. Influence of the Reaction Time

Then, the formation of the various products over time was studied (Figure 7).

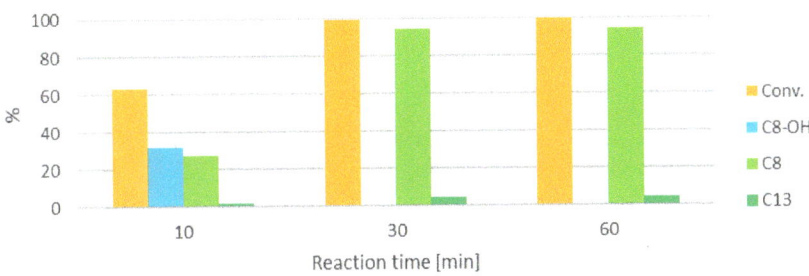

Figure 7. Influence of the reaction time (at 100 °C, furfural:HT 2:1, 0.21 M).

The conversion increases consistently with the reaction time. After 10 min, the major product is C8-OH, which disappears after 30 min to make way for C8. Aldol condensations are reversible reactions, so even after complete conversion has been reached, it could still be possible to obtain C13 from C8, even if formally furfural is absent, especially when one product is more stable than the other. Nevertheless, no such transformation is seen, and the composition of the mixture remains unaltered for 30 min after complete conversion. In addition, this test shows how the condensation products are substantially stable, and do not deteriorate in the reaction conditions.

3.5. Influence of the Temperature

The reaction temperature is the next parameter that was studied (Figure 8).

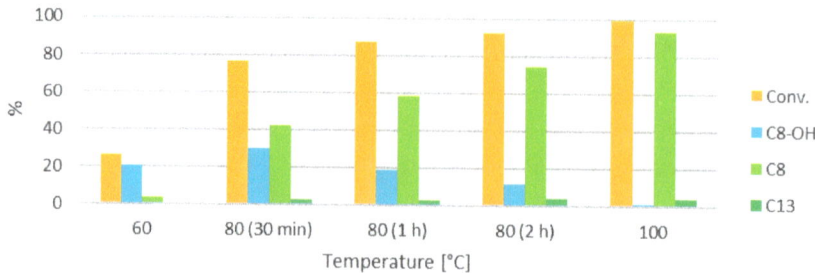

Figure 8. Influence of the reaction temperature (30 min, furfural:HT 2:1, 0.21 M).

The reaction at 60 °C is considerably slower than the reaction at 100 °C. Only traces of condensation products are observed. At 80 °C, an intermediate level of conversion is obtained, and the dehydration also appears to be favored by the higher temperature. However, the reaction at this temperature is still much slower than at 100 °C, and even when the reaction time is increased to 2 h, the conversion does not match the conversion in the experiment at 100 °C for 30 min. This might also be associated with the lower irradiation power. The C8:C13 ratio does not appear to be affected by the lower temperature.

3.6. Choice of the Catalyst

A blank test was performed to confirm the role of the catalyst in the transformation. Indeed, no conversion and formation of product was seen in our reaction conditions. Other HT-derived catalysts were tested: HT was calcined to MMO, which was tested as a catalyst (Figure 9).

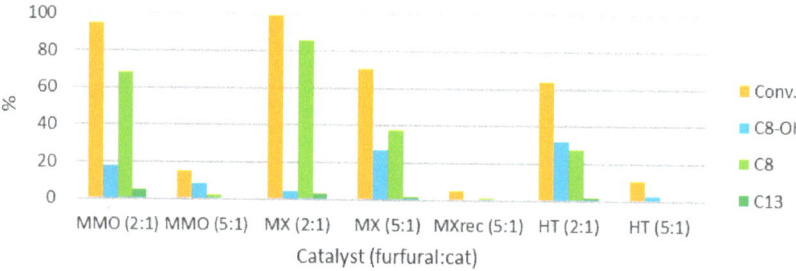

Figure 9. Influence of the catalyst choice (at 100 °C, 10 min, 0.21 M).

MMO proved to be a more active catalyst than HT, in terms of both furfural conversion and dehydration of C8-OH. As seen for HT, the MMO activity decreases if a 5:1 furfural:catalyst ratio is used. MMO was rehydrated in the liquid phase, and the solid collected after this modification, MX, was used as a catalyst for the condensation. MX was the most active of the catalysts tested, with conversion reaching >99% by the end of the heating step of only 10 min, and C8-OH dehydration being almost complete. Remarkably, the performance of this catalyst at a furfural:catalyst ratio of 5:1 was greater than when HT was used at the same ratio, and even at a 2:1 ratio, in the same conditions. Interestingly, the C8:C13 selectivity obtained with MMO as a catalyst was lower than when the parent catalyst was used (13 vs. 22). The selectivity obtained with MX was not significantly different (26). The solid in the mixture of an MX-catalyzed reaction was recovered by filtration, washed with abundant acetone, and then used as a catalyst in a subsequent reaction. This catalyst (MXrec) proved to be basically inactive; differently from the parent catalyst, which is a white solid, MXrec is a brown powder. Apparently, after reaction, organic matter is deposited on the catalyst surface and deactivates the catalyst.

MMO and MX were also tested at lower temperatures (Figure 10).

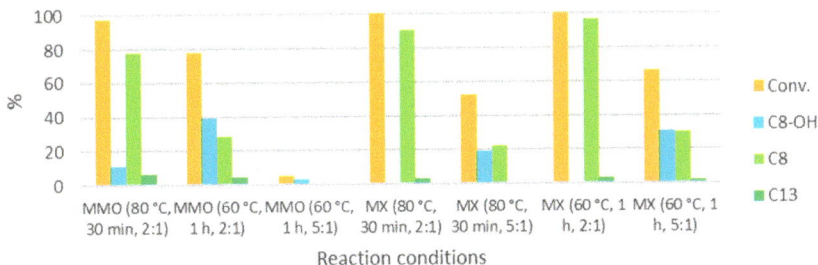

Figure 10. Performances of the various catalysts at lower temperatures (0.21 M).

The conversion obtained with MMO at 80 °C is comparable to the conversion obtained at 100 °C, although the reaction time is much longer (30 min vs. 10). The same applies to the C8:C13 selectivity (13 in both cases). Decreasing the reaction temperature to 60 °C leads to a somewhat lower conversion, even with a reaction time as long as 1 h. Also, in this case, the activity decreases dramatically when the furfural:MMO ratio is decreased to 5:1, as was observed for the reaction at 100 °C. When using MX at 80 °C for 30 min the activity decreases, but with a 5:1 ratio, about half of the furfural can still be converted. When the same reaction is performed at 60 °C for 1 h, the conversion increases slightly. Using a 2:1 furfural:MX ratio leads to complete conversions and C8-OH dehydrations at these lower temperatures.

Differently from what has previously been reported with conventional heating (almost complete conversion at 100 °C in 2 h, excluding the pre-heating, with a catalyst analogous to MX at a 3.25:1 furfural:MX ratio [27]), in our MW-assisted process it was possible to obtain a 70% conversion with a 5:1 furfural:MX ratio with 10 min irradiation in the heating step to the same temperature. Likely, it is the interaction of microwaves with the solid catalysts that is the source of this activity boost. The reaction is also much faster than the reactions catalyzed by the most common heterogeneous catalysts (e.g., Mg–Al and Mg–Zr) in the aqueous phase [18].

4. Conclusions

We developed a microwave-assisted protocol to convert furfural and acetone very quickly into their aldol reaction (C8-OH) and condensation products (C8 and C13). The catalysts used are easy to prepare and to separate from the reaction mixture, they are not corrosive, and only contain earth-abundant elements. Our neat process is selective for the preparation of C8, which is achieved with satisfactory C8:C13 selectivity. This selectivity can be increased even further if the furfural concentration is decreased, whereas shifting the selectivity towards C13 requires the use of a solvent (or of a different catalyst). MX is by far the most active catalyst of the ones tested, followed by MMO and HT. Although this reactivity trend has been observed in many other reports, HT activity has not. HTs are usually reported to be inactive, or they are not even considered for catalytic testing. The formation of hotspots in the HT structure caused by MW irradiation may account for the atypical activity of this catalyst. The reaction also works well at temperatures lower than 100 °C but, of course, a much longer reaction time is required.

Supplementary Materials: The following are available online at http://www.mdpi.com/2073-4352/10/9/833/s1: Figure S1: Parameter profiles in a typical microwave experiment; Figure S2: Mass spectra of the various chemical species detected in the final reaction mixtures by GC-MS.

Author Contributions: Investigation: A.T. and M.L.; writing—original draft preparation: A.T.; writing—review and editing: F.M. and M.C.; supervision: F.M. All authors have read and agreed to the published version of the manuscript.

Funding: A.T. thanks AGAUR (Generalitat de Catalunya) and ESF (European Union) for his postgraduate scholarship (2018 FI_B 01124). F.M. and M.C. thank the Ministerio de Economía y Competitividad for financial support (RTI2018-098310-B-I00).

Acknowledgments: S. Dominguez and the technicians of the Servei de Recursos Científics i Tècnics of URV are gratefully acknowledged for assistance in the characterizations.

Conflicts of Interest: The authors declare no conflict of interest. The funders had no role in the design of the study; in the collection, analyses, or interpretation of data; in the writing of the manuscript; or in the decision to publish the results.

References

1. Capellán-Pérez, I.; Mediavilla, M.; De Castro, C.; Carpintero, Ó.; Miguel, L.J. Fossil fuel depletion and socio-economic scenarios: An integrated approach. *Energy* **2014**, *77*, 641–666. [CrossRef]
2. Levi, P.G.; Cullen, J.M. Mapping Global Flows of Chemicals: From Fossil Fuel Feedstocks to Chemical Products. *Environ. Sci. Technol.* **2018**, *52*, 1725–1734. [CrossRef] [PubMed]
3. Bogdanov, D.; Farfan, J.; Sadovskaia, K.; Aghahosseini, A.; Child, M.; Gulagi, A.; Oyewo, A.S.; De Souza Noel Simas Barbosa, L.; Breyer, C. Radical transformation pathway towards sustainable electricity via evolutionary steps. *Nat. Commun.* **2019**, *10*, 1077. [CrossRef] [PubMed]
4. Falkowski, P.; Scholes, R.J.; Boyle, E.; Canadell, J.; Canfield, D.; Elser, J.; Gruber, N.; Hibbard, K.; Högberg, P.; Linder, S.; et al. The global carbon cycle: A test of our knowledge of earth as a system. *Science* **2000**, *290*, 291–296. [CrossRef] [PubMed]
5. Corma, A.; Iborra, S.; Velty, A. Chemical routes for the transformation of biomass into chemicals. *Chem. Rev.* **2007**, *107*, 2411–2502. [CrossRef] [PubMed]
6. Höök, M.; Tang, X. Depletion of fossil fuels and anthropogenic climate change—A review. *Energy Policy* **2013**, *52*, 797–809. [CrossRef]
7. Ahorsu, R.; Medina, F.; Constantí, M. Significance and Challenges of Biomass as a Suitable Feedstock for Bioenergy and Biochemical Production: A Review. *Energies* **2018**, *11*, 3366. [CrossRef]
8. Wang, W.-C.; Tao, L. Bio-jet fuel conversion technologies. *Renew. Sustain. Energy Rev.* **2016**, *53*, 801–822. [CrossRef]
9. West, R.M.; Liu, Z.Y.; Peter, M.; Dumesic, J.A. Liquid alkanes with targeted molecular weights from biomass-derived carbohydrates. *ChemSusChem* **2008**, *1*, 417–424. [CrossRef]
10. Gruber, H.; Groß, P.; Rauch, R.; Reichhold, A.; Zweiler, R.; Aichernig, C.; Müller, S.; Ataimisch, N.; Hofbauer, H. Fischer-Tropsch products from biomass-derived syngas and renewable hydrogen. *Biomass Conv. Biorefin.* **2019**, *48*, 22. [CrossRef]
11. Zang, H.; Wang, K.; Zhang, M.; Xie, R.; Wang, L.; Chen, E.Y.-X. Catalytic coupling of biomass-derived aldehydes into intermediates for biofuels and materials. *Catal. Sci. Technol.* **2018**, *8*, 1777–1798. [CrossRef]
12. Sutton, A.D.; Waldie, F.D.; Wu, R.; Schlaf, M.; Silks, L.A.P.; Gordon, J.C. The hydrodeoxygenation of bioderived furans into alkanes. *Nat. Chem.* **2013**, *5*, 428–432. [CrossRef] [PubMed]
13. Chang, H.; Motagamwala, A.H.; Huber, G.W.; Dumesic, J.A. Synthesis of biomass-derived feedstocks for the polymers and fuels industries from 5-(hydroxymethyl)furfural (HMF) and acetone. *Green Chem.* **2019**, *21*, 5532–5540. [CrossRef]
14. Mamman, A.S.; Lee, J.-M.; Kim, Y.-C.; Hwang, I.T.; Park, N.-J.; Hwang, Y.K.; Chang, J.-S.; Hwang, J.-S. Furfural: Hemicellulose/xylosederived biochemical. *Biofuels Bioprod. Biorefin. Innov. Sustain. Econ.* **2008**, *2*, 438–454. [CrossRef]
15. Mariscal, R.; Maireles-Torres, P.; Ojeda, M.; Sádaba, I.; López Granados, M. Furfural: A renewable and versatile platform molecule for the synthesis of chemicals and fuels. *Energy Environ. Sci.* **2016**, *9*, 1144–1189. [CrossRef]
16. Kelly, G.J.; King, F.; Kett, M. Waste elimination in condensation reactions of industrial importance. *Green Chem.* **2002**, *4*, 392–399. [CrossRef]
17. Huber, G.W.; Chheda, J.N.; Barrett, C.J.; Dumesic, J.A. Production of liquid alkanes by aqueous-phase processing of biomass-derived carbohydrates. *Science* **2005**, *308*, 1446–1450. [CrossRef]
18. Faba, L.; Díaz, E.; Ordóñez, S. Aqueous-phase furfural-acetone aldol condensation over basic mixed oxides. *Appl. Catal. B Environ.* **2012**, *113–114*, 201–211. [CrossRef]
19. Desai, D.S.; Yadav, G.D. Green Synthesis of Furfural Acetone by Solvent-Free Aldol Condensation of Furfural with Acetone over La_2O_3–MgO Mixed Oxide Catalyst. *Ind. Eng. Chem. Res.* **2019**, *58*, 16096–16105. [CrossRef]

20. Smoláková, L.; Dubnová, L.; Kocík, J.; Endres, J.; Daniš, S.; Priecel, P.; Čapek, L. In-situ characterization of the thermal treatment of Zn-Al hydrotalcites with respect to the formation of Zn/Al mixed oxide active in aldol condensation of furfural. *Appl. Clay Sci.* **2018**, *157*, 8–18. [CrossRef]
21. Kikhtyanin, O.; Chlubná, P.; Jindrová, T.; Kubička, D. Peculiar behavior of MWW materials in aldol condensation of furfural and acetone. *Dalton Trans.* **2014**, *43*, 10628–10641. [CrossRef]
22. Nishimura, S.; Takagaki, A.; Ebitani, K. Characterization, synthesis and catalysis of hydrotalcite-related materials for highly efficient materials transformations. *Green Chem.* **2013**, *15*, 2026. [CrossRef]
23. Lari, G.M.; De Moura, A.B.L.; Weimann, L.; Mitchell, S.; Mondelli, C.; Pérez-Ramírez, J. Design of a technical Mg–Al mixed oxide catalyst for the continuous manufacture of glycerol carbonate. *J. Mater. Chem. A* **2017**, *5*, 16200–16211. [CrossRef]
24. Cueto, J.; Faba, L.; Díaz, E.; Ordóñez, S. Performance of basic mixed oxides for aqueous-phase 5-hydroxymethylfurfural-acetone aldol condensation. *Appl. Catal. B Environ.* **2017**, *201*, 221–231. [CrossRef]
25. Abelló, S.; Medina, F.; Tichit, D.; Pérez-Ramírez, J.; Groen, J.C.; Sueiras, J.E.; Salagre, P.; Cesteros, Y. Aldol condensations over reconstructed Mg-Al hydrotalcites: Structure-activity relationships related to the rehydration method. *Chem. Eur. J.* **2005**, *11*, 728–739. [CrossRef]
26. Abelló, S.; Vijaya-Shankar, D.; Pérez-Ramírez, J. Stability, reutilization, and scalability of activated hydrotalcites in aldol condensation. *Appl. Catal. A General* **2008**, *342*, 119–125. [CrossRef]
27. Hora, L.; Kelbichová, V.; Kikhtyanin, O.; Bortnovskiy, O.; Kubička, D. Aldol condensation of furfural and acetone over MgAl layered double hydroxides and mixed oxides. *Catal. Today* **2014**, *223*, 138–147. [CrossRef]
28. Horikoshi, S.; Schiffmann, R.F.; Fukushima, J.; Serpone, N. *Microwave Chemical and Materials Processing*; Springer: Singapore, 2018; ISBN 978-981-10-6465-4.
29. Kokel, A.; Schäfer, C.; Török, B. Application of microwave-assisted heterogeneous catalysis in sustainable synthesis design. *Green Chem.* **2017**, *19*, 3729–3751. [CrossRef]
30. Zeitsch, K.J. The Discoloration of Furfural. In *the Chemistry and Technology of Furfural and Its Many by-Products*; Elsevier: Amsterdam, The Netherlands, 2000; pp. 28–33. ISBN 9780444503510.
31. Scanlon, J.T.; Willis, D.E. Calculation of Flame Ionization Detector Relative Response Factors Using the Effective Carbon Number Concept. *J. Chromatogr. Sci.* **1985**, *23*, 333–340. [CrossRef]
32. Mokhtar, M.; Inayat, A.; Ofili, J.; Schwieger, W. Thermal decomposition, gas phase hydration and liquid phase reconstruction in the system Mg/Al hydrotalcite/mixed oxide: A comparative study. *Appl. Clay Sci.* **2010**, *50*, 176–181. [CrossRef]
33. Barriga, C.; Gaitán, M.; Pavlovic, I.; Ulibarri, M.A.; Hermosı́n, M.C.; Cornejo, J. Hydrotalcites as sorbent for 2,4,6-trinitrophenol: Influence of the layer composition and interlayer anion. *J. Mater. Chem.* **2002**, *12*, 1027–1034. [CrossRef]
34. Abelló, S.; Medina, F.; Tichit, D.; Pérez-Ramírez, J.; Rodríguez, X.; Sueiras, J.E.; Salagre, P.; Cesteros, Y. Study of alkaline-doping agents on the performance of reconstructed Mg–Al hydrotalcites in aldol condensations. *Appl. Catal. A Gen.* **2005**, *281*, 191–198. [CrossRef]
35. Galindo, R.; López-Delgado, A.; Padilla, I.; Yates, M. Hydrotalcite-like compounds: A way to recover a hazardous waste in the aluminium tertiary industry. *Appl. Clay Sci.* **2014**, *95*, 41–49. [CrossRef]
36. Xu, C.; Gao, Y.; Liu, X.; Xin, R.; Wang, Z. Hydrotalcite reconstructed by in situ rehydration as a highly active solid base catalyst and its application in aldol condensations. *RSC Adv.* **2013**, *3*, 793–801. [CrossRef]
37. Thommes, M.; Kaneko, K.; Neimark, A.V.; Olivier, J.P.; Rodriguez-Reinoso, F.; Rouquerol, J.; Sing, K.S.W. Physisorption of gases, with special reference to the evaluation of surface area and pore size distribution (IUPAC Technical Report). *Pure Appl. Chem.* **2015**, *87*, 1051–1069. [CrossRef]

© 2020 by the authors. Licensee MDPI, Basel, Switzerland. This article is an open access article distributed under the terms and conditions of the Creative Commons Attribution (CC BY) license (http://creativecommons.org/licenses/by/4.0/).

Article

Green Synthesis of Hydrocalumite (CaAl-OH-LDH) from Ca(OH)₂ and Al(OH)₃ and the Parameters That Influence Its Formation and Speciation

Bianca R. Gevers * and Frederick J.W.J. Labuschagné

Institute of Applied Materials, Department of Chemical Engineering, University of Pretoria, Pretoria 0002, South Africa; johan.labuschagne@up.ac.za
* Correspondence: bianca.gevers@tuks.co.za

Received: 14 July 2020; Accepted: 1 August 2020; Published: 3 August 2020

Abstract: Hydrocalumite is a layered double hydroxide (LDH) that is finding increased application in numerous scientific fields. Typically, this material is produced through environmentally polluting methods such as co-precipitation, sol-gel synthesis and urea-hydrolysis. Here, the hydrothermal green (environmentally friendly) synthesis of hydrocalumite (CaAl-OH) from Ca(OH)₂ and Al(OH)₃ in water and the parameters that influence its formation are discussed. The parameters investigated include the reaction temperature, reaction time, molar calcium-to-aluminium ratio, the morphology/crystallinity of reactants used, mixing and the water-to-solids ratio. Hydrocalumite formation was favoured in all experiments, making up between approximately 50% and 85% of the final crystalline phases obtained. Factors that were found to encourage higher hydrocalumite purity include a low water-to-solids ratio, an increase in the reaction time, sufficient mixing, the use of amorphous Al(OH)₃ with a high surface area, reaction at an adequate temperature and, most surprisingly, the use of a calcium-to-aluminium ratio that stoichiometrically favours katoite formation. X-ray diffraction (XRD) and Rietveld refinement were used to determine the composition and crystal structures of the materials formed. Scanning electron microscopy (SEM) was used to determine morphological differences and Fourier-transform infrared analysis with attenuated total reflectance (FTIR-ATR) was used to identify possible carbonate contamination, inter alia. While the synthesis was conducted in an inert environment, some carbonate contamination could not be avoided. A thorough discussion on the topic of carbonate contamination in the hydrothermal synthesis of hydrocalumite was given, and the route to improved conversion as well as the possible reaction pathway were discussed.

Keywords: HC; hydrothermal synthesis; layered double hydroxide; AFm phase; calcium hemicarboaluminate; cement phases; cement hydration; C_3AH_6; $C_4A\tilde{C}H_{11}$; katoite

1. Introduction

Hydrocalumite (HC) is a CaAl-LDH with the general formula

$$[Ca_{1-x}Al_x(OH)_2][X_{x/q}^{q-} \cdot nH_2O] \tag{1}$$

where x is the ratio of trivalent to total cations in the layered double hydroxide (LDH) lattice, X is the interlayer anion, q its charge and n the amount of water present in the interlayer [1]. As a result of its corrugated-iron-like structure, the ratio of Ca:Al is limited to 2:1 ($x = 0.3\bar{3}$) [2]. In nature, HC occurs with chloride and hydroxide anions in the interlayer, but many other anions can be intercalated.

HC is used in numerous scientific fields, including catalysis [3–5], sensors [6,7], medical applications [8,9], environmental remediation [10–12], agriculture [13], polymers [6,14,15] and occurs in cementitious phases during curing [16]. As with other LDHs for study in applications, this material is typically produced using co-precipitation—the most widely used synthesis technique for LDHs [1,17]. While co-precipitation has many advantages, such as simplicity, speed and high tailorability of the materials produced, it is also one of the most polluting synthesis methods available to produce LDHs, causing large amounts of salt-rich waste water.

Co-precipitation synthesis utilises a mixture of metal salts (typically metal chlorides, nitrates or sulphates) and a base (frequently NaOH or KOH) which are added dropwise to a beaker, typically containing the anion to be intercalated. While many derivatives of this method exist, the basic concept remains. During synthesis, LDH precipitates out of solution. After synthesis, this precipitate must be filtered from the resulting slurry to further process the material. During filtering, large amounts of water are used to wash the filtrate liquor—that would otherwise dry with the LDH filter cake and contaminate the final product—out of the material. Depending on the chemicals used, the resulting filtrate can be rich in sodium or potassium and chlorides, nitrates or sulphates and, of course, excess intercalant ions. In addition to large-scale environmental pollution associated with these untreated waste-streams (if released to the environment), the chemicals required for the synthesis of LDHs are expensive. Considering worsening environmental pollution and the drive to reduce the impact of the chemical industry on the environment, it has become evermore important to find alternative syntheses for these materials.

There exist a multitude of synthesis routes to produce LDHs. Urea hydrolysis and sol-gel syntheses are often used. These methods use urea (for urea hydrolysis) or metal alkoxides, alcohol and acids as chelating agents (for sol-gel synthesis). While these are not environmentally friendly alternatives to co-precipitation, attempts have been made to make sol-gel synthesis more environmentally-friendly [18]. In addition to these, there exist less frequently used methods that can be used in an environmentally friendly manner, such as hydrothermal or mechanochemical synthesis. In both methods, metal hydroxides or metal oxides are common starting materials that are mixed with water and processed at elevated temperatures and pressures (hydrothermal method) or milled together (mechanochemical synthesis). Use of the metal hydroxides and oxides as starting materials is hereby key to producing less polluting waste streams. In previous work, we have shown that the hydrothermal process can produce very pure phases of MgAl-LDH, even with a recycle-based system that reduces waste-streams [19]. It has been shown that mechanochemical synthesis (wet-milling) of $Ca(OH)_2$ and $Al(OH)_3$ could lead to an HC-phase content similar to that achieved in this work [20].

HC has been synthesised using a hydrothermal method in previous works, with a great interest in the effects of the presence of $CaO/Ca(OH)_2$ and $Al(O)OH/Al(OH)_3/Al_2O_3$ in presence of water and CO_2 on the curing of cement [16,21–23]. In fact, some of the earliest reports concerning the study of the material and its characterisation used a hydrothermal synthesis [22,24,25]. HC is especially well suited to hydrothermal synthesis because of the metal oxides and hydroxides that can be used at mild conditions to produce the desired phase. In these early reports of hydrothermal HC synthesis and related studies, HC (also frequently referred to in cement and concrete literature as tetracalcium monocarboaluminate, a carbonate intercalated CaAl-LDH) was synthesised using several approaches. Ref. [25] synthesised HC by reacting $Ca(OH)_2$, $Al(OH)_3$ and $CaCO_3$ in water for one month at 2 kbar and at 100 °C. Ref. [26] prepared HC using different sources of aluminium (gibbsite and boehmite) and $CaCO_3$ (with different surface areas and particle sizes), and $Ca(OH)_2$ at varying temperatures and reacting the mixture for 24 h to 48 h. They also did a small study on the effect of temperatures between 70 °C and 90 °C, and found that the largest fraction of HC is formed at 80 °C. It was found that the reaction temperature significantly affects the product formed, being HC or katoite ($Ca_3Al_2(OH)_{12}$). In our own study of the effect of temperatures between 30 °C and 90 °C on the hydrothermal synthesis of HC using $Ca(OH)_2$ and highly crystalline $Al(OH)_3$, similar results were obtained [27]. Ref. [28] investigated the crystal structure and phase transitions in HC from −115 °C to 45 °C. They used $Ca(OH)_2$, $Al(OH)_3$, $CaCl_2 \cdot 6H_2O$ and $CaCO_3$ to create single crystals at 120 °C and 2 kbar in two months. They prepared powder HC samples by reacting $Ca_3Al_2O_6$, $CaCl_2 \cdot 6H_2O$ and $CaCO_3$ in water at room temperature and under inert atmosphere for four weeks. Ref. [29] synthesised HC using CaO, Al_2O_3 and K_2CO_3 within an hour at 100 °C with microwave irradiation assistance. There also exist several thermodynamic studies [21–24] and studies concerning pressure-induced reactions [16] in the system $CaO-Al_2O_3-H_2O$. These studies have shown that the species formed can be similar to HC [24]. Ref. [3] used a carbonate-free approach for the synthesis of HC by reacting $Ca(OH)_2$ and Al(O)OH in water under an inert atmosphere at 80 °C for 3 h. However, it was shown that the calcium hydroxide used for synthesis was contaminated with calcite. No FTIR analysis was conducted to determine the interlayer anions, but preparation of the HC-like compounds with nitric acid led to the formation of a pure HC-like phase similar to nitrate intercalated HC.

From the above literature it is evident that several attempts on the hydrothermal synthesis of HC have been made; however, most of these attempts were either time-intensive or time- and energy-intensive, because of the long reaction times, temperatures and pressures required. Further, most work has focused on the synthesis of a carbonate-intercalated LDH. To create a $CaAl-CO_3$-LDH, a carbonate source is required during synthesis. In others works, this was achieved using $CaCO_3$ or K_2CO_3. However, as carbonate uptake is favoured, it is difficult to reverse. As many HC applications have anion-exchange reactions as a subsequent step or end goal, the presence of carbonate is frequently undesirable. Removing the interlayer carbonate, usually involves calcination—a process that may alter the structure and morphology of the LDH, which can be undesirable. Thus, this contribution focuses on the preparation of CaAl-OH-LDH by using $Ca(OH)_2$, $Al(OH)_3$ and water under low-temperature and atmospheric-pressure conditions. The purpose of this paper is to investigate whether it is possible to achieve a high purity HC using hydrothermal synthesis with this minimal number of chemicals and at which synthesis conditions this is possible, if at all. Several parameters can influence the hydrothermal synthesis of HC, these being temperature, time, molar calcium-to-aluminium ratio, chemical morphology/crystallinity of the reactants, mixing and the water-to-solids ratio used. The influence of these on the conversion of reactants to HC will be discussed in the following text.

2. Results

The method chosen leads to a self-regulated-pH synthesis through a dissolution-precipitation mechanism. The only externally variable parameters during the reaction are those that will be discussed in this paper (temperature, time, molar calcium-to-aluminium ratio, mixing, chemical morphology/crystallinity and water-to-solids ratio) and the use of other reactants, such as oxides, etc. which will not be discussed here. In addition to these, the chosen system was self-contained. During the synthesis, the system remained in an inert atmosphere to mitigate atmospheric carbonate contamination. The synthesis pH was sampled at regular intervals and was recorded continuously during some syntheses.

2.1. The Standard LDH and Its Carbonate Form

One LDH was common between all sets of experiments, with it synthesised at 80 °C for 3 h, a stirring speed of 750 rpm, a water-to-solids ratio of 80:20, at a molar calcium-to-aluminium ratio of 4:2 and using amorphous aluminium hydroxide (SA).

The standard sample was synthesised in triplicate to determine the repeatability of the synthesis method used and the trustworthiness of the Rietveld refinement applied to the wet-phase XRD data. The XRD and FTIR-ATR results for the standard LDHs are shown in Figure 1a,b, respectively.

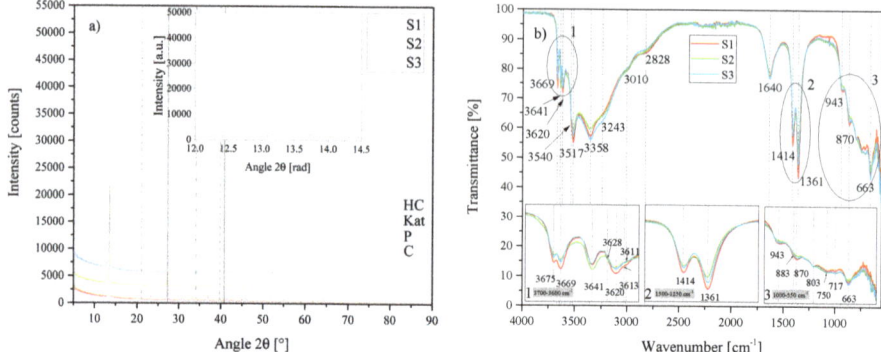

Figure 1. (a) XRD results of the three standard hydrocalumite (HC) layered double hydroxides (LDHs) S1, S2 and S3 between 5° 2θ and 90° 2θ. Each scan is y-shifted by 2500 counts. The inset depicts the primary LDH peaks and change in position. HC = hydrocalumite, Kat = katoite, P = portlandite and C = calcite. (b) FTIR-ATR scans of the three standard HC LDHs S1, S2 and S3 between 4000 cm^{-1} and 550 cm^{-1}. The three insets show the regions: (1) 3700 cm^{-1}–3600 cm^{-1}, (2) 1500 cm^{-1}–1250 cm^{-1}, (3) 1000 cm^{-1}–550 cm^{-1} in detail. Dashed and solid grey lines indicate the maxima of vibrations.

As a result of cation ordering and interlayer offsets, HC is typically presented in monoclinic form but has also been found in nature in the 6T polytype [2]. Other polytypes also exist, especially in the synthetically produced HCs, although a comprehensive review of all different polytypes of HC is lacking. The standard HCs were best described by HC of the formula [Ca$_4$Al$_2$(OH)$_{12}$][(CO$_3$)·5H$_2$O] in the anorthic (triclinic) crystal system and P1 space group with crystal parameters a = 5.7747 Å, b = 8.4689 Å and c = 9.9230 Å (reference code: 98-005-9327). The XRD patterns of the three standard HC LDHs were very similar with the exception of S2 which had a secondary phase that could not be identified. Contaminants of katoite (Ca$_3$Al$_2$(OH)$_{12}$), portlandite (Ca(OH)$_2$) and calcite (CaCO$_3$) were found in all three LDHs to varying degrees.

FTIR-ATR analysis showed that the standard HC LDHs S1, S2 and S3 all portrayed almost identical vibrational responses at 3675 cm^{-1}, 3669 cm^{-1}, 3641 cm^{-1}, 3620 cm^{-1}, 3540 cm^{-1}, 3517 cm^{-1}, 3358 cm^{-1}, 3243 cm^{-1}, 3010 cm^{-1}, 2828 cm^{-1}, 1640 cm^{-1}, 1414 cm^{-1}, 1361 cm^{-1}, 943 cm^{-1}, 883 cm^{-1}, 870 cm^{-1}, 803 cm^{-1}, 750 cm^{-1}, 717 cm^{-1} and 663 cm^{-1}. Typically, for LDHs, vibrations in the region between 3700 cm^{-1} and 2500 cm^{-1} are ascribed to OH str. vibrations of hydroxides bonded to the metal ions, of interlayer water or of water bonded to carbonate in the interlayer region. Ref. [30] assigned the vibrations around 3500 cm^{-1} and 3305 cm^{-1} to the Al–OH str. and Ca–OH str. vibration in the brucite-like lattice, respectively and the vibrations around 3100 cm^{-1} and between 2915 cm^{-1} and 2935 cm^{-1} to the OH str. vibrations of interlayer water and water bonded to interlayer carbonate, respectively. A similar assignment has been made for MgAl-LDHs, spanning different synthesis methods [31]. Thus, the vibrations at 3517 cm^{-1}, 3358 cm^{-1}, 3243 cm^{-1} and 3010 cm^{-1} were assigned to the OH str. vibrations of Al–OH and Ca–OH bonds, and interlayer H$_2$O and CO$_3^{2-}$ bonded to interlayer H$_2$O, respectively. The vibration at 3641 cm^{-1} could be assigned to the remaining portlandite in the samples (see Figure 2b). Some of the other vibrations between 3700 cm^{-1} and 3500 cm^{-1} have been observed in spectra of gibbsite (Al(OH)$_3$) [32] and have also been ascribed to different OH species [26]. The vibration at 2828 cm^{-1} could be a combination band as mentioned by [31]. On the lower end of the spectrum, the vibration at 1640 cm^{-1} could be present due to both, ν_{asym} vibrations of interlayer anions (such as hydroxides or carbonates) and ν_2 (H$_2$O) [31,32]. The vibration at 1414 cm^{-1} corresponds strongly to calcite (see Figure 2b) and is frequently assigned with 1365 cm^{-1} as the doublet ν_3 (CO$_3^{2-}$) vibration [33]. However, [31] have found this vibration to also correspond to the ν_{sym} vibration of the interlayer anion (for both, CO$_3^{2-}$ and OH$^-$). The vibrations at 943 cm^{-1}, 870 cm^{-1}, 717 cm^{-1} and 663 cm^{-1} could be assigned to ν_{sym} (OH), ν_2 (CO$_3^{2-}$) (interlayer and contaminant

CaCO$_3$), contaminant CaCO$_3$ and ν_4 (CO$_3^{2-}$), respectively. Vibrations at 803 cm^{-1} and 750 cm^{-1} have been observed in gibbsite spectra but remained unassigned [32]. The vibration at 803 cm^{-1} has also previously been assigned to katoite [26].

The best-fit crystal structure of HC—as identified by XRD—contained carbonate in the interlayer, contrary to the carbonate-free synthesis utilised. Although this, of course, is not evidence enough of carbonate intercalation, FTIR also showed the existence of carbonate-related vibrations. An HC LDH containing carbonate (sample ID: CO3) was, therefore, synthesised, the characteristics of which are shown in Figure 2.

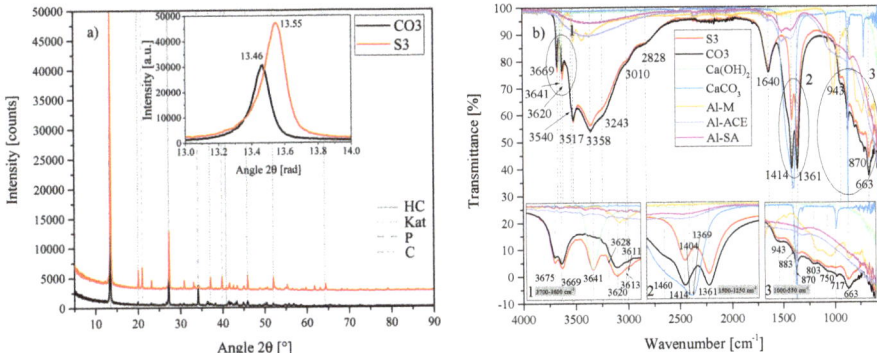

Figure 2. (**a**) XRD results of the HC-CO3-LDH plotted against the standard HC LDH S2 between 5° 2θ and 90° 2θ. Each scan is y-shifted by 2500 counts. The inset depicts the primary LDH peaks and change in position. HC = hydrocalumite, Kat = katoite, P = portlandite and C = calcite. (**b**) FTIR-ATR scans of the HC-CO3-LDH plotted against the standard HC LDH S2 and the starting materials used (Al(OH)$_3$, Ca(OH)$_2$ and CaCO$_3$) between 4000 cm^{-1} and 550 cm^{-1}. The three insets show the regions: (**1**) 3700 cm^{-1} – 3600 cm^{-1}, (**2**) 1500 cm^{-1} – 1250 cm^{-1}, (**3**) 1000 cm^{-1} – 550 cm^{-1} in detail. Dashed and solid grey lines indicate the maxima of vibrations.

The standard and CO3 specimens had the anorthic [Ca$_4$Al$_2$(OH)$_{12}$][(CO$_3$)·5H$_2$O] phase as best fit (Figure 2a). However, the primary reflection of CO3 proved to be shifted to the left of that of S3. The FTIR-ATR results (Figure 2b) showed that some differences exist between these two phases. Synthesis of HC-CO3-LDH was conducted with adjustment for the stoichiometric amount of CO$_3^{2-}$ required through replacement of Ca(OH)$_2$ with CaCO$_3$. No Ca(OH)$_2$ was detected at the end of this synthesis but a large amount of CaCO$_3$ remained. This is clearly evident on the FTIR scans (Figure 2b) through an increase in vibration strength at 1414 cm^{-1}, 870 cm^{-1}, 717 cm^{-1} and 663 cm^{-1}, and the disappearance of the vibration at 3641 cm^{-1}.

Rietveld refinement of S1 and S3 showed that a material consisting of approximately 60% HC could be obtained, the rest being katoite (approximately 28%), portlandite (approximately 10%) and calcite (approximately 2%). No Rietveld refinement could be performed on S2 due to the presence of the unidentifiable phase. CO3 consisted of approx 70% LDH, 10% katoite and 20% calcite. The analysis of S3 will be used for comparative purposes in the remainder of the text.

Figure 3 depicts the micrographs obtained for S1, S2, S3, CO3 and the Ca(OH)$_2$, Al(OH)$_3$ and CaCO$_3$ used in the synthesis.

LDH platelet formation was evident in all of the synthesised materials in Figure 3 and constituted the majority of the identifiable phase with thin, large hexagonal platelets, typically with an elongated facet. Katoite was visually identifiable as faceted garnet-like balls and calcite as cubic crystals. Small particulate matter was present in all samples, sticking on the LDH platelets and other phases and sometimes forming small agglomerates. This could be either Ca(OH)$_2$ or Al(OH)$_3$ left over. While no left-over Al(OH)$_3$ was identified through XRD (unsurprisingly, since an amorphous Al(OH)$_3$ was used in the synthesis) it is possible that the small structures are either portlandite or amorphous Al(OH)$_3$, which proved to be similar in size and appearance. SEM showed that CO3 consisted of smaller platelets than the standard LDHs. It was also more difficult to find remnants of the left-over $CaCO_3$ used, even though this LDH consisted of 72.47% LDH, 9.24% katoite and 18.29% calcite, according to Rietveld refinement. This could be attributed to the $CaCO_3$ consisting of ill-defined oblong shapes that would be difficult to identify on SEM micrographs. It does, however, introduce a difference between the calcite fed with $Ca(OH)_2$ (cubic crystals) and the calcite supplied to CO3.

Figure 3. SEM micrographs of the HC LDHs S1, S2, S3 and CO3 at 1 keV and 2k magnification, and $Ca(OH)_2$, $Al(OH)_3$-SA (amorphous) and $CaCO_3$ at 1 kEV and 10k magnification. The scale bar is indicated under the label. LDH: turquoise hexagon, katoite: turquoise circle, calcite: turquoise asterisk.

pH measurements during the reaction showed that the self-regulated pH remained close to 11 for the duration of the synthesis at 80 °C for S1, S2 and S3. pH fluctuations were, however, common—especially after 1 h to 1.5 h of reaction time when the reaction mixture considerably thickened up—and, overall, the pH decreased slightly with time during the 3 h of synthesis. The pH during synthesis of CO3 was significantly lower than that of S1, S2 and S3. The pH probe (used for sampled pH recording) required substantial agitation to read the pH correctly after 2 h. As a result of these fluctuations and sampling difficulties, even though the results showed very good comparability, the pHs shown in this work are to be taken as an estimate. Figure 4 shows the pH readings obtained for S1, S2 and S3 over 3 h.

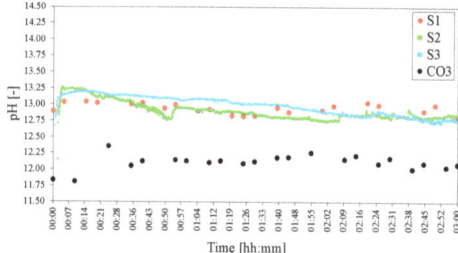

Figure 4. pHs of the standard HC LDHs S1, S2, S3 and CO3. The pH was adjusted to 25 °C to facilitate comparison.

For S1 a sampling pH method was used where the pH probe was cleaned before every measurement. The pH of S2 and S3 was measured continuously. The pH of S2 exhibited greater fluctuations than that of S3. This could be attributed to the encased pH probe used for S2 which was subject to material build-up. Due to the sampling method employed for other experiments, the pH measurements of S1 will be used for comparative purposes in the remainder of the text.

In the following sections, only deviations from the standard S3 discussed in this section—caused by varying one of the studied parameters—will be highlighted. S3 will thus, in the following text, be synonymous with T4, t2, MR2, A1, M2 and WS2.

2.2. Reaction Temperature

The reaction temperature was varied between 20 °C and 90 °C (T1 = 20 °C, T2 = 40 °C, T3 = 60 °C, T4 = 80 °C and T5 = 90 °C). The corresponding XRD (a) and ATR-FTIR (b) results are shown in Figure 5.

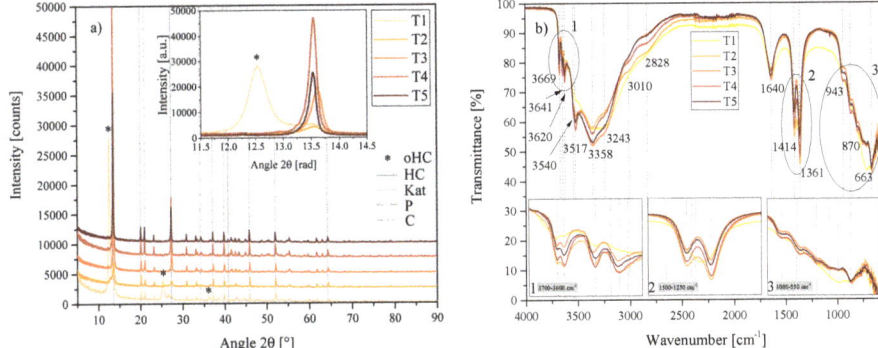

Figure 5. (a) XRD results of the temperature series T1, T2, T3, T4 and T5 between 5° 2θ and 90° 2θ. Each scan is y-shifted by 2500 counts. The inset depicts the primary LDH peaks and change in position. HC = hydrocalumite, oHC = secondary HC-like phase, Kat = katoite, P = portlandite and C = calcite. (b) FTIR-ATR scans of the temperature series T1, T2, T3, T4 and T5 between 4000 cm^{-1} and 550 cm^{-1}. The three insets show the regions: (1) 3700 cm^{-1} – 3600 cm^{-1}, (2) 1500 cm^{-1} – 1250 cm^{-1}, (3) 1000 cm^{-1} – 550 cm^{-1} in detail. Dashed and solid grey lines indicate the maxima of vibrations. T1 = 20 °C, T2 = 40 °C, T3 = 60 °C, T4 = 80 °C and T5 = 90 °C.

The [Ca$_4$Al$_2$(OH)$_{12}$][(CO$_3$)·5H$_2$O] anorthic phase once again proved to be the best fit for the phase produced. However, in this series some additional phases were present, depending on the synthesis temperature. As shown in the inset of Figure 5a, an increase in temperature led to first the formation and then the disappearance of an LDH-like phase with a layered structure, imitating HC. This phase was most prominent in the synthesis at 20 °C and was not present at 60 °C. The phase was best fit by 3 CaO · Al$_2$O$_3$ · 0.5 Ca(OH)$_2$ · 0.5 CaCO$_3$ · 11.5 H$_2$O, a rhombohedral structure. Being an outdated entry, however, the closest fit that could be used for Rietveld refinement is the structure identified by [34] for calcium hemicarboaluminate, a partially carbonated (low carbonate content) phase that is formed at the start of carbonation reactions in cement Ca$_4$Al$_2$(OH)$_{12}$(OH)(CO$_3$)$_{0.5}$·4 H$_2$O. The first phase, a calcium aluminate carbonate hydrate was studied by [35]. At the time it was noted that this compound could be a meta-stable phase that forms in cement but had not been found in Portland cement at the time. In T2, this phase was still present but its reflection shifted slightly to the right—closer resembling calcium hemicarboaluminate—and more of the always-observed HC phase [Ca$_4$Al$_2$(OH)$_{12}$][(CO$_3$)·5H$_2$O] was present. With an increase in temperature, HC formation increased up to 80 °C. Katoite and portlandite were present in all samples, while calcite was only present in T4 and T5. Calcite content increased at 90 °C. Table 1 shows the progression of the phase contents with temperature as determined by Rietveld refinement.

Table 1. Rietveld refinement of the temperature series of LDHs T1 = 20 °C, T2 = 40 °C, T3 = 60 °C, T4 = 80 °C and T5 = 90 °C. HC indicated the phase [$Ca_4Al_2(OH)_{12}$][$(CO_3) \cdot 5 H_2O$], oHC indicates the phase $Ca_4Al_2(OH)_{12}(OH)(CO_3)_{0.5} \cdot 4 H_2O$, Kat is katoite, P is portlandite and C is calcite. All values are given in percentages of the crystalline phases.

	HC	oHC	Kat	P	C
T1	29.48	47.1	12.46	10.95	
T2	30.14	24.22	32.57	13.07	
T3	57.57		28.8	11.37	2.26
T4	61.5		26.52	9.84	2.14
T5	52.18		29.52	10.36	8.24

Some differences in spectra could also be observed with FTIR-ATR, most notably in T1 and T2—which showed a shoulder left of the vibrations at 3669 cm^{-1} and changed intensity of other vibrations as indicated in the insets of Figure 5b and in the region between 3517 cm^{-1} and 2828 cm^{-1}. With an increase in temperature, the 3010 cm^{-1} and 2828 cm^{-1} vibrations decreased in intensity. The 3620 cm^{-1} Al–OH str. vibration was significantly reduced in T1. Further, T1 showed less definition of the 1414 cm^{-1}/1361 cm^{-1} doublet and a greater resemblance of the calcite vibration in this region shown in Figure 2b. T1 also showed less definition in the vibrations between 1000 cm^{-1} and 550 cm^{-1}.

SEM micrographs showed clear differences in the materials obtained and a change in distribution of phases (Figure 6).

Figure 6. SEM micrographs of the HC LDHs T1 = 20 °C, T2 = 40 °C, T3 = 60 °C, T4 = 80 °C and T5 = 90 °C at 1 keV and 2k magnification. The scale bar is indicated under the label.

All samples showed plate-like structures. In T1, these structures were highly agglomerated and packed into stacks of ill-defined platelets. Round, ball-like structures were mixed in-between these platelet agglomerates. It is possible that these were ill-defined katoite structures. T2 showed better definition of the platelet structures, albeit with them still being small and unevenly crystallised, and a large amount of katoite, which was omnipresent between stacks of platelets. No cubic calcite structures were visible in these micrographs. T3, T4 and T5 showed well-defined platelets, with the best-defined platelets being present in T4. Katoite and cubic calcite structures were visible in all three micrographs, corresponding to the Rietveld analysis. Particulate matter was visible in all samples and could again be either unreacted $Ca(OH)_2$ or amorphous $Al(OH)_3$. The secondary calcium aluminate carbonate hydrate phase or calcium hemicarboaluminate phase present in T1 and T2 could not be discerned from other phases.

The pH documented during synthesis varied greatly with temperature. Figure 7 shows the sampled pHs (adjusted to 25 °C) as a function of time for comparative purposes.

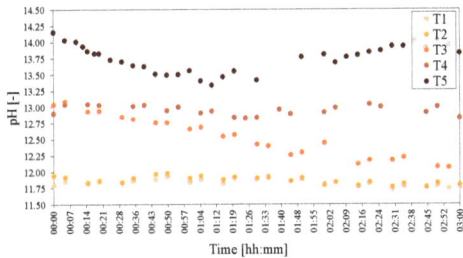

Figure 7. pHs of the temperature series of LDHs T1 = 20 °C, T2 = 40 °C, T3 = 60 °C, T4 = 80 °C and T5 = 90 °C. The pH was adjusted to 25 °C to facilitate comparison.

The recorded pH rose with an increase in temperature. The pHs at 20 °C and 40 °C were almost identical. Increasing the temperature to 60 °C led to the start of the reaction at a higher pH, lowering with the progression of time. At 80 °C the pH stayed constant (only decreasing slightly with time), while at 90 °C, the pH was very high at the start of reaction and dipped before rising again after 2 h of reaction time had passed.

2.3. Reaction Time

The impact of three reaction times on HC formation was tested (t1 = 1 h, t2 = 3 h and t3 = 6 h), the XRD and FTIR-ATR results of which are shown in Figure 8a and Figure 8b, respectively.

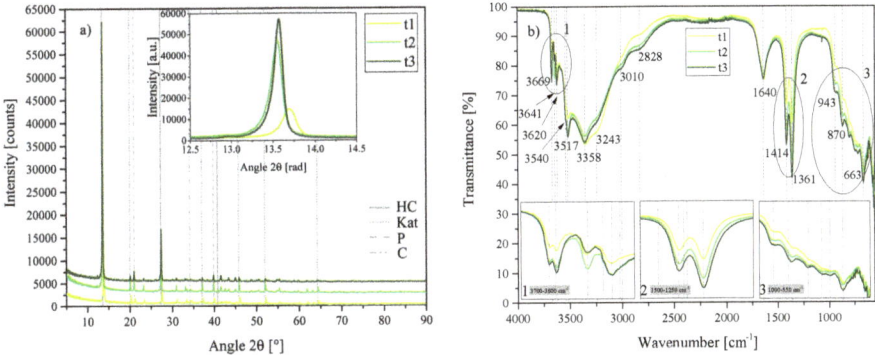

Figure 8. (a) XRD results of the time series t1, t2 and t3 between 5° 2θ and 90° 2θ. Each scan is y-shifted by 2500 counts. The inset depicts the primary LDH peaks and change in position. HC = hydrocalumite, Kat = katoite, P = portlandite and C = calcite. (b) FTIR-ATR scans of the time series t1, t2 and t3 between 4000 cm^{-1} and 550 cm^{-1}. The three insets show the regions: (1) 3700 cm^{-1} − 3600 cm^{-1}, (2) 1500 cm^{-1} − 1250 cm^{-1}, (3) 1000 cm^{-1} − 550 cm^{-1} in detail. Dashed and solid grey lines indicate the maxima of vibrations. t1 = 1 h, t2 = 3 h and t3 = 6 h.

XRD showed a large similarity between the phases formed; however, the LDH phase formed in t1 was shifted to the right (see inset of Figure 5a, more closely resembling the primary reflection position of the phase formed in T2. The best-fit LDH crystal structure for all three, t1, t2 and t3, was again anorthic $[Ca_4Al_2(OH)_{12}][(CO_3) \cdot 5H_2O]$. A small additional phase with a similar crystal structure to chloride-intercalated HC was detected in t3 ($[Ca_4Al_2(OH)_{12}][(CO_3)_{0.5}Cl_4 \cdot 8H_2O]$). Overall, Rietveld refinement showed that an increase in reaction time led to an increase in LDH phase and decrease in katoite and portlandite present. Calcite content also decreased slightly with an increase in reaction time. Table 2 depicts the Rietveld refinement results obtained for the different reaction times.

Table 2. Rietveld refinement of the time series of LDHs t1 = 1 h, t2 = 3 h and t3 = 6 h. HC indicates the phase $[Ca_4Al_2(OH)_{12}][(CO_3) \cdot 5H_2O]$, HC2 indicates the phase $[Ca_4Al_2(OH)_{12}][(CO_3)_{0.5}Cl_4 \cdot 8H_2O]$, Kat is katoite, P is portlandite and C is calcite. All values are given in percentages of the crystalline phases.

	HC	HC2	Kat	P	C
t1	49.72		37.39	10.47	2.42
t2	61.5		26.52	9.84	2.14
t3	77.97	2.04	12.66	6.2	1.13

FTIR scans of the three LDHs were similar; most notably showing an increase in vibration in the doublet at $1414\,cm^{-1}/1361\,cm^{-1}$, corresponding to a decreased amount of calcite and possible increased intercalation of carbonate into the interlayer. Most vibrations increased in strength with an increase in reaction time, except the vibration at $3243\,cm^{-1}$, which showed a stronger vibration in t1. Further, the vibration at $3641\,cm^{-1}$ (linked to unreacted $Ca(OH)_2$) was notably stronger in t2, even though this LDH contained less $Ca(OH)_2$ according to Rietveld refinement.

Morphologically, the three materials were very similar; the only easily noticeable differences being the amount of particulate matter clinging to the other phases, improved crystallisation of the platelets and a weak correlation between the amount of katoite and calcite observed, and Rietveld refinement results (Figure 9).

Figure 9. SEM micrographs of the time series of HC LDHs t1 = 1 h, t2 = 3 h and t3 = 6 h at 1 keV and 2k magnification. The scale bar is indicated under the label.

Recording the pH for the duration of the 6 h t3 experiment showed that the pH approximately levels out after 3 h (Figure 10). Small differenced between the pH measurements were observed between t1, t2 and t3 in their respective time-frames.

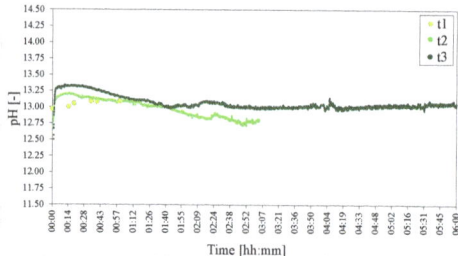

Figure 10. pHs of the time series of LDHs t1 = 1 h, t2 = 3 h and t3 = 6 h. The pH was adjusted to 25 °C to facilitate comparison.

2.4. Molar Calcium-to-Aluminium Ratio

The reaction of $Ca(OH)_2$ and $Al(OH)_3$ to form HC stands in competition with the formation of katoite $(Ca_3Al_2(OH)_{12})$ [26,27]. Three molar calcium-to-aluminium ratios were, thus, tested to determine the effect of this ratio on the formation of HC. MR1 (3:2) should favour katoite formation (based on stoichiometry), while MR2 (4:2) should favour HC formation and MR3 (6:2) would provide an excess of $Ca(OH)_2$ to the reaction. Out of all results, the result of this series was most surprising, inverting expectations. XRD (shown in Figure 11) showed that, once again, similar phases formed, just with different ratios and that the $[Ca_4Al_2(OH)_{12}][(CO_3) \cdot 5\,H_2O]$ phase again provided the best fit for the crystal structure of the LDH formed.

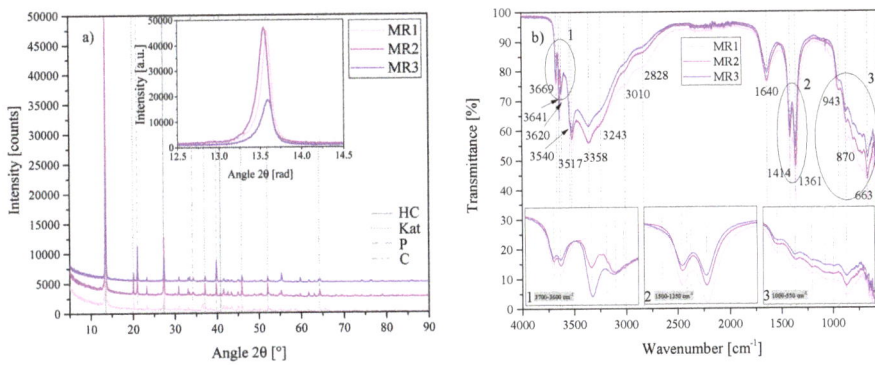

Figure 11. (a) XRD results of the molar calcium-to-aluminium ratio series MR1, MR2 and MR3 between 5° 2θ and 90° 2θ. Each scan is y-shifted by 2500 counts. The inset depicts the primary LDH peaks and change in position. HC = hydrocalumite, Kat = katoite, P = portlandite and C = calcite. (b) FTIR-ATR scans of the molar calcium-to-aluminium ratio series MR1, MR2 and MR3 between 4000 cm^{-1} and 550 cm^{-1}. The three insets show the regions: (1) 3700 cm^{-1} – 3600 cm^{-1}, (2) 1500 cm^{-1} – 1250 cm^{-1}, (3) 1000 cm^{-1} – 550 cm^{-1} in detail. Dashed and solid grey lines indicate the maxima of vibrations. MR1 = 3:2, MR2 = 4:2 and MR3 = 6:2.

The primary MR3 LDH peak was slightly shifted to the right of MR1 and MR2 (inset of Figure 11a). No significant amounts of other LDH phases were formed in this series. However, Rietveld refinement revealed that MR1 led to the greatest purity HC phase (>80%), while MR3 resulted in a purity of less than 50%. The Rietveld refinement results of this series are shown in Table 3.

Table 3. Rietveld refinement of the molar calcium-to-aluminium series of LDHs MR1 = 3:2, MR2 = 4:2 and MR3 = 6:2. HC indicates the phase $[Ca_4Al_2(OH)_{12}][(CO_3) \cdot 5\,H_2O]$, Kat is katoite, P is portlandite and C is calcite. All values are given in percentages of the crystalline phases.

	HC	Kat	P	C
MR1	84.91	14.59		0.51
MR2	61.5	26.52	9.84	2.14
MR3	46.19	22.34	28.16	3.31

FTIR-ATR results (Figure 11b) showed good correlation between the Rietveld results and the vibrational intensities of each sample. The vibration at 3641 cm^{-1}, linked to unreacted $Ca(OH)_2$, showed excellent correlation between the Rietveld and FTIR results, increasing in intensity with an increase in left-over portlandite. The twin-peak at 1414 cm^{-1}/1361 cm^{-1} correlated well with the shift in primary LDH reflection position and possible carbonate content. In the region between 1000 cm^{-1} and 550 cm^{-1} the vibration strength decreased with an increase in molar ratio.

Morphologically, clear differences were visible between the amount of particulate matter present (Ca(OH)$_2$ or amorphous Al(OH)$_3$) in each sample (Figure 12). MR3 contained the largest amount of particulate matter, correlating well with the Rietveld refinement results. MR3 also contained visibly more katoite and calcite, as expected. The platelets showed some differences too. MR1 produced smaller, rugged platelets with many broken edges and the agglomeration of different platelet sizes. MR2 produced much smoother, better defined and larger platelets than MR1 and MR3. The platelets of MR3 were less rugged than those of MR1, but due to the coverage in particulate matter, proper description of their definition was challenging.

Figure 12. SEM micrographs of the series of molar calcium-to-aluminium ratio MR1 = 3:2, MR2 = 4:2 and MR3 = 6:2 at 1 keV and 2k magnification. The scale bar is indicated under the label.

The pHs of MR1 and MR3 were very similar and dropped slightly with time in comparison with MR2. MR3 started with a slightly elevated pH (as expected due to the large amount of Ca(OH)$_2$ present). Conversely, MR1 had a lower starting pH, as expected, due to the lower amount of Ca(OH)$_2$ present and MR2 lay between the two at the start—as portrayed in Figure 13. With progression in time, the pHs of MR1 and MR3 fell slightly below that of MR2.

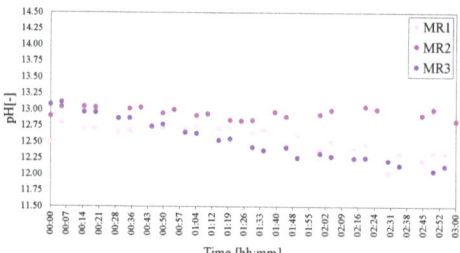

Figure 13. pHs of the molar calcium-to-aluminium series of LDHs MR1 = 3:2, MR2 = 4:2 and MR3 = 6:2. The pH was adjusted to 25 °C to facilitate comparison.

2.5. Chemical Morphology/Crystallinity of the Al(OH)$_3$

Three different Al(OH)$_3$ sources were tested in this work. The first (A1), from Sigma Aldrich (SA) was used as the standard source in all experiments. This Al source proved to be amorphous by XRD and have a surface area of 59.35 m^2 g^{-1}. The second Al source (A2) from ACE Chemicals (ACE) proved to be minimally crystalline boehmite with a surface area of 69.12 m^2 g^{-1} (due to the low crystallinity it is possible that some of the phase consisted of other amorphous crystal structures or polymorphs). The third Al source (A3) from Merck (M) proved to be highly crystalline gibbsite with a surface area of 0.75 m^2 g^{-1}. Figure 14 shows the XRD results for these three sources and isotherms obtained for the determination of the Brunauer–Emmett–Teller (BET) surface area.

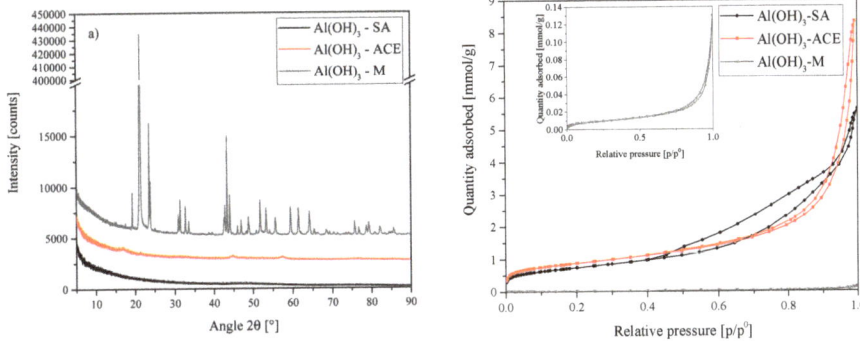

Figure 14. (**a**) XRD results of the Al sources used for the synthesis of A1, A2 and A3. SA = amorphous, ACE = boehmite and M = gibbsite—Sigma Aldrich (SA), ACE Chemicals (ACE) and Merck (M). Each scan is y-shifted by 2500 counts. (**b**) Isotherms obtained for each Al source. Sigma Aldrich (SA), ACE Chemicals (ACE) and Merck (M).

The Al sources used thus had very different crystallinities and crystal structures. SEM further showed that the morphologies of the three Al sources were very different (see Figure 15).

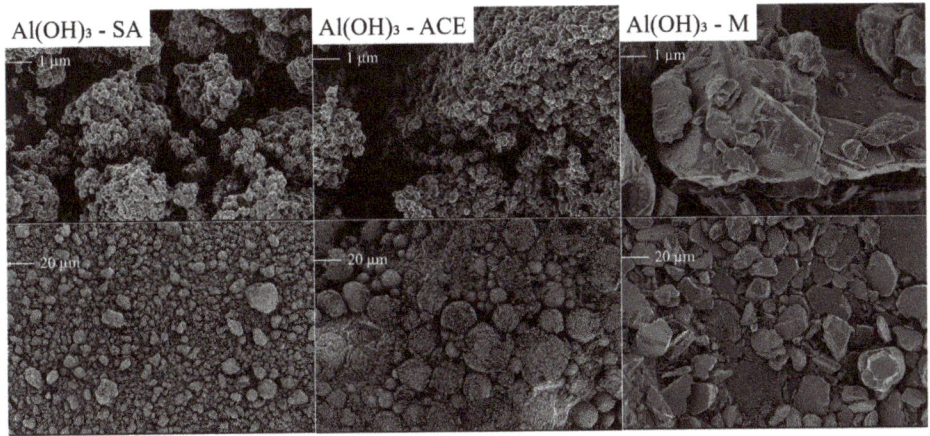

Figure 15. SEM micrographs of the three different Al(OH)$_3$ sources used from Sigma Aldrich (SA), ACE Chemicals (ACE) and Merck (M) taken at 1 keV and 10k (top) and 500 times (bottom) magnification. The scale bar is indicated on each micrograph.

The SA and ACE reagents had similarly sized particles but different "macro" morphologies. Where SA consisted of a carpet-like covering of small dot-like particles (agglomerated into random shapes), ACE consisted of the same size particles but agglomerated into a myriad of variably sized balls. These balls were damaged in the sample preparation process, breaking up/squashing and revealing some of the internal structure. The M source, on the other hand, consisted of large, thick chunks of gibbsite.

These three materials caused the formation of three very different reaction outcomes. The outcome of A1 (the standard sample S3) has already been discussed. In comparison, A2 consisted of two HC phases, being mainly the typical one with a crystal structure similar to $[Ca_4Al_2(OH)_{12}][(CO_3) \cdot 5 H_2O]$ but also a small fraction of the HC phase with a crystal structure similar to $[Ca_4Al_2(OH)_{12}][(CO_3)_{0.5}Cl \cdot 4.8H_2O]$, as already present in t3. These two phases are clearly distinguishable on the XRD inset in Figure 16a depicting the primary LDH reflections.

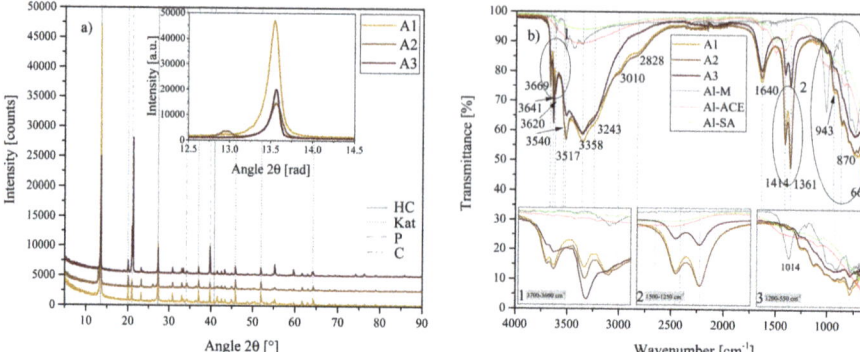

Figure 16. (a) XRD results of the chemical morphology/crystallinity of Al(OH)$_3$ series A1, A2 and A3 between 5° 2θ and 90° 2θ. Each scan is y-shifted by 2500 counts. The inset depicts the primary LDH peaks and change in position. HC = hydrocalumite, HC2 = hydrocalumite phase 2, Kat = katoite, P = portlandite, C = calcite and G = gibbsite. (b) FTIR-ATR scans of the chemical morphology/crystallinity of Al(OH)$_3$ A1, A2 and A3 between 4000 cm^{-1} and 550 cm^{-1}. The three insets show the regions: (1) 3700 cm^{-1} – 3600 cm^{-1}, (2) 1500 cm^{-1} – 1250 cm^{-1}, (3) 1200 cm^{-1} – 550 cm^{-1} in detail. Dashed and solid grey lines indicate the maxima of vibrations. A1 = Al(OH)$_3$-SA, A2 = Al(OH)$_3$-ACE and A3 = Al(OH)$_3$-SA.

A3 consisted of only one phase of LDH but at a lower overall purity—with large amounts of Ca(OH)$_2$ remaining unreacted as shown through Rietveld refinement in Table 4. A2 produced the most katoite and A3 contained no calcite, but large amounts of unreacted gibbsite. Both A2 and A3 showed slight right-shifts in their primary LDH reflection in comparison to A1.

Table 4. Rietveld refinement of the morphology/crystallinity series of LDHs A1 = SA, A2 = ACE and A3 = M. HC indicates the phase [Ca$_4$Al$_2$(OH)$_{12}$][(CO$_3$)·5H$_2$O], HC2 is [Ca$_4$Al$_2$(OH)$_{12}$][(CO$_3$)$_{0.5}$Cl$_4$·8H$_2$O], Kat is katoite, P is portlandite, C is calcite and G is gibbsite. All values are given in percentages of the crystalline phases.

	HC	HC2	Kat	P	C	G
A1	61.5		26.52	9.84	2.14	
A2	43.51	11	34.04	8.25	3.2	
A3	27.9		23.48	28.97		19.65

There were also differences between the FTIR-ATR scans of A1, A2 and A3 depicted in Figure 16b. While the scans for A1 and A2 were quite similar, A3 deviated from almost all major vibrations with either a decreased or increased vibration. The large amount of unreacted portlandite in A3 was clearly visible at 3641 cm^{-1}. Most other vibrations, however, were weaker in A3—most notably those indicating the OH str. vibration of water bonded to interlayer carbonate (3010 cm^{-1}) and the combination band at 2828 cm^{-1} as well as the twin peak at 1414 cm^{-1}/1361 cm^{-1} associated with both, interlayer carbonate and interlayer hydroxide anions. The scan for A3 gained an additional band at 1014 cm^{-1} which correlates to the strongest vibration of the Al(OH)$_3$-M source itself. The remaining vibrations between 1000 cm^{-1} and 550 cm^{-1} were less strongly defined than those of A1 and A2. There also existed some stark differences between the FTIR scans of the three Al(OH)$_3$ sources, which are also depicted in Figure 16b. Al(OH)$_3$-M showed defined vibrations in the 3700 cm^{-1} to 3300 cm^{-1} region and between 1200 cm^{-1} and 550 cm^{-1}, while Al(OH)$_3$-SA and Al(OH)$_3$-ACE were very similar and had much less strongly defined features. Both consisted of a broad band in the 3700 cm^{-1} to 3300 cm^{-1} region and immolated well the 1000 cm^{-1} to 550 cm^{-1} region. Both materials, however, had a doublet feature blue-shifted from the 1414 cm^{-1}/1361 cm^{-1} doublet. This feature has been linked to adsorbed carbonate species [36–38] and will be discussed in-depth later in the text.

Morphological examination of these three materials showed that A1 and A2 formed similar structures but that A3 formed much larger HC platelets (Figure 17). A far greater quantity of particulate matter was also present in A3, as expected from the sample consisting of almost 50% of portlandite and gibbsite. No large chunks of gibbsite (as initially present from the reactant) were visible, though, indicating that these chunks of gibbsite must have broken up during synthesis.

Figure 17. SEM micrographs of the chemical morphology/crystallinity of Al(OH)$_3$ series A1 = SA, A2 = ACE and A3 = M taken at 1 keV and 2k magnification. The scale bar is indicated below the label.

Even though the phase compositions were very different for A1, A2 and A3, the synthesis pHs observed were almost identical throughout the reaction, as shown in Figure 18—indicating that Al(OH)$_3$ crystallinity and surface area, and the subsequent dissoluted Al-species, do not contribute to the mixture pH significantly.

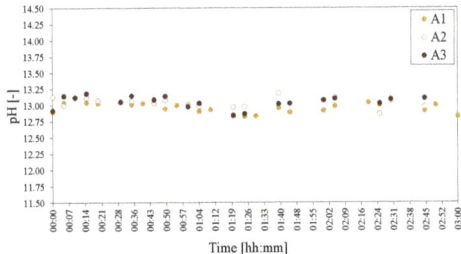

Figure 18. pHs of the morphology/crystallinity of Al(OH)$_3$ series of LDHs A1 = SA, A2 = ACE and A3 = M. The pH was adjusted to 25 °C to facilitate comparison.

2.6. Mixing

The mixing during synthesis was varied by altering the rotational speed of the magnetic stirrer used. Three rotational speeds were used: M1 = 500 rpm, M2 = 750 rpm and M3 = 1000 rpm. During the synthesis, these rotational speeds led to mixing at full incorporation at all times (1000 rpm), sufficient mixing to keep the system visibly agitated but cause a thin condensate film forming (750 rpm) and mixing that kept the reaction mixture agitated but led to the formation of a thicker condensate film on top of the mixture (500 rpm). Little difference was observed in the outcome of the reaction in XRD and FTIR-ATR results, as shown in Figure 19a,b.

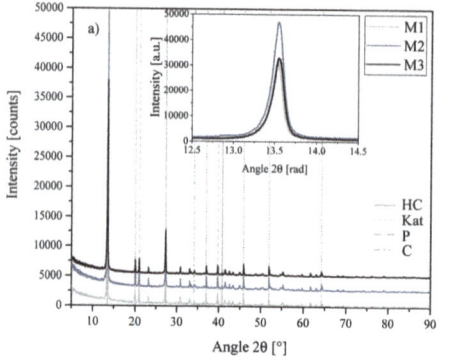
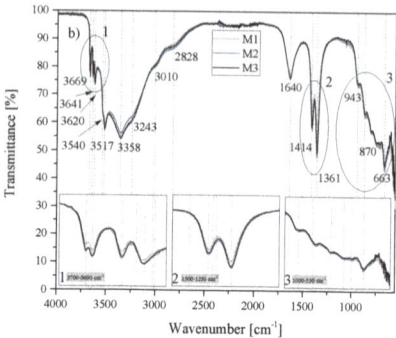

Figure 19. (a) XRD results of the mixing series M1, M2 and M3 between 5° 2θ and 90° 2θ. Each scan is y-shifted by 2500 counts. The inset depicts the primary LDH peaks and change in position. HC = hydrocalumite, Kat = katoite, P = portlandite and C = calcite. (b) FTIR-ATR scans of the mixing series M1, M2 and M3 between 4000 cm^{-1} and 550 cm^{-1}. The three insets show the regions: (1) 3700 cm^{-1} – 3600 cm^{-1}, (2) 1500 cm^{-1} – 1250 cm^{-1}, (3) 1000 cm^{-1} – 550 cm^{-1} in detail. Dashed and solid grey lines indicate the maxima of vibrations. M1 = 500 rpm, WS2 = 750 rpm and WS3 = 1000 rpm.

Rietveld refinement further confirmed the similarity between these three reaction outcomes. As shown in Table 5, only small variations in HC, katoite, portlandite and calcite content were observed. Unsurprisingly, SEM (Figure 20) also showed very little difference in the morphologies of M1, M2 and M3.

Table 5. Rietveld refinement of the mixing series of LDHs M1 = 500 rpm, M2 = 750 rpm and M3 = 1000 rpm. HC indicates the phase [Ca$_4$Al$_2$(OH)$_{12}$][(CO$_3$)·5 H$_2$O], Kat is katoite, P is portlandite and C is calcite. All values are given in percentages of the crystalline phases.

	HC	Kat	P	C
M1	58.29	28.77	11.61	1.33
M2	61.5	26.52	9.84	2.14
M3	59.83	27.74	10.36	2.07

Figure 20. SEM micrographs of the mixing series M1 = 500 rpm, M2 = 750 rpm and M3 = 1000 rpm taken at 1 keV and 2k magnification. The scale bar is indicated below the label.

However, surprisingly, even though there was almost no difference seen between M1, M2 and M3 in terms of the other results, the pH of M1 and M3 decreased notably with time in comparison to that of M2 (Figure 21).

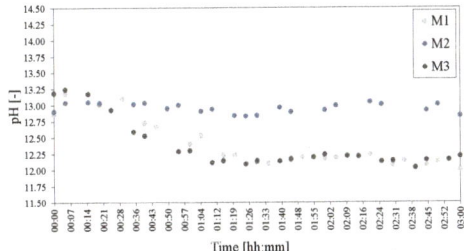

Figure 21. pHs of the mixing series of LDHs M1 = 500 rpm, M2 = 750 rpm and M3 = 1000 rpm. The pH was adjusted to 25 °C to facilitate comparison.

2.7. Water-to-Solids Ratio

The final parameter investigated was the effect of the water-to-solids ratio on the formation of HC. Three water-to-solids ratios were tested, being WS1 = 70:30, WS2 = 80:20 and WS3 = 90:10. During the reaction, these three ratios behaved very differently. Where, usually, reaction mixtures with an 80:20 ratio thickened considerably after 1 h to 1.5 h, WS1 had already thickened considerably after 30 min. After 2 h, this mixture became difficult to agitate even with the pH probe during measurement and was mixing with great difficulty until the 3 h of reaction time had passed. WS3, on the contrary, never thickened to a considerable degree, fully incorporating all mixture contents continuously.

The XRD and FTIR-ATR results depicted in Figure 22 show some differences between the three materials formed.

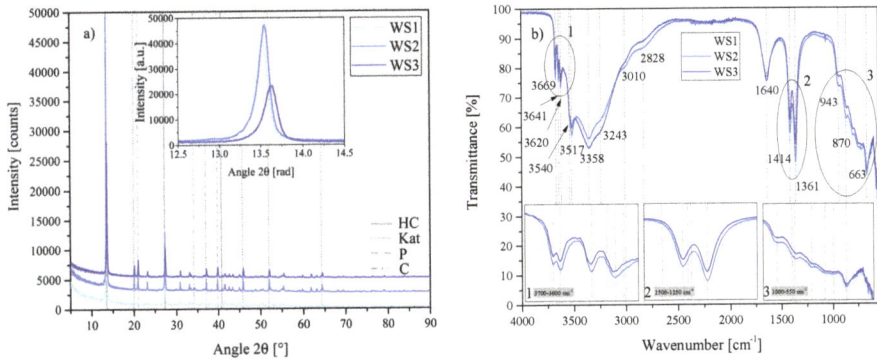

Figure 22. (a) XRD results of the water-to-solids ratio series WS1, WS2 and WS3 between 5° 2θ and 90° 2θ. Each scan is y-shifted by 2500 counts. The inset depicts the primary LDH peaks and change in position. HC = hydrocalumite, Kat = katoite, P = portlandite and C = calcite. (b) FTIR-ATR scans of the water-to-solids ratio series WS1, WS2 and WS3 between 4000 cm^{-1} and 550 cm^{-1}. The three insets show the regions: (1) 3700 cm^{-1} – 3600 cm^{-1}, (2) 1500 cm^{-1} – 1250 cm^{-1}, (3) 1000 cm^{-1} – 550 cm^{-1} in detail. Dashed and solid grey lines indicate the maxima of vibrations. WS1 = 70:30, WS2 = 80:20 and WS3 = 90:10.

The primary LDH peak of HC was shifted somewhat to the right for WS3 with respect to WS1 and WS2 (inset in Figure 22a). $[Ca_4Al_2(OH)_{12}][(CO_3) \cdot 5H_2O]$ remained the best fitting crystal structure analogue in all three materials, however. FTIR-ATR showed little difference between the three materials, although there was a clear reduction in the 3641 cm^{-1} peak associated with unreacted $Ca(OH)_2$. Intensity of vibration also decreased in the 1414 cm^{-1}/1361 cm^{-1} doublet with an increase in water-to-solids ratio; as did the vibrations between 1000 cm^{-1} and 550 cm^{-1}. Rietveld refinement, however, showed very large differences between these three materials. WS1 led to a HC purity greater than 70%, while WS3 achieved only approximately 55%. There were also large differences in the amount of katoite formed and the amount of portlandite left unreacted, as well as the amount of calcite remaining, as shown in Table 6.

Table 6. Rietveld refinement of the water-to-solids ratio series of LDHs WS1 = 70:30, WS2 = 80:20 and WS3 = 90:10. HC indicates the phase $[Ca_4Al_2(OH)_{12}][(CO_3)\cdot 5H_2O]$, Kat is katoite, P is portlandite and C is calcite. All values are given in percentages of the crystalline phases.

	HC	Kat	P	C
WS1	73.48	14.64	1.91	9.98
WS2	61.5	26.52	9.84	2.14
WS3	54.59	32.39	13.02	

Even with this large difference in the amount of each phase present, crystallisation of the LDH phases appeared similar on the SEM micrographs, shown in Figure 23. However, less particulate matter was present in WS1 and more in WS3, in good correlation with the decreased amount of portlandite present. An increased amount of katoite was also present in WS2 and WS3. The large amount of calcite determined through the Rietveld analysis for WS1 was not observed in its micrograph.

Figure 23. SEM micrographs of the water-to-solids ratio series WS1 = 70:30, WS2 = 80:20 and WS3 = 90:10 taken at 1 keV and 2k magnification. The scale bar is indicated below the label.

The pHs recorded for WS1, WS2 and WS3 showed that significant differences existed between the starting pH of the reaction mixture and the behaviour during synthesis with time (Figure 24). WS1, as expected due to the larger amount of reactants present, had a very high initial pH but then exhibited a considerable reduction in pH over time. WS3, on the other hand started off with a much lower pH, which remained almost constant throughout the reaction. WS1 and WS3 had opposing viscosities, the one being very thick, the other very thin. Yet they still ended on a similarly low pH after 3 h.

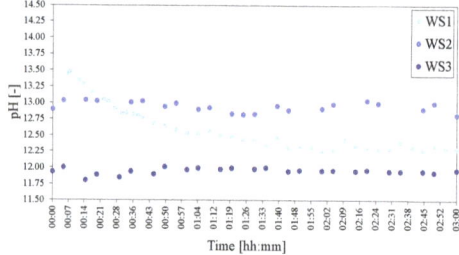

Figure 24. pHs of the water-to-solids ratio series of LDHs WS1 = 70:30, WS2 = 80:20 and WS3 = 90:10. The pH was adjusted to 25 °C to facilitate comparison.

3. Discussion

The results of this work have shown that the synthesis of an LDH phase (with high purity, max. = 84.91%) is possible using only $Ca(OH)_2$, $Al(OH)_3$ and water at mild synthesis conditions (atmospheric pressure and temperatures less than boiling point) as well as in a short period of time (3 h). This section serves to discuss the following points:

- carbonate contamination;
- increasing conversion to HC;
- and a hint towards the reaction mechanism of the formation of HC in a hydrothermal process.

3.1. Carbonate Contamination

The synthesis of carbonate-free HC from metal oxides and hydroxides has not been attempted with this aim in literature before, to our knowledge. However, some authors have used only calcium and aluminium oxides and hydroxides for their HC synthesis which could facilitate the formation of a CaAl-OH-LDH phase [3,23–25]. In most of these studies involving the formation of HC, especially when attempting to form nitrate or chloride forms or studying the thermodynamics, carbonate contamination was limited. When carbonate contamination occurred, it was typically said to have occurred during synthesis, filtering or drying. In the studies where carbonate intercalation was not desired, great care was taken to carry out the synthesis in an inert environment (glove boxes or covered systems), use freshly precipitated $Ca(OH)_2$ and/or dried materials under nitrogen or in a desiccator to minimise contact with CO_2 [3,23–25].

In this study, an inert environment was accomplished by synthesising the LDHs under a slight over-pressure induced by constant nitrogen flow. The materials were filtered in open atmosphere but only required approximately 5 min to filter. XRD and FTIR-ATR analysis was carried out wet and within 0.5 h and 1 h after synthesis, respectively. The only possible CO_2 contamination from the atmosphere could have occurred during the 5 min of filtering (after which the filtered samples were sealed from air immediately and only unsealed for the wet XRD analysis and again for wet FTIR-ATR analysis). To put this into perspective, if the carbonate contamination were to come from CO_2 in air, at the atmospheric conditions in Pretoria, South Africa, where these experiments were conducted, 6.69 m^3 of air would have needed to pass through the sample and every available mol of CO_2 absorbed within the 5 min of filtration time or surface adsorbed during XRD and subsequently intercalated. As this is essentially impossible, the interlayer carbonate, if present to a significant degree, thus had to come from a different source.

For the synthesis of the materials, a fresh bottle of analytical grade $Ca(OH)_2$ was used. Unfortunately, after synthesis and our surprise of this possible CO_2 contamination, it was found that the $Ca(OH)_2$ used nevertheless consisted to 3.76% of calcite. This constitutes only 13.36% of the amount of calcite stoichiometrically required to form a fully carbonate intercalated CaAl-LDH, though. Taking into account theoretical requirements of carbonate to form a fully carbonate intercalated CaAl-LDH, one could thus expect that the LDH consisted of 13.36% carbonate in the interlayer. However, almost every material synthesised contained some calcite at the end of the synthesis. In prior studies [25,26,28], $CaCO_3$ was used as a carbonate source during synthesis and most of these studies produced very pure $CaAl-CO_3$-LDHs. Considering the approximately dried material yield, our materials consisted of between 0.5 g to 1 g of calcite after synthesis (approximately 2% as determined from Rietveld refinement). This is approximately as much as initially fed with the $Ca(OH)_2$, indicating that a small amount of carbonate could have been taken up by the CaAl-OH-LDHs, but a large fraction remained as calcite.

Our Rietveld refinement results showed that between 1% and 3% calcite typically remained in the material after synthesis. However, there were some outliers. Some materials (MR1, WS3, T1, T2 and A3) contained no or less than 1% calcite, while others (WS1 and T5) consisted of close to 10% calcite and CO_3 even of 20% calcite. MR1, due to the use of less $Ca(OH)_2$ during synthesis to achieve the required molar calcium-to-aluminium ratio would contain less calcite to begin with. This result is thus not entirely surprising. WS3 contained less material in general to achieve the desired water-to-solids ratio. In this case, one could argue, that the calcite could have dissolved better due to a higher shear synthesis induced by the thinner reaction mixture (the same would apply to MR1 and A3). However, FTIR-ATR showed a decreased vibrational strength in the doublet 1414 cm^{-1}/1361 cm^{-1} vibration, which in most literature is designated to interlayer carbonate. T1 and T2 both showed no remaining calcite, however, the primary phase present in these materials was found to most likely be a calcium aluminate carbonate hydrate of the form 3 CaO · Al$_2$O$_3$ · 0.5Ca(OH)$_2$ · 0.5CaCO$_3$ · 11.5H$_2$O, which could explain where some of the CaCO$_3$ was used. Interestingly, the calcium aluminate hydrate phases presented by [23] at 20 °C and 30 °C were not observed. The only difference between A3 and its other series materials (A1 and A2) was the use of a highly crystalline gibbsite as Al source. While interlayer carbonate was present in this material as identified through FTIR-ATR analysis, it had a much lower vibrational response at 1414 cm^{-1}/1361 cm^{-1} and 663 cm^{-1}, three vibrations that are typically linked to interlayer carbonate—especially the vibration at 663 cm^{-1}, where this material had by far the least intense vibrational response of all materials.

This leads to the discussion of the importance of the blue-shifted doublet (from the doublet 1414 cm^{-1}/1361 cm^{-1}) vibrations of the Al(OH)$_3$-SA and Al(OH)$_3$-ACE sources indicated in the results section. To our amazement, the possibility of carbonate contamination in HC through adsorption of CO$_2$ onto the surface of Al(OH)$_3$ has not been indicated in any previous research, to our knowledge. We were also unable to find any FTIR spectra of Al sources used in previous research for the synthesis of HC. It is, however, well established in the catalytic field that aluminium oxides and hydroxides adsorb carbonate species. CO$_2$ adsorption onto boehmite and other aluminium phases was studied by [38] at different activation temperatures and pressures of CO$_2$. While their findings of the adsorbed species do not match our vibrational responses, the amorphous nature of the Al(OH)$_3$ sources used in this text could still allow for good explanation of the reason why CO$_2$ adsorbed onto the surface of Al(OH)$_3$-SA and Al(OH)$_3$-ACE and not onto the crystalline gibbsite phase. Amorphicity of materials typically goes hand in hand with some degree of disorder in these systems, thus making it more likely to adsorb CO$_2$ onto coordinatively unsaturated Al atoms with octahedral coordination that are very active toward CO$_2$ adsorption. Another explanation could come from the presence of small amounts of sodium present in the amorphous Al(OH)$_3$. While Al does not form carbonates itself, the presence of Na easily leads to the formation of dawsonite (NaAl(OH)$_2$CO$_3$) phases. Prior to the mid-1990s, several authors reported vibrational bands between 1520 cm^{-1}–1570 cm^{-1} and/or 1350 cm^{-1}–1410 cm cm^{-1} that were linked to CO$_2$ adsorption on γ-alumina phases. Ref. [36] linked these to carbonate vibrations of the bidentate form which is enhanced through the reaction of moisture, CO$_2$ and Na to dawsonite-like phases that are vibrationally active in this region. The amorphous Al(OH)$_3$ used in this work, unfortunately, did not have an elemental analysis provided by the manufacturer, but some Na was present in Al(OH)$_3$-M. If small amounts of Na were present, this is a possible reason for these vibrations. The vibrational intensity also scaled with the surface area of the Al sources used. [37] also found these vibrations on their amorphous Al(OH)$_3$ but ascribed them to the monodentate carbonate form. They performed experiments reacting amorphous aluminium hydroxide with sodium bicarbonate solutions of different concentrations. The wet environment could be the reason for the difference in carbonate orientation as later described by [39] for the adsorption of CO$_2$ onto hydrotalcites in the presence of water vapour. Nevertheless, the different amounts of sodium bicarbonate solution used by [37] could give an indication towards the amount of carbonate adsorbed on the amorphous Al(OH)$_3$. The absorbance spectrum of their sample prepared in 0.1 M sodium bicarbonate solution (corresponding to 0.26 g C/kg Al(OH)$_3$ on the surface) fit the spectra obtained for Al(OH)$_3$-SA and Al(OH)$_3$-ACE best. If this was the amount of carbonate adsorbed on the surface of Al(OH)$_3$-SA, this would make the starting materials consist of an additional 1.56% of carbonate. Thus equalling a total carbonate (not calcium carbonate) percentage of approximately 2% supplied with the raw materials during synthesis.

If this was the amount of carbonate present in the system initially, this would not, however, explain the large amounts of calcite remaining in WS1, T5 and CO3. The 20% content of calcite in CO3 is easily explained by the calcite fed to the system to prepare this carbonate-intercalated form. The relatively high percentage of calcite in WS1 can be reasoned through the large amount of starting material present to achieve the desired water-to solids ratio. In T5 the only difference between this material and that synthesised with S3 is the reaction temperature (90 °C). A much higher pH was also recorded during this synthesis, which can easily be explained by the effect of temperature on the dissolution of the solid phases. There exist two possibilities for this increased amount in calcite. Either the high temperature reaction favours the formation of $CaCO_3$ (and insolubility of this phase) instead of a carbonate intercalated LDH at this elevated temperature and pH—thus depicting a close approximation to the "true" amount of carbonate in the system—or there was simply more carbonate present to start with and it is also less soluble at these conditions. $CaCO_3$ solubility is known to decrease at increasing temperatures and also at increasing pHs (at 25 °C and 1 bar) [40]. As all $Ca(OH)_2$ and $Al(OH)_3$ was taken from the same bottle, the last explanation thus seems unlikely, albeit not impossible.

While carbonate contamination definitely occurred (due to the variety of reasons mentioned), the phases formed in this work most likely consisted mainly of CaAl-OH-LDH as desired and as can be inferred by considering the difference in FTIR spectra of A3 and A1 in terms of the presence of adsorbed carbonate species on the Al sources and overall carbonate supplied to the system.

3.2. Increasing Conversion to HC

Considering the results presented in this work, it was evident that certain reaction parameters had a larger influence on the reaction outcome than others. Generally speaking, a higher purity HC was obtained by lowering the water-to-solids ratio, increasing the reaction time, having sufficient mixing, using an amorphous $Al(OH)_3$ source with high surface area, using an adequate reaction temperature (80 °C for the highest purity within 3 h) and most surprisingly, by using a calcium-to-aluminium ratio stoichiometrically favouring katoite formation. It is possible that this occurred because of an overall slightly lower pH during synthesis. pH sampling results showed that the materials with the highest HC purity (WS1 and MR1, 73.48% and 84.91%, respectively) followed a similar pH behaviour during synthesis, starting with a high pH and ending on a lower pH. We expect that this occurred due to the conversion to the LDH phase. It is possible that the high initial pH facilitates better dissolution of the phases (especially $Al(OH)_3$) for reaction. Unpublished results of the same molar-ratio-experiment (but conducted for a shorter period of time) indicated that the best purity was achieved by using the stoichiometric ratio for hydrocalumite formation, contradicting the results found here [41]. Time, especially in this regard, thus seems a very important factor.

During one of our previous studies, entirely different results were obtained for the temperature series by using a more crystalline $Al(OH)_3$ source and a shorter reaction time [27]. While the $Al(OH)_3$ used in that study was also sourced from ACE Chemicals, it showed to be a highly crystalline material, akin to the Merck source used in this work. Reaction kinetics were thus seemingly very different and the LDH-like phases formed in T1 and T2 remained undetected. The morphology, surface area, crystallinity and even crystal structure of the starting materials most likely remains one of the main contributing factors towards a successful synthesis in a short period of time. As seen by comparison of even our own results, the choice of starting materials can change the reaction outcome completely. Conversion to better crystallised and more morphologically even HCs seems to be achievable by using a higher crystallinity Al source. This is possibly observed due to lower reaction rates, a subsequent slower crystal growth and hence higher crystallinity and larger platelets formed.

In closing the discussion in this section, the presence of the particulate matter in LDHs formed and XRD identified phases deserve some discussion. As shown through the morphological evaluation of the $Ca(OH)_2$ and $Al(OH)_3$ phases used, it was difficult to discern these using SEM. Due to the high amount of portlandite remaining in many samples, it is highly likely that this particulate matter is a mixture of the two phases and that an increase in reaction time would increase the yield of the HC phase. As the cyrstal structure of the amorphous $Al(OH)_3$ could not be determined, it is also possible that slightly too little aluminium was fed to the system, which could have an effect on the final outcome with sufficient reaction time. During XRD, most scans contained very, very small amounts of phases that remained unidentified. It is possible that small amounts of especially the Cl-intercalated

HC form exist in some more samples. However, due to shifted peaks and/or other effects, these phases could not be assigned. Further, any Rietveld refinement results (as mentioned in the text) were only applicable to the crystalline phases formed. The fraction of amorphous material present in each material remains undetermined.

3.3. A Hint Towards the Reaction Mechanism of the Formation of HC in a Hydrothermal Process

Finally, a short discussion on what could potentially be inferred regarding the reaction mechanism through which HC phases form using hydrothermal synthesis. The temperature experiments showed that, at least at low temperatures, an alternate LDH-like phase was formed. Some research suggests that this compound be a meta-stable phase that forms in cement (although it had not been found in Portland cement at the time) [35]. It is possible that this is a sort of pre-cursor to the HC phase, at least in low temperature systems. This seems likely considering observations in cement literature. Ref. [42] investigated the time dependent formation of calcium carboaluminate phases. They found that calcium hemicarboaluminate transforms into calcium monocarboaluminate given time and reaches high conversion after 100 days of reaction in the cement phase. Only the calcium monocarboaluminate phase is present in well-hydrated fully cured cement. Further, the calcium hemicarboaluminate form occurs early in the hydration process of cement, even if large quantities of calcium carbonate are present. Ref. [43] described how these calcium carboaluminate phases only really constitute a large fraction of the cement phase after about a day of reaction.

No study of the hydrothermal formation mechanism of HC exists (to our knowledge), especially at elevated temperatures. At elevated temperatures, these calcium hemi/monocarboaluminate phases could not be identified, but there seems to be a strong relationship to the formation/depletion of katoite during synthesis. In our previous results [27], these and those of [26], katoite was present to a large degree at a synthesis temperature of 90 °C. Katoite was also present in larger amounts at lower temperatures than 80 °C. The higher katoite content at 90 °C could indicate that this is the more stable phase past 80 °C. No elevated temperatures are required, however, to form katoite, which made up a large fraction of the phases formed even at 20 °C. The largest fraction of katoite was present at 40 °C and 90 °C, though, possibly indicating that HC is the favoured phase between these two temperatures. The time experiments also revealed a great amount regarding the progression of the phase contents with time. At 6 h, only approximately a quarter of the katoite present after 1 h was observed. This seems to be a strong suggestion that, given enough time, katoite is converted into HC during synthesis and could thus be a precursor to the HC phase.

4. Materials and Methods

Chemically pure $Ca(OH)_2$ and $Al(OH)_3$ were sourced from ACE Chemicals (analytical reagent grade) and Sigma Aldrich (SA) (reagent grade). Two other $Al(OH)_3$ sources were used in the experiments. They were sourced from ACE Chemicals (chemical purity grade) and from Merck Chemicals (95% purity). Distilled and dissolved gas free (boiled prior to use and cooled/heated to the desired temperature under nitrogen flow) water was used in all experiments.

Experiments were performed using a bench top reactor set up as shown in Figure 25. The figure also depicts all variables that were investigated (time: t, temperature: T, mixing: M, molar ratio: MR, aluminium source: A and water-to-solids ratio: WS). An inert N_2 environment was maintained in all experiments at slight over-pressure. The dry reactant powders were added to preheated water at the desired temperature under constant stirring. The mixture was kept in suspension by magnetic stirring at the desired speed and reacted for the desired time at the desired temperature. pH measurements were taken intermittently. The samples were filtered using vacuum filtration, immediately sealed and analysed as a wet paste within 0.5 h with XRD and 1 h with ATR-FTIR.

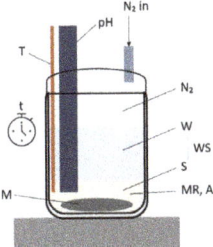

Figure 25. Setup used for the synthesis of hydrocalumite.

The synthesis of HC using Ca(OH)$_2$, Al(OH)$_3$ and H$_2$O stands in competition with the formation of katoite Ca$_3$Al$_2$(OH)$_{12}$ (a member of the hydrogrossular family) as shown Equation (2).

$$a\text{Ca(OH)}_2 + b\text{Al(OH)}_3 + c\text{H}_2\text{O} \longrightarrow d\text{Ca}_3\text{Al}_2(\text{OH})_{12} + c\text{H}_2\text{O} \quad (2)$$
$$\longrightarrow e\text{Ca}_4\text{Al}_2(\text{OH})_{12}A_{2/n} \cdot x\text{H}_2\text{O} + (c-x)\text{H}_2\text{O}$$

All lower case letters represent stoichiometric coefficients. A denotes the desired intercalated anion. Table 7 shows the experimental conditions used for each experiment group.

Table 7. Experimental conditions used for each of the experiments performed. The experiment IDs and colour codes defined in the table were used for each experiment in the text. Purple: molar Ca:Al ratio, red: temperature, green: time, blue: water : solids ratio, brown: Al source, grey: mixing. Note: S1, S2, S3 = MR2, T4, t2, WS2, A1 and M2. *Ca(OH)$_2$ used in CO3 was partially substituted with CaCO$_3$ as described in the text to achieve stoichiometric carbonate intercalation.

ID	Molar Ca:Al Ratio [-]	Temperature [°C]	Time [h]	Water: Solids Ratio [-]	Al Source [-]	Mixing [rpm]
S1, S2, S3, CO3*	2	80	3	80:20	SA	750
MR1	1.5	80	3	80:20	SA	750
MR2	2	80	3	80:20	SA	750
MR3	3	80	3	80:20	SA	750
T1	2	25	3	80:20	SA	750
T2	2	40	3	80:20	SA	750
T3	2	60	3	80:20	SA	750
T4	2	80	3	80:20	SA	750
T5	2	90	3	80:20	SA	750
t1	2	80	1	80:20	SA	750
t2	2	80	3	80:20	SA	750
t3	2	80	6	80:20	SA	750
WS1	2	80	3	70:30	SA	750
WS2	2	80	3	80:20	SA	750
WS3	2	80	3	90:10	SA	750
A1	2	80	3	80:20	SA	750
A2	2	80	3	80:20	ACE	750
A3	2	80	3	80:20	M	750
M1	2	80	3	80:20	SA	500
M2	2	80	3	80:20	SA	750
M3	2	80	3	80:20	SA	1000

XRD measurements were performed on a Panalytical X'Pert PRO X-ray diffractometer in $\theta - \theta$ configuration, using Fe filtered Co-Kα radiation (1.789), an X'Celerator detector and variable divergence- and fixed receiving slits. The data were collected in the angular range of $5° \leq 2\theta \leq 90°$ with a step size and time of $0.008° 2\theta$ and 13 s, respectively. Phases were identified using X'Pert Highscore plus software. Molar fractions of phases present were determined using Rietveld refinement. The samples were analysed wet in order to minimise carbonate contamination and changes in the crystal structure due to drying effects. This process was chosen to closely simulate in-situ XRD.

ATR-FTIR spectra were obtained using a Perkin Elmer 100 Spectrophotometer. Samples were pressed in place with a force arm. Spectra were obtained in the range of 550–4000 cm^{-1} each with 32 scans at a resolution of 2 cm^{-1}.

SEM micrographs were obtained using a Zeiss Ultra PLUS FEG SEM at 1 keV. Samples were coated with 1.4 nm carbon prior to analysis. The LDH samples were dried in a desiccator prior to study with SEM.

The BET surface areas of the materials were determined using isotherms recorded at 77.35 K with a Micromeritics TriStar II 3020. The samples were degassed at 80 °C for 1.5 h prior to the analysis.

5. Conclusions

In conclusion, HC formation was favoured in all experiments, making up between approximately 50% and 85% of the final crystalline phases obtained. The formation of HC stood in competition with the formation of katoite, with constituted a large fraction of the remaining phases. It is expected that only small amounts of carbonate were present in the CaAl-OH-LDH and that any carbonate contamination came from the $Ca(OH)_2$ (as calcite) and from the $Al(OH)_3$ as surface adsorbed carbonate species rather than from the air as previously suggested. At high temperatures, $CaCO_3$ formation seemed to be favoured instead of carbonate intercalation. The low solubility of carbonate species at elevated temperatures could be contributing factor to the low amount of carbonate intercalated.

The largest effect on HC purity was seen using a low water-to-solids ratio, increasing the reaction time, having sufficient mixing, using an amorphous $Al(OH)_3$ with a high surface area, using an adequate reaction temperature and most surprisingly, by using a calcium-to-aluminium ratio stoichiometrically favouring katoite formation instead of HC formation. The morphology, surface area and crystallinity of the starting materials played a significant role. pH effects caused by the amount of reactants supplied could also have played a role in the increased purity observed—possibly by facilitating better dissolution of the $Al(OH)_3$ phase—especially for low water-to-solids ratios and the stoichiometrically unfavoured molar calcium-to-aluminium ratios.

Finally, it was possible to obtain a hint regarding the reaction mechanism at elevated temperatures. At lower temperatures, it is possible that the formation of HC follows through the formation of calcium aluminate carbonate hydrate phases in conjunction with katoite, while at high temperatures, katoite formation seems to precede the formation of HC.

Author Contributions: Conceptualisation, B.R.G. and F.J.W.J.L.; methodology, B.R.G.; validation, B.R.G.; formal analysis, B.R.G.; investigation, B.R.G.; resources, F.J.W.J.L.; data curation, B.R.G.; writing—original draft preparation, B.G.; writing—review and editing, B.R.G. and F.J.W.J.L.; visualisation, B.R.G.; supervision, F.J.W.J.L.; project administration, F.J.W.J.L.; funding acquisition, F.J.W.J.L. All authors have read and agreed to the published version of the manuscript.

Funding: This research was funded by Techsparks (Pty) Ltd and the Technology and Human Resources for Industry Programme (THRIP) administered by the Department of Trade and Industry, South Africa, (grant number THRIP/133/31/03/2016). The APC was funded by the University of Pretoria, South Africa.

Acknowledgments: Particular thanks are extended to Wiebke Gröte (X-ray diffraction analyst) at the Department of Geology, University of Pretoria, South Africa, for making these time sensitive wet XRD measurements possible and performing the subsequent Rietveld refinement. We also thank David Viljoen for his assistance in the laboratory to allow for a timely publication of the results and editing the manuscript for language.

Conflicts of Interest: The authors declare no conflict of interest. The funders had no role in the design of the study, in the collection, analyses, or interpretation of data or in the writing of the manuscript. They consented to publish the results.

References

1. Forano, C.; Hibino, T.; Leroux, F.; Taviot-Guého, C. Chapter 13.1 Layered Double Hydroxides. In *Handbook of Clay Science*; Bergaya, F., Theng, B.K., Lagaly, G., Eds.; Elsevier: Amsterdam, The Netherlands, 2006; Volume 1, pp. 1021–1095. Developments in Clay Science.

2. Mills, S.J.; Christy, A.G.; Génin, J.M.R.; Kameda, T.; Colombo, F. Nomenclature of the hydrotalcite supergroup: Natural layered double hydroxides. *Mineral. Mag.* **2012**, *76*, 1289–1336. [CrossRef]
3. Sánchez-Cantú, M.; Camargo-Martínez, S.; Pérez-Díaz, L.M.; Hernández-Torres, M.E.; Rubio-Rosas, E.; Valente, J.S. Innovative method for hydrocalumite-like compounds' preparation and their evaluation in the transesterification reaction. *Appl. Clay Sci.* **2015**, *114*, 509–516. [CrossRef]
4. Sankaranarayanan, S.; Antonyraj, C.A.; Kannan, S. Transesterification of edible, non-edible and used cooking oils for biodiesel production using calcined layered double hydroxides as reusable base catalysts. *Bioresour. Technol.* **2012**, *109*, 57–62. [CrossRef]
5. Sipos, P.; Pálinkó, I. As-prepared and intercalated layered double hydroxides of the hydrocalumite type as efficient catalysts in various reactions. *Catal. Today* **2018**, *306*, 32–41. [CrossRef]
6. Shamitha, C.; Mahendran, A.; Anandhan, S. Effect of polarization switching on piezoelectric and dielectric performance of electrospun nanofabrics of poly(vinylidene fluoride)/Ca–Al LDH nanocomposite. *J. Appl. Polym. Sci.* **2020**, *137*, 48697. [CrossRef]
7. Soussou, A.; Gammoudi, I.; Kalboussi, A.; Grauby-Heywang, C.; Cohen-Bouhacina, T.; Baccar, Z.M. Hydrocalumite Thin Films for Polyphenol Biosensor Elaboration. *IEEE Trans. NanoBiosci.* **2017**, *16*, 650–655. [CrossRef]
8. Saha, S.; Ray, S.; Acharya, R.; Chatterjee, T.K.; Chakraborty, J. Magnesium, zinc and calcium aluminium layered double hydroxide-drug nanohybrids: A comprehensive study. *Appl. Clay Sci.* **2017**, *135*, 493–509. [CrossRef]
9. Shahabadi, N.; Razlansari, M.; Zhaleh, H.; Mansouri, K. Antiproliferative effects of new magnetic pH-responsive drug delivery system composed of Fe_3O_4, CaAl layered double hydroxide and levodopa on melanoma cancer cells. *Mater. Sci. Eng. C* **2019**, *101*, 472–486. [CrossRef] [PubMed]
10. Dutta, K.; Pramanik, A. Synthesis of a novel cone-shaped CaAl-layered double hydroxide (LDH): Its potential use as a reversible oil sorbent. *Chem. Commun.* **2013**, *49*, 6427–6429. [CrossRef] [PubMed]
11. Xu, Y.; Xu, X.; Hou, H.; Zhang, J.; Zhang, D.; Qian, G. Moisture content-affected electrokinetic remediation of Cr(VI)-contaminated clay by a hydrocalumite barrier. *Environ. Sci. Pollut. Res.* **2016**, *23*, 6517–6523. [CrossRef] [PubMed]
12. Chrysochoou, M.; Dermatas, D. Evaluation of ettringite and hydrocalumite formation for heavy metal immobilization: Literature review and experimental study. *J. Hazard. Mater.* **2006**, *136*, 20–33. [CrossRef] [PubMed]
13. Bernardo, M.P.; Moreira, F.K.; Ribeiro, C. Synthesis and characterization of eco-friendly Ca-Al-LDH loaded with phosphate for agricultural applications. *Appl. Clay Sci.* **2017**, *137*, 143–150. [CrossRef]
14. Das, S.; Roy, S. A newly designed softoxometalate $[BMIm]_2[DMIm][\alpha\text{-}PW_{12}O_{40}]$@hydrocalumite that controls the chain length of polyacrylic acid in the presence of light. *RSC Adv.* **2016**, *6*, 37583–37590. [CrossRef]
15. Labuschagne, F.J.W.J.; Molefe, D.M.; Focke, W.W.; Van Der Westhuizen, I.; Wright, H.C.; Royeppen, M.D. Heat stabilising flexible PVC with layered double hydroxide derivatives. *Polym. Degrad. Stab.* **2015**, *113*, 46–54. [CrossRef]
16. Moon, J.H.; Oh, J.E.; Balonis, M.; Glasser, F.P.; Clark, S.M.; Monteiro, P.J. Pressure induced reactions amongst calcium aluminate hydrate phases. *Cem. Concr. Res.* **2011**, *41*, 571–578. [CrossRef]
17. Rives, V. *Layered Double Hydroxides: Present and Future*; Nova Science Publishers: New York, NY, USA, 2001; pp. 39–92.
18. Muráth, S.; Somosi, Z.; Kukovecz, Á.; Kónya, Z.; Sipos, P.; Pálinkó, I. Novel route to synthesize CaAl- and MgAl-layered double hydroxides with highly regular morphology. *J. Sol-Gel Sci. Technol.* **2019**, *89*, 844–851. [CrossRef]
19. Labuschagné, F.J.; Wiid, A.; Venter, H.P.; Gevers, B.R.; Leuteritz, A. Green synthesis of hydrotalcite from untreated magnesium oxide and aluminum hydroxide. *Green Chem. Lett. Rev.* **2018**, *11*, 18–28. [CrossRef]
20. Ferencz, Z.; Kukovecz, Á.; Kónya, Z.; Sipos, P.; Pálinkó, I. Optimisation of the synthesis parameters of mechanochemically prepared CaAl-layered double hydroxide. *Appl. Clay Sci.* **2015**, *112–113*, 94–99. [CrossRef]
21. Peppler, R.; Wells, L. The system of lime, alumina, and water from 50-degrees to 250-degrees C. *J. Res. Natl. Bur. Stand.* **1954**, *52*, 75–92. [CrossRef]
22. Majumdar, A.J.; Roy, R. The system $CaO-Al_2O_3-H_2O$. *J. Am. Ceram. Soc.* **1956**, *40*, 434–442. [CrossRef]

23. Lothenbach, B.; Pelletier-Chaignat, L.; Winnefeld, F. Stability in the system CaO–Al_2O_3–H_2O. *Cem. Concr. Res.* **2012**, *42*, 1621–1634. [CrossRef]
24. Buttler, F.G.; Glasser, D.; Taylor, H.F.W. Studies on $4CaO \cdot Al_2O_3 \cdot 13 H_2O$ and the Related Natural Mineral Hydrocalumite. *J. Am. Ceram. Soc.* **1958**, *42*, 121–126. [CrossRef]
25. Renaudin, G.; Francois, M.; Evrard, O. Order and disorder in the lamellar hydrated tetracalcium monocarboaluminate compound. *Cem. Concr. Res.* **1999**, *29*, 63–69. [CrossRef]
26. Gabrovšek, R.; Vuk, T.; Kaučič, V. The preparation and thermal behavior of calcium monocarboaluminate. *Acta Chim. Slov.* **2008**, *55*, 942–950.
27. Gevers, B.R.; Labuschagné, F.J. Temperature effects on the dissolution-precipitation synthesis of hydrocalumite. In *AIP Conference Proceedings*; The American Institute of Physics: College Park, MD, USA, 2019; Volume 2055. [CrossRef]
28. Mesbah, A.; Rapin, J.P.; François, M.; Cau-Dit-Coumes, C.; Frizon, F.; Leroux, F.; Renaudin, G. Crystal structures and phase transition of cementitious Bi-anionic AFm-(Cl^-, CO_3^{2-}) compounds. *J. Am. Ceram. Soc.* **2011**, *94*, 262–269. [CrossRef]
29. Abd-El-Raoof, F.; Tawfik, A.; Komarneni, S.; Ahmed, S.E. Hydrotalcite and hydrocalumite as resources from waste materials of concrete aggregate and Al-dross by microwave-hydrothermal process. *Constr. Build. Mater.* **2019**, *207*, 10–16. [CrossRef]
30. Mora, M.; López, M.I.; Jiménez-Sanchidrián, C.; Ruiz, J.R. Ca/Al mixed oxides as catalysts for the meerwein-ponndorf-verley reaction. *Catal. Lett.* **2010**, *136*, 192–198. [CrossRef]
31. Kagunya, W.; Baddour-Hadjean, R.; Kooli, F.; Jones, W. Vibrational modes in layered double hydroxides and their calcined derivatives. *Chem. Phys.* **1998**, *236*, 225–234. [CrossRef]
32. Schroeder, P. Infrared spectroscopy in clay science. *Teach. Clay Sci.* **2002**, *11*, 181–206.
33. Kloprogge, J.T.; Ruan, H.D.; Frost, R.L. Thermal decomposition of bauxite minerals: Infrared emission spectroscopy of gibbsite, boehmite and diaspore. *J. Mater. Sci.* **2002**, *37*, 1121–1129. [CrossRef]
34. Runčevski, T.; Dinnebier, R.E.; Magdysyuk, O.V.; Pöllmann, H. Crystal structures of calcium hemicarboaluminate and carbonated calcium hemicarboaluminate from synchrotron powder diffraction data. *Acta Crystallogr. Sect. B: Struct. Sci.* **2012**, *68*, 493–500. [CrossRef] [PubMed]
35. Carlson, E.T.; Berman, H.A. Some observations on the calcium aluminate carbonate hydrates. *J. Res. Nation. Bureau Stand. Sect. A Phys. Chem.* **1960**, *64A*, 333–341. [CrossRef]
36. Lee, D.H.; Condrate, R.A. An FTIR spectral investigation of the structural species found on alumina surfaces. *Mater. Lett.* **1995**, *23*, 241–246. [CrossRef]
37. Su, C.; Suarez, D.L. In situ infrared speciation of adsorbed carbonate on aluminum and iron oxides. *Clays Clay Miner.* **1997**, *45*, 814–825. [CrossRef]
38. Morterra, C.; Emanuel, C.; Cerrato, G.; Magnacca, G. Infrared study of some surface properties of boehmite (γ-AlO_2H). *J. Chem. Soc. Faraday Trans.* **1992**, *88*, 339–348. [CrossRef]
39. Coenen, K.; Gallucci, F.; Mezari, B.; Hensen, E.; van Sint Annaland, M. An in-situ IR study on the adsorption of CO_2 and H_2O on hydrotalcites. *J. CO_2 Util.* **2018**, *24*, 228–239. [CrossRef]
40. Coto, B.; Martos, C.; Peña, J.L.; Rodríguez, R.; Pastor, G. Effects in the solubility of $CaCO_3$: Experimental study and model description. *Fluid Phase Equilibria* **2012**, *324*, 1–7. [CrossRef]
41. Gevers, B.R.; Labuschagné, F.J.W.J. Parameters influencing the formation of katoite. **2020** unpublished.
42. Ipavec, A.; Gabrovšek, R.; Vuk, T.; Kaučič, V.; Maček, J.; Meden, A. Carboaluminate phases formation during the hydration of calcite-containing Portland cement. *J. Am. Ceram. Soc.* **2011**, *94*, 1238–1242. [CrossRef]
43. Bizzozero, J.; Scrivener, K.L. Limestone reaction in calcium aluminate cement-calcium sulfate systems. *Cem. Concr. Res.* **2015**, *76*, 159–169. [CrossRef]

© 2020 by the authors. Licensee MDPI, Basel, Switzerland. This article is an open access article distributed under the terms and conditions of the Creative Commons Attribution (CC BY) license (http://creativecommons.org/licenses/by/4.0/).

Article

On the Reconstruction Peculiarities of Sol–Gel Derived Mg$_{2-x}$M$_x$/Al$_1$ (M = Ca, Sr, Ba) Layered Double Hydroxides

Ligita Valeikiene [1], Marina Roshchina [2], Inga Grigoraviciute-Puroniene [1], Vladimir Prozorovich [2], Aleksej Zarkov [1], Andrei Ivanets [2] and Aivaras Kareiva [1,*]

[1] Institute of Chemistry, Faculty of Chemistry and Geosciences, Vilnius University, Naugarduko 24, LT-03225 Vilnius, Lithuania; ligita.valeikiene@chgf.vu.lt (L.V.); inga.grigoraviciute@gmail.com (I.G.-P.); aleksej.zarkov@chf.vu.lt (A.Z.)
[2] Institute of General and Inorganic Chemistry of National Academy of Sciences of Belarus, st. Surganova 9/1, 220072 Minsk, Belarus; che.roschina@bsu.by (M.R.); vladimirprozorovich@gmail.com (V.P.); andreiivanets@yandex.by (A.I.)
* Correspondence: aivaras.kareiva@chgf.vu.lt; Tel.: +37061567428

Received: 6 May 2020; Accepted: 30 May 2020; Published: 2 June 2020

Abstract: In this study, the reconstruction peculiarities of sol–gel derived Mg$_{2-x}$M$_x$/Al$_1$ (M = Ca, Sr, Ba) layered double hydroxides were investigated. The mixed metal oxides (MMO) were synthesized by two different routes. Firstly, the MMO were obtained directly by heating Mg(M)–Al–O precursor gels at 650 °C, 800 °C, and 950 °C. These MMO were reconstructed to the Mg$_{2-x}$M$_x$/Al$_1$ (M = Ca, Sr, Ba) layered double hydroxides (LDHs) in water at 50 °C for 6 h (pH 10). Secondly, in this study, the MMO were also obtained by heating reconstructed LDHs at the same temperatures. The synthesized materials were characterized using X-ray powder diffraction (XRD) analysis and scanning electron microscopy (SEM). Nitrogen adsorption by the Brunauer, Emmett, and Teller (BET) and Barrett, Joyner, and Halenda (BJH) methods were used to determine the surface area and pore diameter of differently synthesized alkaline earth metal substituted MMO compounds. It was demonstrated for the first time that the microstructure of reconstructed MMO from sol–gel derived LDHs showed a "memory effect".

Keywords: layered double hydroxides; sol–gel processing; alkaline earth metals; mixed metal oxides; reconstruction effect; surface properties

---

1. Introduction

Layered double hydroxides ([M$^{2+}_{1-x}$M$^{3+}_x$(OH)$_2$]$^{x+}$(A^{y-})$_{x/y}$·zH$_2$O, where M^{2+} and M^{3+} are divalent and trivalent metal cations, respectively, and A^{y-} is a intercalated anion, LDHs) are widely used in catalysis, in ion-exchange processes, as catalyst support precursors, adsorbents, anticorrosion inhibitors, anion exchangers, flame retardants, polymer stabilizers, and in pharmaceutical applications, optics, in separation science and photochemistry [1–7]. The most common preparation technique of LDHs is the co-precipitation method starting from the soluble salts of the metals [8–10]. The second synthetic technique also widely used for the preparation of LDHs is anion exchange [11–13]. Recently, for the preparation of Mg$_3$Al$_1$ LDHs, we developed the indirect sol–gel synthesis route [14–17]. In this synthetic approach, the synthesized Mg–Al–O precursor gels were converted to the mixed metal oxides (MMO) by heating the gels at 650 °C. The LDHs were fabricated by the reconstruction of MMO in deionized water at 80 °C. The proposed sol–gel synthesis route for LDHs showed some benefits over the co-precipitation and anion-exchange methods such as simplicity, high homogeneity, and good crystallinity of the end synthesis products, effectiveness, cost efficiency, and suitability for the synthesis of different LDH compositions.

Recently, this newly developed sol–gel synthesis method has been successfully applied for the synthesis of the transition metal substituted layered double $Mg_{3-x}M_x/Al_1$ (M = Mn, Co, Ni, Cu, Zn) [18]. Calcined at temperatures higher than 600–650 °C, the M/Mg/Al LDHs form $M_xMg_{1-x}Al_2O_4$ solid solutions having the spinel structure and various cations distributions [19–23]. It was reported that these spinel structure compounds obtained at high temperatures cannot be reconstructed to the LDHs [24–29].

The investigation of mixed oxides derived from calcined LDHs prepared by direct and indirect methods is an interesting topic, since the reformation conditions could have an effect not only on the composition of a solid but also on the morphology of oxides and consequently on the properties. Usually, the calcined LDHs materials or mixed metal oxides have high surface areas. During the calcination, the dehydroxylation of LDHs with different chemical composition gave rise to the crystal deformation and interstratified structure of metal oxides, resulting in the development of mesopores and enhancement of specific surface area and enhanced sorption capacity [30–34]. Interestingly, the obtained MMO sometimes can preserve the morphology of the LDH precursor and show also the efficient recycling of the spent adsorbent [35,36]. It has been found that the nature of the partially introduced cation into the M^{2+} position influences the conditions of thermal decomposition of LDHs and also the structural and morphological features of the formed mixed metal oxides [37]. The obtained data can be used to synthesize the oxide supports with desired adsorption and other physical properties. In this study, the alkaline earth metal substituted $Mg_{2-x}M_x/Al_1$ (M = Ca, Sr, Ba) layered double hydroxides were synthesized by an indirect sol–gel method. The aim of this study was to decompose the sol–gel-derived LDHs at different temperatures and investigate the possible reconstruction of obtain mixed metal oxides to LDHs. The surface area and porosity as important characteristics of these alkaline earth metal substituted MMO materials were investigated in this study as well.

2. Experimental

Aluminium nitrate nonahydrate ($Al(NO_3)_3 \cdot 9H_2O$, 98.5%, Chempur, Plymouth, MI, USA), magnesium nitrate hexahydrate ($Mg(NO_3)_2 \cdot 6H_2O$), 99,0% Chempur, Plymouth, MI, USA), calcium nitrate tetrahydrate ($Ca(NO_3)_2 \cdot 4H_2O$, 99%, Chempur, Plymouth, MI, USA), strontium nitrate ($Sr(NO_3)_2$, 99.0%, Chempur, Plymouth, MI, USA) and barium nitrate ($Ba(NO_3)_2$, 99.0%, Chempur, Plymouth, MI, USA) were used as metal sources in the preparation of $Mg_{2-x}M_x/Al_1$ (M = Ca, Sr, Ba) layered double hydroxides. In the sol–gel processing, citric acid monohydrate ($C_6H_8O_7 \cdot H_2O$, 99.5%, Chempur, Plymouth, MI, USA) and 1,2-ethanediol ($C_2H_6O_2$, 99.8%, Chempur, Plymouth, MI, USA) were used as complexing agents. Ammonia solution (NH_3, 25%, Chempur, Plymouth, MI, USA) was used to change pH of the solution.

For the synthesis of $Mg_{2-x}M_x/Al_1$ (M = Ca, Sr, Ba; x is a molar part of substituent metal) LDHs, the stoichiometric amounts of starting materials were dissolved in distilled water under continuous stirring. Citric acid was added to the above solution, and the obtained mixture was stirred for an additional 1 h at 80 °C. Then, 2 mL of 1,2-ethanediol was added to the resulting solution. The transparent gels were obtained by the complete evaporation of the solvent under continuous stirring at 150 °C. The synthesized precursor gels were dried at 105 °C for 24 h. The mixed metal oxides (MMO) were obtained by heating the gels at 650 °C, 800 °C, and 950 °C for 4 h. The $Mg_{2-x}M_x/Al_1$ (M = Ca, Sr, Ba) LDHs were obtained by reconstruction of the MMO in water at 50 °C for 6 h under stirring and by changing the pH of the solution to 10 with ammonia.

X-ray diffraction (XRD) analysis was performed using a MiniFlex II diffractometer (Rigaku, The Woodlands, TX, USA) (Cu Kα radiation) in the 2θ range from 10° to 70° (step of 0.02°) with the exposition time of 2 min per step. The morphological features of MMO samples were estimated using a scanning electron microscope (SEM) Hitachi SU-70, Tokyo, Japan. Nitrogen adsorption by the Brunauer, Emmett, and Teller (BET) and Barret method was used to determine the surface area and pore diameter of the materials (Tristar II, Norcross, GA, USA). The pore-size distribution was evaluated by the Barrett–Joyner–Halenda (BJH) procedure. Prior to analysis, the calcined samples were outgassed at 523 K for 5 h.

3. Results and Discussion

To study the reconstruction peculiarities of sol–gel derived $Mg_{2-x}M_x/Al_1$ (M = Ca, Sr, Ba) layered double hydroxides (LDHs), the precursor gels were firstly annealed at 650 °C, 800 °C, and 950 °C for 4 h. The XRD patterns of mixed metal oxides (MMO) obtained by heating the Mg_2/Al_1 LDHs precursor gels at different temperatures are presented in Figure 1. The XRD pattern of the sample heated at 650 °C had two XRD peaks, which show the formation of a mixed metal oxide (MMO) phase with an MgO-like structure (JCPDS No. 96-100-0054) [14]. Thermal treatment of the precursor gels at 800 °C resulted in the formation of two phases, namely MMO and a low-crystallinity spinel phase with the composition of $MgAl_2O_4$ (JCPDS No. 96-154-0776). After heating at 950 °C, evidently, the highly crystalline $MgAl_2O_4$ phase has formed along with the MgO phase.

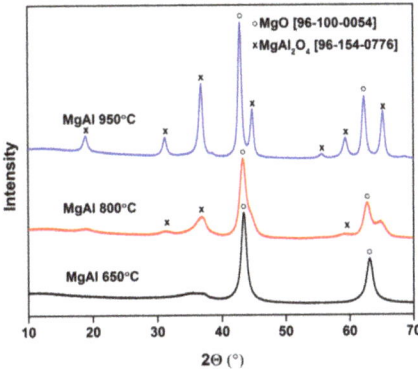

Figure 1. XRD patterns of mixed metal oxides (MMO) obtained by heating the Mg_2/Al_1 precursor gels at 650 °C, 800 °C, and 950 °C.

The XRD patterns of the Mg/Al LDH synthesized by the indirect sol–gel method (reconstruction of sol–gel derived MMO) show the formation of layered double hydroxides independent of the annealing temperature of the precursor gels (see Figure 2).

Three basal reflections typical of an LDH structure were observed: at 2θ of about 10° (003), 23° (006), and 35° (009) [14,15]. Besides, the spinel phase obtained at 800 °C and 950 °C remain almost unchanged during the reconstruction process. These results are in good agreement with those previously published elsewhere [38].

The XRD patterns of synthesis products with the same substitutional level of Ca, Sr, and Ba obtained at 800 °C and 950 °C are almost identical and revealed in all cases with the formation of crystalline magnesium oxide, magnesium spinel phase $MgAl_2O_4$ and an appropriate spinel of alkaline earth metal ($CaAl_2O_4$, $SrAl_2O_4$ and $BaAl_2O_4$). Again, during the partial reconstruction process, the phase purity of sol–gel derived $Mg_{2-x}M_x/Al_1$ LDHs evidently is dependent on the nature of introduced metal. As was expected, the spinel phases obtained at 800 °C and 950 °C remained almost unchanged during the partial reconstruction process. Moreover, during the reconstruction process, a negligible amount of metal carbonates ($CaCO_3$, $SrCO_3$, and $BaCO_3$) have formed as well. The XRD patterns of mixed metal oxides (MMO) obtained by heating the $Mg_{2-x}M_x/Al_1$ (M = Ca, Sr, Ba) precursor gels at different temperatures and reconstructed LDHs are presented in Figures 3–5, respectively. The XRD analysis results confirmed that the phase purity of alkaline earth substituted LDHs obtained by an indirect sol–gel synthesis approach is highly dependent on both the annealing temperature of the precursor gels and that of the alkaline earth metal.

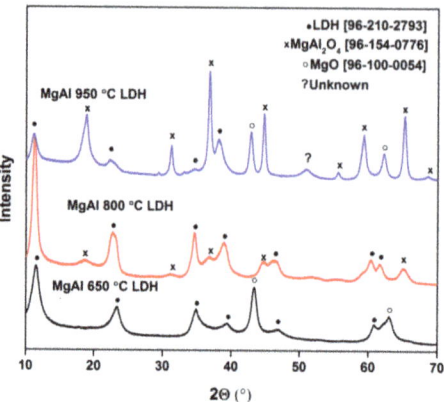

Figure 2. XRD patterns of sol–gel derived Mg$_2$/Al$_1$ layered double hydroxides (LDHs, reconstructed from MMO). The annealing temperature of the precursor gels was 650 °C, 800 °C, and 950 °C.

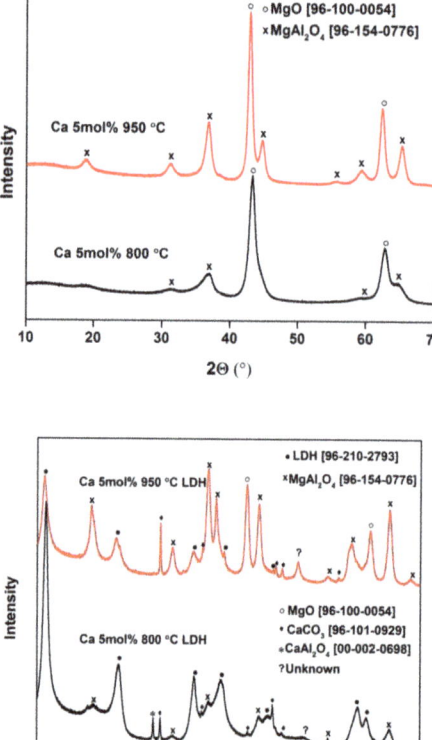

Figure 3. XRD patterns of mixed metal oxides (MMO) obtained by heating the Mg$_{1.95}$Ca$_{0.05}$/Al$_1$ precursor gels at 800 °C and 950 °C (**top**) and reconstructed Mg$_{1.95}$Ca$_{0.05}$/Al$_1$ LDHs (**bottom**).

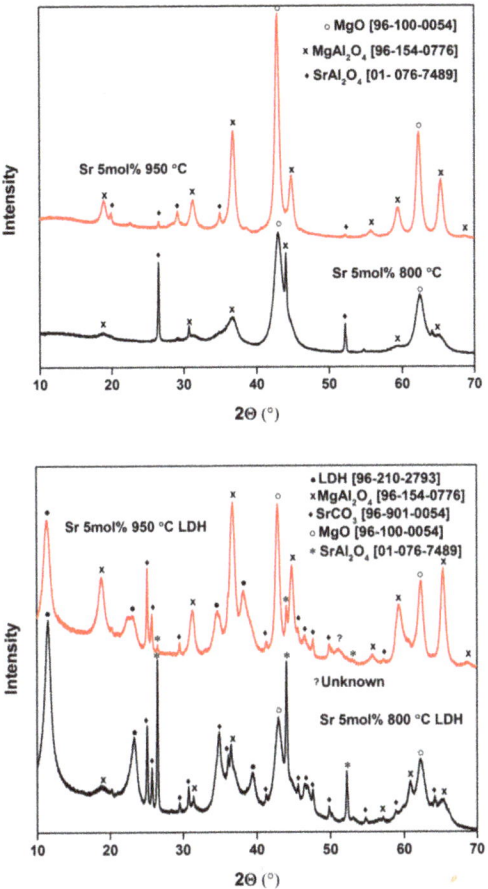

Figure 4. XRD patterns of mixed metal oxides (MMO) obtained by heating the $Mg_{1.95}Sr_{0.05}/Al_1$ precursor gels at 800 °C and 950 °C (**top**) and reconstructed $Mg_{1.95}Sr_{0.05}/Al_1$ LDHs (**bottom**).

The obtained mixed metal LDH samples were repeatedly heated at different temperatures to obtain MMO and compare the phase composition, morphology, and surface properties with obtained ones after initial annealing. The XRD patterns of non-substituted and Ca, Sr, and Ba containing MMO obtained after the heating of LDHs are shown in Figures 6 and 7, respectively. Evidently, the XRD patterns of mixed metal oxides (MMO) obtained by heating the Mg_2/Al_1 precursor gels (see Figure 1) and obtained by heating the Mg_2/Al_1 LDHs (Figure 6) are very similar, confirming the same phase composition. However, the reflections of just obtained MMO are more intense in comparison with ones presented in the repeatedly obtained MMO from LDHs. Obviously, the second time obtained Ca and Sr substituted MMO samples contain much more side phases (see Figures 3, 4 and 7). However, this is not the case for the Ba-substituted MMO samples. Both synthesis products obtained from precursor gels and by heating LDHs were composed of several crystalline phases.

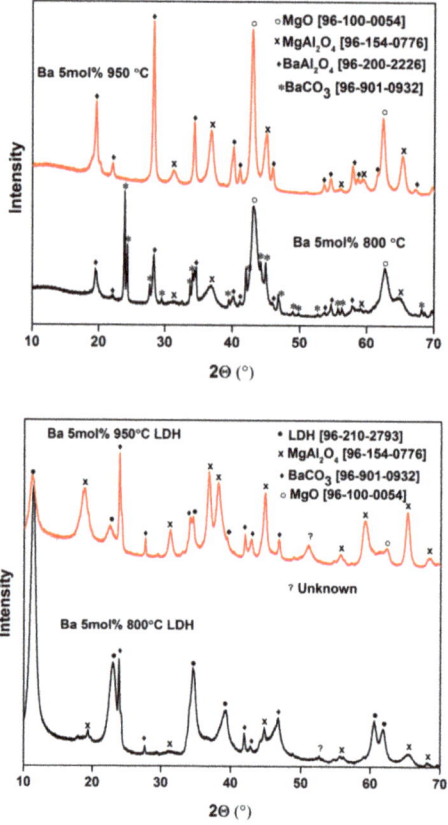

Figure 5. XRD patterns of mixed metal oxides (MMO) obtained by heating the $Mg_{1.95}Ba_{0.05}/Al_1$ precursor gels at 800 °C and 950 °C (**top**) and reconstructed $Mg_{1.95}Ba_{0.05}/Al_1$ LDHs (**bottom**).

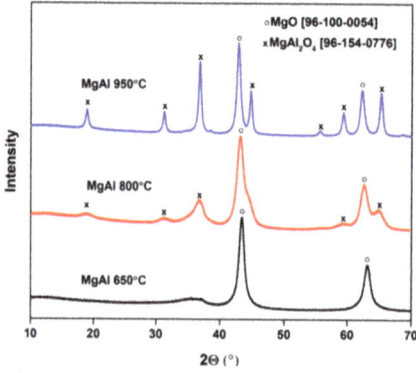

Figure 6. XRD patterns of mixed metal oxides (MMO) obtained by heating the Mg_2/Al_1 LDHs at different temperatures.

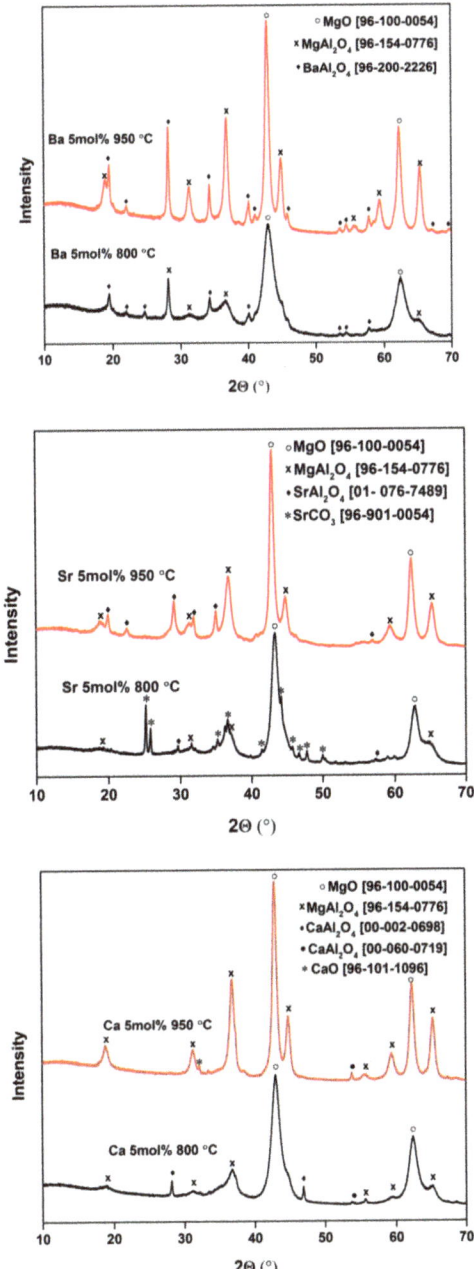

Figure 7. XRD patterns of mixed metal oxides (MMO) obtained by heating the $Mg_{1.95}Ca_{0.05}/Al_1$ LDHs (**bottom**), $Mg_{1.95}Sr_{0.05}/Al_1$ LDHs (**middle**), and $Mg_{1.95}Ba_{0.05}/Al_1$ LDHs (**top**) at different temperatures.

The interplanar spacings and lattice parameters of sol–gel derived $Mg_{2-x}M_x/Al_1$ (M = Ca, Sr and Ba) LDHs with standard deviations in parentheses are presented in Table 1.

Table 1. The interplanar spacings and lattice parameters of sol–gel derived $Mg_{1.95}M_{0.05}/Al_1$ (M = Ca, Sr and Ba) LDHs.

Compound	d(003), Å	d(006), Å	d(110), Å	c, Å	a, Å
Mg_2Al_1	7.601(5)	3.810(3)	1.511(3)	22.818(3)	3.040(2)
$Mg_{1.95}Ca_{0.05}/Al_1$	7.690(4)	3.813(4)	1.512(2)	23.027(4)	3.041(3)
$Mg_{1.95}Sr_{0.05}/Al_1$	7.697(3)	3.826(3)	1.517(3)	23.034(5)	3.039(4)
$Mg_{1.95}Ba_{0.05}/Al_1$	7.695(6)	3.828(4)	1.521(1)	23.176(5)	3.041(4)

Surprisingly, the calculated values of parameter a are not increasing monotonically with the increasing ionic radius of metal in $Mg_{1.95}M_{0.05}/Al_1$. On the other hand, the amount of substituent is rather small, and the obtained LDHs were not fully monophasic.

Figures 8–11 show the morphological features of non-substituted and alkaline earth metal-substituted LDHs and MMO obtained by heating the precursor gels or Mg_2/Al_1 LDHs. The SEM micrographs of Mg–Al MMO obtained by heating Mg–Al–O precursor gel (Figure 8) confirm that the surface of synthesized compounds is composed of large monolithic particles at about 15–20 µm in size independent of the annealing temperature (800 °C and 950 °C). The surface of these monoliths is randomly covered with smaller needle-like particles, and some pores also could be detected. The SEM micrographs of reconstructed from MMO Mg_2/Al_1 LDH samples showed different morphological features. The formation of round particles (3–15 µm) could be observed, and these particles are composed of nanosized plate-like crystallites. The most interesting observation is that the surface morphology of MMO samples obtained by heating Mg_2/Al_1 LDH specimens show "memory effect". In this case, the surface morphology of MMO is almost identical to the morphology of primary Mg_2/Al_1 LDHs. On the other hand, the morphological features of differently obtained MMO (MMO obtained by heating Mg–Al–O precursor gel and MMO obtained by heating Mg_2/Al_1 LDHs) differ considerably (see Figure 8).

The SEM micrographs of MMO obtained by heating the $Mg_{1.95}Ca_{0.05}/Al_1$ precursor gels, sol–gel derived $Mg_{1.95}Ca_{0.05}/Al_1$ LDHs, and MMO obtained by heating the $Mg_{1.95}Ca_{0.05}/Al_1$ LDHs are presented in Figure 9. The surface of Ca containing MMO obtained by heating the $Mg_{1.95}Ca_{0.05}/Al_1$ precursor gels is composed of large monolithic particles (≥20 µm). Apparently, the different morphological features could be determined for the reconstructed $Mg_{2-x}Ca_x/Al_1$ LDH samples. The plate-like crystals with sizes of 5–15 µm composed of nanosized plate-like crystallites have formed. An almost identical microstructure was observed for the MMO specimens obtained after heating $Mg_{2-x}Ca_x/Al_1$ LDH samples. SEM micrographs of strontium containing MMO and related $Mg_{2-x}Sr_x/Al_1$ LDHs are presented in Figure 10. The surface microstructure of Sr-containing MMO obtained by heating the $Mg_{1.95}Sr_{0.05}/Al_1$ precursor gels is very similar to the Ca-containing ones. However, on the surface of plate-like crystals of reconstructed $Mg_{2-x}Sr_x/Al_1$ LDH samples, additionally spherical particles (approximately 1 µm) were determined. These spherical particles as a "memory effect" remain on the surface of already heat-treated Sr containing LDHs. Again, the microstructure of investigated samples was not dependent on the annealing temperature. Interestingly, the barium containing $Mg_{2-x}Ba_x/Al_1$ LDH samples showed the formation of smaller LDH particles (2–5 µm) (Figure 11). The formation of plate-like crystals of MMO with the size of 7.5–12.5 µm was observed by heating these LDHs at elevated temperatures.

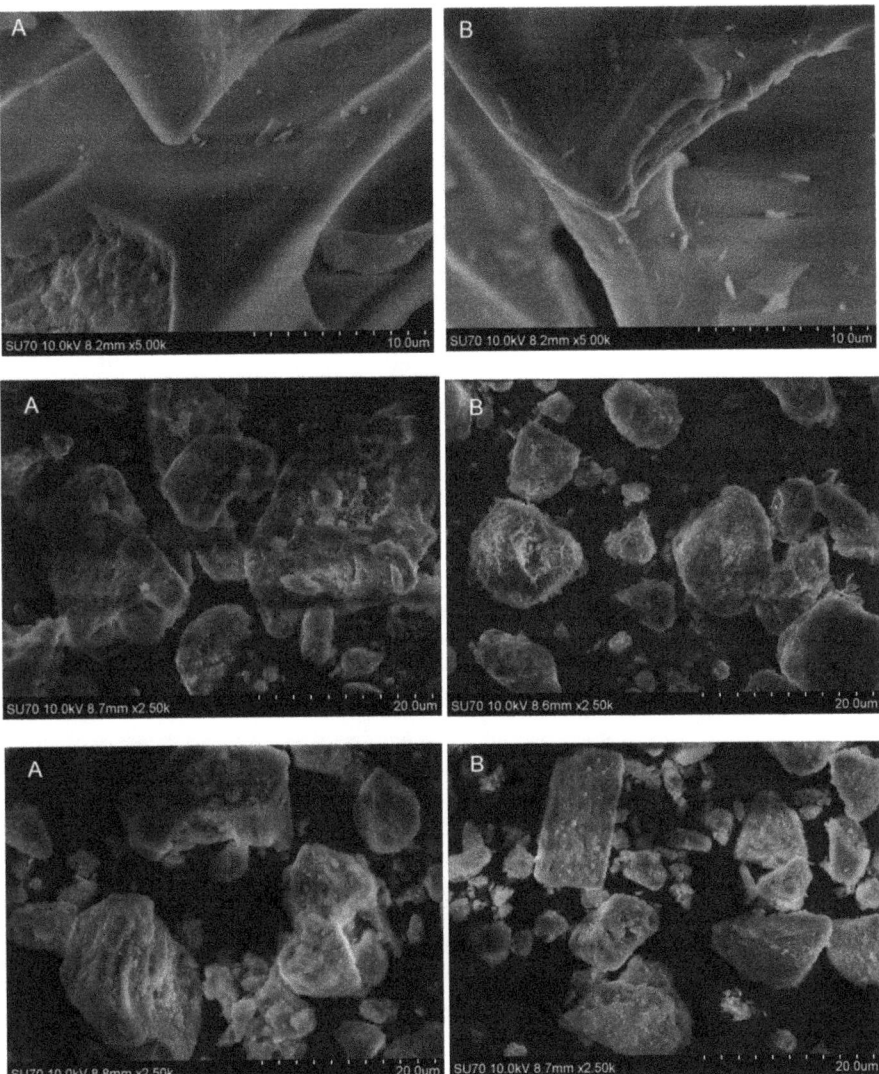

Figure 8. SEM micrographs of MMO obtained by heating the Mg_2/Al_1 precursor gels (**top**), sol–gel derived Mg_2/Al_1 LDHs (**middle**) and MMO obtained by heating the Mg_2Al_1 LDHs (**bottom**). Annealing temperatures: 800 °C (**A**) and 950 °C (**B**).

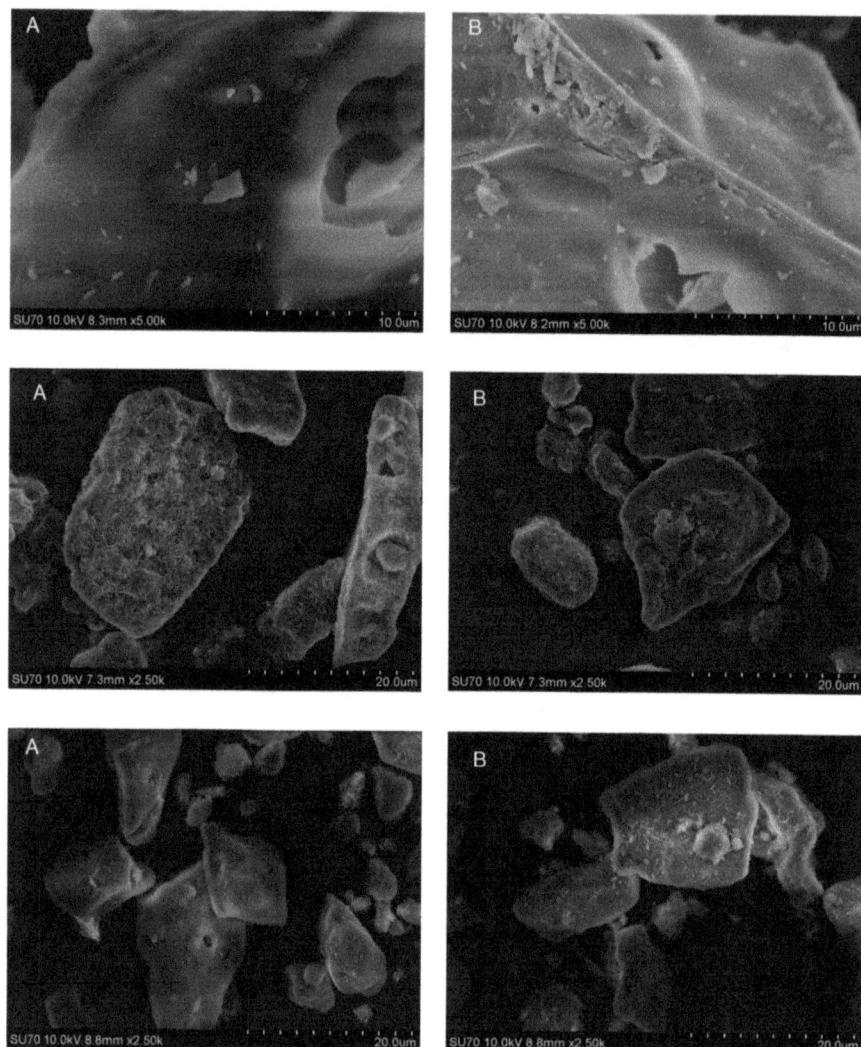

Figure 9. SEM micrographs of MMO obtained by heating the $Mg_{1.95}Ca_{0.05}/Al_1$ precursor gels (**top**), sol–gel derived $Mg_{1.95}Ca_{0.05}/Al_1$ LDHs (**middle**) and MMO obtained by heating the $Mg_{1.95}Ca_{0.05}/Al_1$ LDHs (**bottom**). Annealing temperatures: 800 °C (**A**) and 950 °C (**B**).

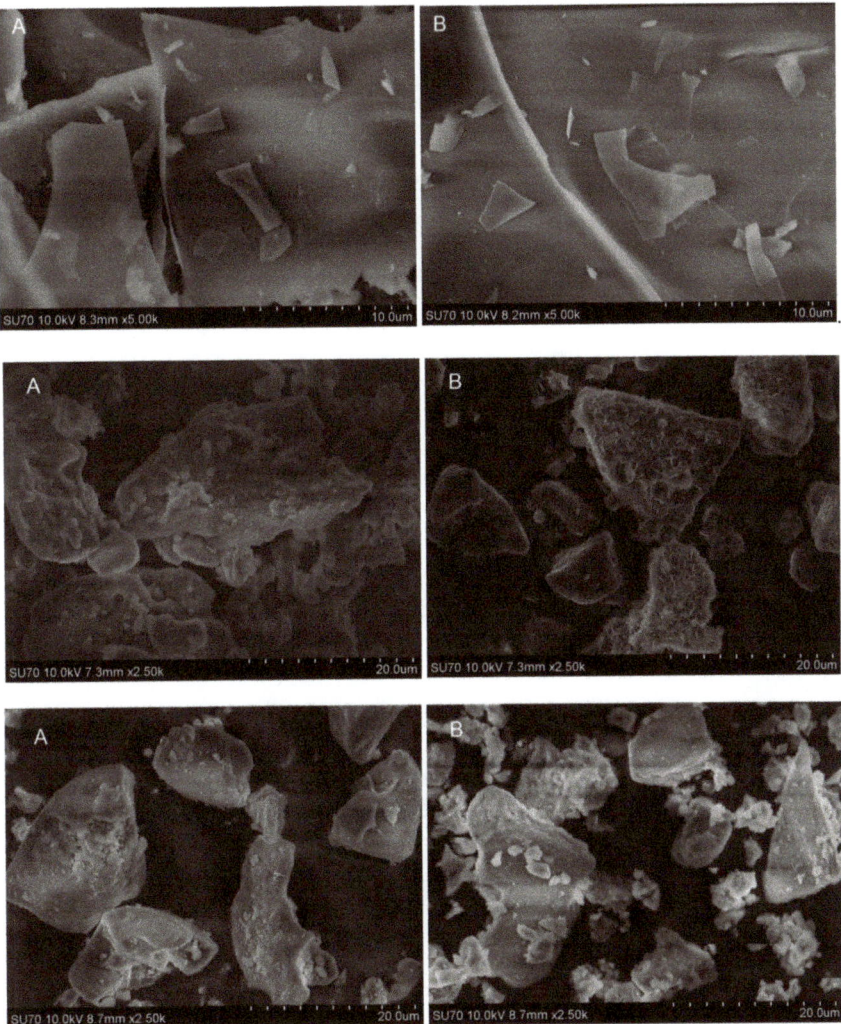

Figure 10. SEM micrographs of MMO obtained by heating the $Mg_{1.95}Sr_{0.05}/Al_1$ precursor gels (**top**), sol–gel derived $Mg_{1.95}Sr_{0.05}/Al_1$ LDHs (**middle**), and MMO obtained by heating the $Mg_{1.95}Sr_{0.05}/Al_1$ LDHs (**bottom**). Annealing temperatures: 800 °C (**A**) and 950 °C (**B**).

The results received by the BET method on Mg–Al MMO obtained by heating Mg–Al–O precursor gel and Mg_2/Al_1 LDHs are presented in Figure 12. Interestingly, these results of MMO obtained from Mg_2/Al_1 LDHs are comparable with those determined for the Mg_3/Al_1 LDH samples [18]. These samples exhibit type IV isotherms independent of the annealing temperature. At higher pressure values, the H1 hystereses are seen. This type of hysteresis is characteristic for the mesoporous (pore size in the range of 2–50 nm) materials. However, in the case of the MMO obtained by heating Mg–Al–O precursor gel, the steep increase at relatively low pressures let us predict the type of H4 isotherms, especially for the MMO samples obtained at lower temperature (800 °C). The surface area of these MMO samples evidently depends on the synthesis temperature.

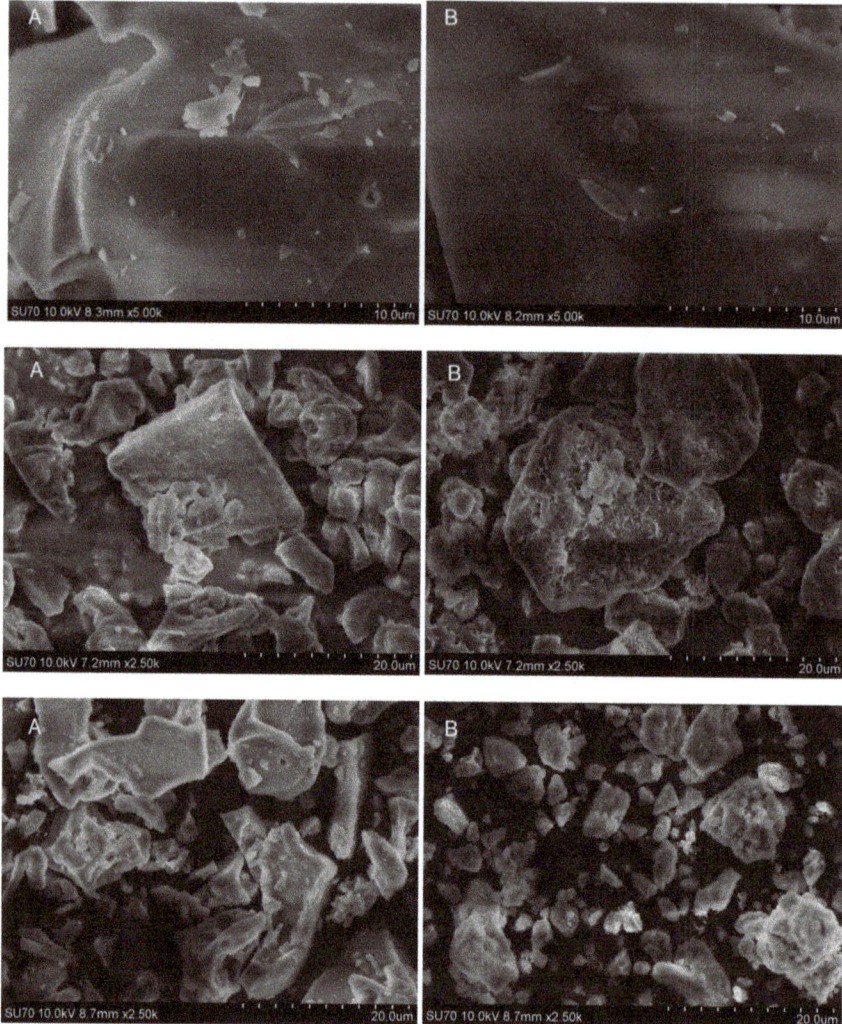

Figure 11. SEM micrographs of MMO obtained by heating the $Mg_{1.95}Ba_{0.05}/Al_1$ precursor gels (**top**), sol–gel derived $Mg_{1.95}Ba_{0.05}/Al_1$ LDHs (**middle**) and MMO obtained by heating the $Mg_{1.95}Ba_{0.05}/Al_1$ LDHs (**bottom**). Annealing temperatures: 800 °C (**A**) and 950 °C (**B**).

Thus, the isotherms and hystereses are dependent on both synthesis pathway and annealing temperature. The nitrogen adsorption–desorption results obtained for the mixed metal oxides containing Ca, Sr, and Ba (Figure 13) demonstrated that the N_2 adsorption–desorption isotherms show very similar trends.

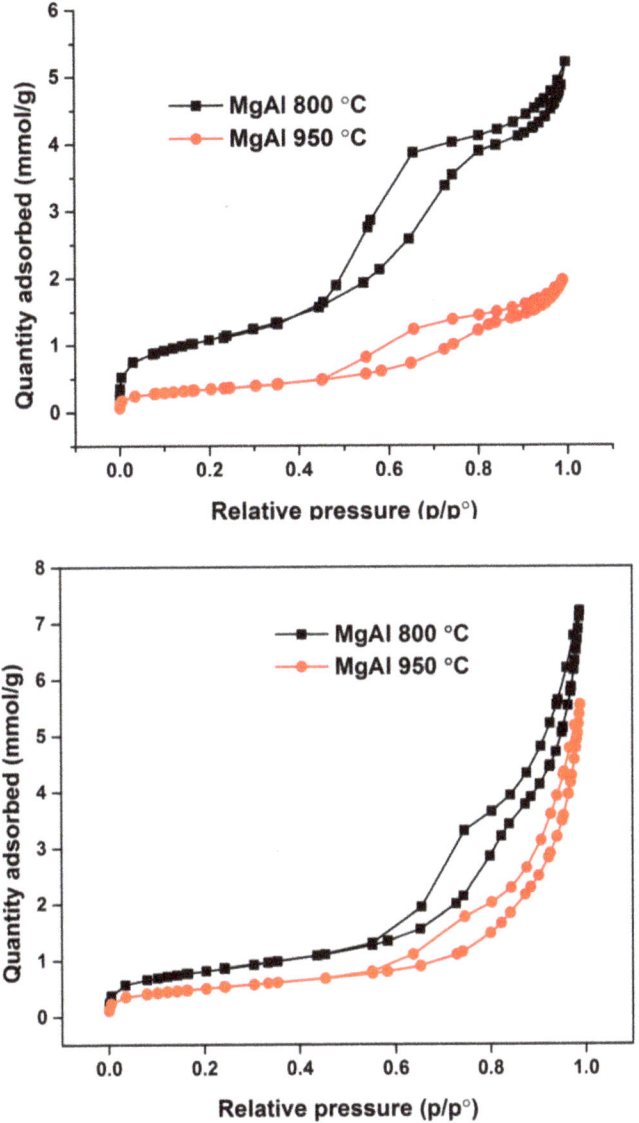

Figure 12. Nitrogen adsorption–desorption isotherms of mixed metal oxides (MMO) obtained by heating the Mg_2/Al_1 precursor gels (**top**) and obtained by heating the Mg_2Al_1 LDHs (**bottom**) at 800 °C and 950 °C.

However, in the case of barium-substituted MMO samples synthesized at 800 °C, the determined N_2 adsorption–desorption isotherms exhibited same type of isotherms independent of the synthesis method.

The results of the BET analysis of MMO samples are summarized in Table 2.

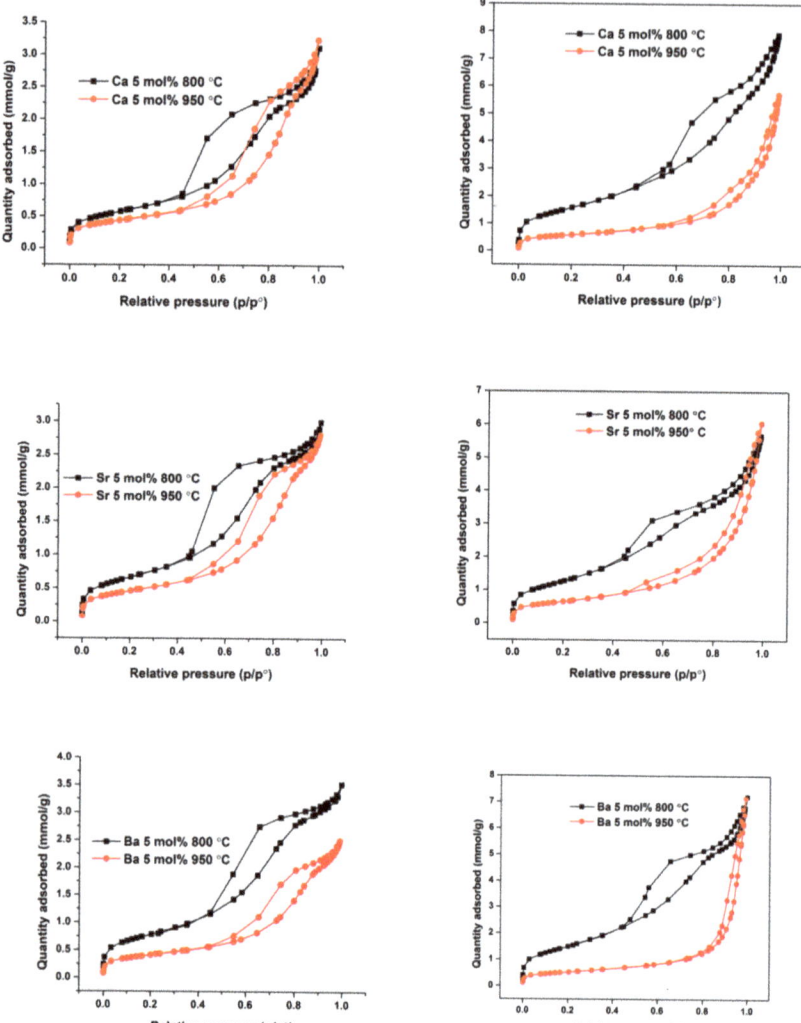

Figure 13. Nitrogen adsorption–desorption isotherms of mixed metal oxides (MMO) obtained by heating the $Mg_{1.95}Ca_{0.05}/Al_1$ precursor gels (**top, left**), by heating the $Mg_{1.95}Ca_{0.05}/Al_1$ LDHs (**top, right**), by heating the $Mg_{1.95}Sr_{0.05}/Al_1$ precursor gels (**middle, left**), by heating the $Mg_{1.95}Sr_{0.05}/Al_1$ LDHs (**middle, right**), by heating the $Mg_{1.95}Ba_{0.05}/Al_1$ precursor gels (**bottom, left**), and by heating the $Mg_{1.95}Ba_{0.05}/Al_1$ LDHs (**bottom, right**) at 800 °C and 950 °C.

Figure 14 shows the pore size distributions obtained by the BJH method for the MMO specimens obtained by heating Mg–Al–O precursor gels and Mg_2/Al_1 LDHs. Both samples demonstrate narrow pore size distributions (PSD) almost at the mesoporous level, but very close to micropores domain.

Table 2. Brunauer, Emmett and Teller (BET) surface area of sol–gel derived $Mg_{1.95}M_{0.05}/Al_1$ (M = Ca, Sr and Ba) MMO.

Precursor Compound	Temperature	BET Surface Area m^2/g
Mg_2Al_1 precursor gels	800 °C	87.470
Mg_2Al_1	800 °C	65.450
Mg_2Al_1 precursor gels	950 °C	27.749
Mg_2Al_1	950 °C	40.528
$Mg_{1.95}Ca_{0.05}/Al_1$ precursor gels	800 °C	46.461
$Mg_{1.95}Ca_{0.05}/Al_1$	800 °C	129.16
$Mg_{1.95}Ca_{0.05}/Al_1$ precursor gels	950 °C	34.791
$Mg_{1.95}Ca_{0.05}/Al_1$	950 °C	46.078
$Mg_{1.95}Sr_{0.05}/Al_1$ precursor gels	800 °C	53.847
$Mg_{1.95}Sr_{0.05}/Al_1$	800 °C	104.543
$Mg_{1.95}Sr_{0.05}/Al_1$ precursor gels	950 °C	36.292
$Mg_{1.95}Sr_{0.05}/Al_1$	950 °C	51.961
$Mg_{1.95}Ba_{0.05}/Al_1$ precursor gels	800 °C	63.217
$Mg_{1.95}Ba_{0.05}/Al_1$	800 °C	122.486
$Mg_{1.95}Ba_{0.05}/Al_1$ precursor gels	950 °C	32.498
$Mg_{1.95}Ba_{0.05}/Al_1$	950 °C	40.576

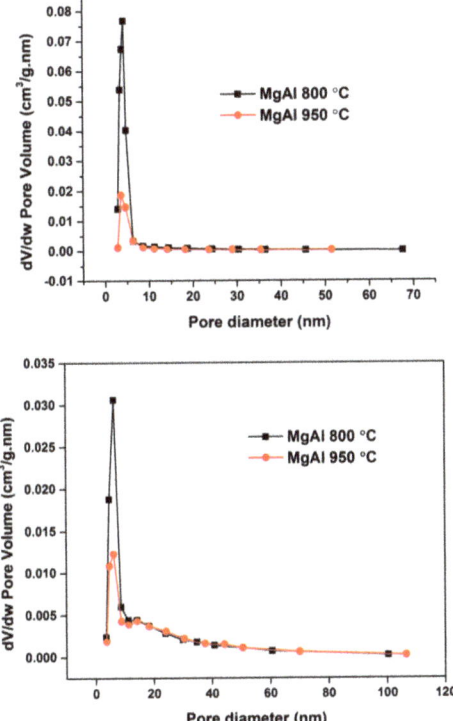

Figure 14. The pore size distribution of mixed metal oxides (MMO) obtained by heating the Mg_2/Al_1 precursor gels (**top**) and obtained by heating the Mg_2/Al_1 LDHs (**bottom**) at 800 °C and 950 °C.

Surprisingly, the PSD width does not depend neither on the synthetic procedure nor on the annealing temperature. The determined average pore diameter in the mesopore region is approximately 3.0–5.5 nm. The PSD results obtained for the mixed metal oxides containing Ca, Sr, and Ba are shown in Figure 15.

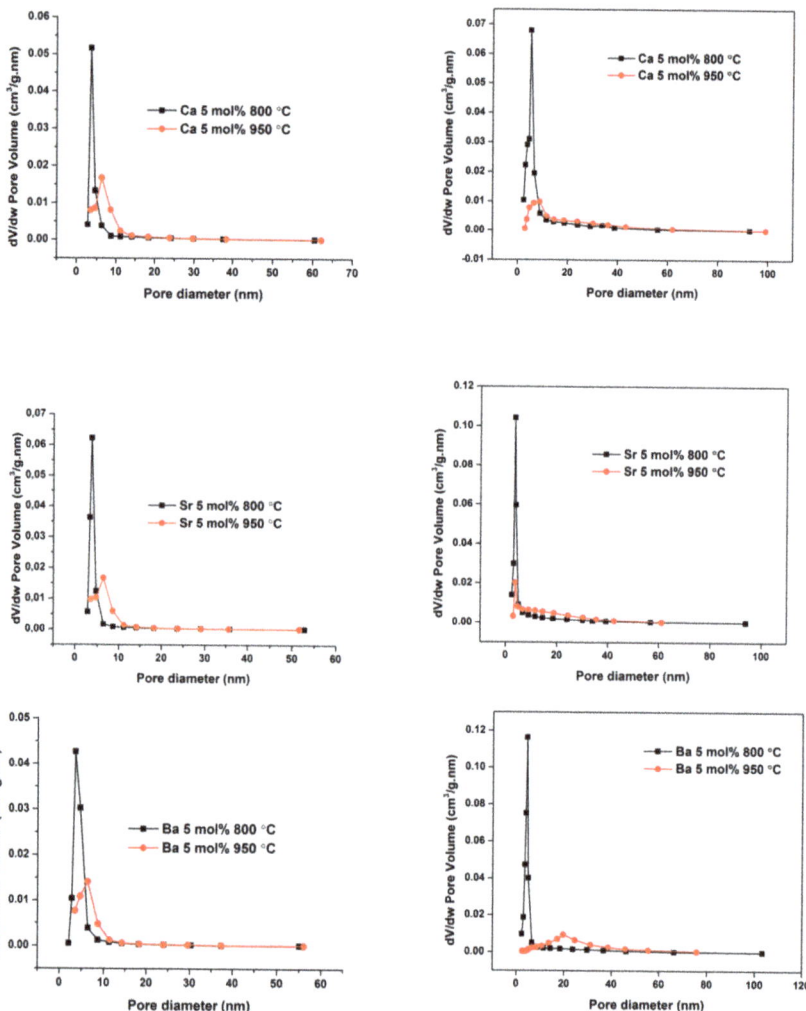

Figure 15. The pore size distribution of mixed metal oxides (MMO) obtained by heating the $Mg_{1.95}Ca_{0.05}/Al_1$ precursor gels (**top, left**), by heating the $Mg_{1.95}Ca_{0.05}/Al_1$ LDHs (**top, right**), by heating the $Mg_{1.95}Sr_{0.05}/Al_1$ precursor gels (**middle, left**), by heating the $Mg_{1.95}Sr_{0.05}/Al_1$ LDHs (**middle, right**), by heating the $Mg_{1.95}Ba_{0.05}/Al_1$ precursor gels (**bottom, left**), and by heating the $Mg_{1.95}Ba_{0.05}/Al_1$ LDHs (**bottom, right**) at 800 °C and 950 °C.

As seen, various surface properties could be detected for the MMO samples synthesized by two different methods. The pore size distribution of directly obtained MMO by heating Mg–Al–O precursor gels depends on the heating temperature and less on the nature of alkaline earth metal. The determined average pore diameter in the mesopore region is approximately 2.5–8 nm, 2.5–7 nm, and 2.5–9 nm for the Ca-MMO, Sr-MMO, and Ba-MMO samples, respectively, synthesized at 800 °C. The pore size distribution is visible wider for the MMO synthesized at 950 °C (approximately 3–10.15 nm, 3–12.5 nm, and 3.5–10.1 nm for the Ca-MMO, Sr-MMO, and Ba-MMO samples, respectively). As seen from Figure 15, the pore size distributions obtained by the BJH method for the MMO specimens synthesized from the reconstructed Mg_2/Al_1 LDHs depends on both synthesis temperature and nature of substituent. The most narrow pore size distribution was determined for Sr-containing MMO (2.5–3.5 nm for the sample heat-treated at 800 °C). On the other hand, the sample with 5% mol of Ba and prepared at 950 °C has very broad pore size distribution. In general, the gain in the volume of mesopores is clearly visible for the MMO samples synthesized at lower temperature. However, the pore diameter, wall thickness, and pore size distribution depend on the used synthesis method, heating temperature, and nature of alkali earth metal in the MMO host matrix, indicating that these MMO could have the potential for the application as catalysts, catalyst supports, and adsorbents.

4. Conclusions

In this study, the reconstruction peculiarities of sol–gel derived $Mg_{2-x}M_x/Al_1$ (M = Ca, Sr, Ba) layered double hydroxides (LDHs) were investigated. For the synthesis of $Mg_{2-x}M_x/Al_1$ (M = Ca, Sr, Ba) LDHs, the indirect sol–gel synthesis method has been used. Citric acid and 1,2-ethanediol were used as complexing agents in sol–gel processing [39]. The mixed metal oxides (MMO) were synthesized by two different routes in this work. Firstly, the MMO were obtained directly by heating Mg(M)–Al–O precursor gels at 650 °C, 800 °C, and 950 °C. The XRD pattern of the MMO sample obtained by heating Mg–Al–O precursor gels at 650 °C showed the formation of monophasic MMO. However, with increasing annealing temperature up to 800 °C or 950 °C and upon the substitution of Mg by Ca, Sr, and Ba, highly crystalline spinel ($MgAl_2O_4$, $CaAl_2O_4$, $SrAl_2O_4$ and $BaAl_2O_4$) phases have also formed. All MMO samples were successfully reconstructed to the $Mg_{2-x}M_x/Al_1$ (M = Ca, Sr, Ba) layered double hydroxides (LDHs) in water at 50 °C for 6 h (pH 10). However, the spinel phases were not reconstructed and remained as impurity phases. Moreover, during the reconstruction process, a negligible amount of metal carbonates ($CaCO_3$, $SrCO_3$, and $BaCO_3$) have formed as well. Secondly, the MMO were also obtained by heating the reconstructed LDHs at the same temperatures and the phase composition, morphology, and surface properties of MMO were compared with obtained ones after initial annealing. It was demonstrated that the second time obtained Ca and Sr-substituted MMO samples contained more side phases. However, this was not the case for the Ba-substituted MMO samples, since both synthesis products obtained from precursor gels and by heating LDHs were composed of several crystalline phases. It was demonstrated for the first time that the microstructure of reconstructed MMO from sol–gel derived LDHs showed a "memory effect", i.e., the microstructural features of MMO were almost identical as was determined for LDHs. Besides, the microstructure of investigated samples was not dependent on the annealing temperature and substitution. The synthesized Mg(M)–Al MMO samples exhibited type IV isotherms independent of the annealing temperature. At higher pressure values, the H1 hystereses were detected, which are characteristic for the mesoporous (pore size in the range of 2–50 nm) materials. It was found that the pore size distributions obtained by the BJH method for the MMO specimens synthesized from the reconstructed Mg_2/Al_1 LDHs depended on both the synthesis temperature and nature of the substituent. The most narrow pore size distribution was determined for Sr-containing MMO (2.5–3.5 nm for the sample heat-treated at 800 °C). On the other hand, the sample with 5% mol of Ba and prepared at 950 °C had very broad pore size distribution. The pore diameter, wall thickness, and pore size distribution was found to be dependent on used synthesis method, heating temperature, and nature of alkali earth metal in the MMO host matrix.

Author Contributions: Formal Analysis, L.V., A.Z., A.I., and A.K.; Investigation, L.V., M.R., I.G.-P., A.Z., and V.P.; Resources, A.Z., A.I., and A.K.; Data Curation, L.V.; Writing—Original Draft Preparation, L.V., I.G.-P., and A.K.; Writing—Review and Editing, A.K.; Visualization, A.Z. and V.P.; Supervision, A.I. and A.K. All authors have read and agreed to the published version of the manuscript.

Funding: This work was supported by a Research grant N°CAMAT (No. S-LB-19-2) from the Research Council of Lithuania and Belarusian Republican Found for Fundamental Research (No. 19LITG-007).

Conflicts of Interest: The authors declare that they have no conflict of interest.

References

1. Li, F.; Duan, X. Applications of layered double hydroxides. In *Layered Double Hydroxides (Structure and Bonding)*; Mingos, D.M.P., Ed.; Springer-Verlag: Berlin/Heidelberg, Germany, 2006; pp. 193–223.
2. Sokol, D.; Klemkaite-Ramanauske, K.; Khinsky, A.; Baltakys, K.; Beganskiene, A.; Baltusnikas, A.; Pinkas, J.; Kareiva, A. Reconstruction effects on surface properties of Co/Mg/Al layered double hydroxide. *Mater. Sci. (Medžiagotyra)* **2017**, *23*, 144–149. [CrossRef]
3. Mishra, G.; Dash, B.; Pandey, S. Layered double hydroxides: A brief review from fundamentals to application as evolving biomaterials. *Appl. Clay Sci.* **2018**, *1539*, 172–186. [CrossRef]
4. Vieira, D.E.L.; Sokol, D.; Smalenskaite, A.; Kareiva, A.; Ferreira, M.G.S.; Vieira, J.M.; Brett, C.M.A.; Salak, A.N. Cast iron corrosion protection with chemically modified Mg-Al layered double hydroxides synthesized using a novel approach. *Surf. Coat. Technol.* **2019**, *375*, 158–163. [CrossRef]
5. Smalenskaite, A.; Pavasaryte, L.; Yang, T.C.K.; Kareiva, A. Undoped and Eu^{3+} doped magnesium-aluminium layered double hydroxides: Peculiarities of intercalation of organic anions and investigation of luminescence properties. *Materials* **2019**, *12*, 736. [CrossRef] [PubMed]
6. Li, Y.L.; Ma, J.; Yuan, Y.X. Enhanced adsorption of chromium by stabilized Ca/Al-Fe layered double hydroxide decorated with ferric nanoparticles. *Sci. Adv. Mater.* **2020**, *12*, 441–448. [CrossRef]
7. Zhang, Z.D.; Qin, J.Y.; Zhang, W.C.; Pan, Y.T.; Wang, D.Y.; Yang, R.J. Synthesis of a novel dual layered double hydroxide hybrid nanomaterial and its application in epoxy nanocomposites. *Chem. Eng. J.* **2020**, *381*, 122777. [CrossRef]
8. Sato, T.; Fujita, H.; Endo, T.; Shimada, M.; Tsunashima, A. Synthesis of hydrotalcite-like compounds and their physic-chemical properties. *Reactiv. Solids* **1988**, *5*, 219–228. [CrossRef]
9. Klemkaite, K.; Prosycevas, I.; Taraskevicius, R.; Khinsky, A.; Kareiva, A. Synthesis and characterization of layered double hydroxides with different cations (Mg, Co, Ni, Al), decomposition and reformation of mixed metal oxides to layered structures. *Centr. Eur. J. Chem.* **2011**, *9*, 275–282. [CrossRef]
10. Salak, A.N.; Tedim, J.; Kuznetsova, A.I.; Ribeiro, J.L.; Vieira, L.G.; Zheludkevich, M.L.; Ferreira, M.G.S. Comparative X-ray diffraction and infrared spectroscopy study of Zn-Al layered double hydroxides: Vanadate vs. nitrate. *Chem. Phys.* **2012**, *397*, 102–108. [CrossRef]
11. Meyn, M.; Beneke, K.; Lagaly, G. Anion exchange reactions of layered double hydroxides. *Inorg. Chem.* **1990**, *29*, 5201–5206. [CrossRef]
12. Newman, S.P.; Jones, W. Synthesis, characterization and applications of layered double hydroxides containing organic guests. *New J. Chem.* **1998**, *22*, 105–115. [CrossRef]
13. Olfs, H.W.; Torres-Dorante, L.O.; Eckelt, R.; Kosslick, H. Comparison of different synthesis routes for Mg-Al layered double hydroxides (LDH): Characterization of the structural phases and anion exchange properties. *Appl. Clay Sci.* **2009**, *43*, 459–464. [CrossRef]
14. Smalenskaite, A.; Vieira, D.E.L.; Salak, A.N.; Ferreira, M.G.S.; Katelnikovas, A.; Kareiva, A. A comparative study of co-precipitation and sol-gel synthetic approaches to fabricate cerium-substituted Mg/Al layered double hydroxides with luminescence properties. *Appl. Clay Sci.* **2017**, *143*, 175–183. [CrossRef]
15. Sokol, D.; Salak, A.N.; Ferreira, M.G.S.; Beganskiene, A.; Kareiva, A. Bi-substituted Mg_3Al-CO_3 layered double hydroxides. *J. Sol-Gel Sci. Technol.* **2018**, *85*, 221–230. [CrossRef]
16. Smalenskaite, A.; Salak, A.N.; Kareiva, A. Induced neodymium luminescence in sol-gel derived layered double hydroxides. *Mendeleev Commun.* **2018**, *28*, 493–494. [CrossRef]
17. Smalenskaite, A.; Kaba, M.M.; Grigoraviciute-Puroniene, I.; Mikoliunaite, L.; Zarkov, A.; Ramanauskas, R.; Morkan, I.A.; Kareiva, A. Sol-gel synthesis and characterization of coatings of Mg-Al layered double hydroxides (LDHs). *Materials* **2019**, *12*, 3738. [CrossRef]
18. Valeikiene, L.; Paitian, R.; Grigoraviciute-Puroniene, I.; Ishikawa, K.; Kareiva, A. Transition metal substitution effects in sol-gel derived $Mg_{3-x}M_x/Al_1$ (M = Mn, Co, Ni, Cu, Zn) layered double hydroxides. *Mater. Chem. Phys.* **2019**, *237*, 121863. [CrossRef]

19. Kovanda, F.; Grygar, T.; Dornicak, V. Thermal behaviour of Ni-Mn layered double hydroxide and characterization of formed oxides. *Solid State Sci.* **2003**, *5*, 1019–1026. [CrossRef]
20. Liu, X.W.; Wu, Y.L.; Xu, Y.; Ge, F. Preparation of Mg/Al bimetallic oxides as sorbents: Microwave calcination, characterization, and adsorption of Cr(VI). *J. Solid State Chem.* **2016**, *79*, 122–132. [CrossRef]
21. Vicente, P.; Perez-Bernal, M.E.; Ruano-Casero, R.J.; Ananias, D.; Almeida Paz, F.A.; Rocha, J.; Rives, V. Luminescence properties of lanthanide-containing layered double hydroxides. *Microp. Mesop. Mater.* **2016**, *226*, 209–220. [CrossRef]
22. Kryshtab, T.; Calderon, H.A.; Kryvko, A. Microstructure characterization of metal mixed oxides. *MRS Adv.* **2017**, *2*, 4025–4030. [CrossRef]
23. Bugris, V.; Adok-Sipiczki, M.; Anitics, T.; Kuzmann, E.; Homonnay, Z.; Kukovecz, A.; Konya, Z.; Sipos, P.; Palinko, I. Thermal decomposition and reconstruction of CaFe-layered double hydroxide studied by X-ray diffractometry and Fe-57 Mossbauer spectroscopy. *J. Molec. Struct.* **2015**, *1090*, 19–24. [CrossRef]
24. Millange, F.; Walton, R.I.; O'Hare, D. Time-resolved in situ X-ray diffraction study of the liquid-phase reconstruction of Mg-Al-carbonate hydrotalcite-like compounds. *J. Mater. Chem.* **2000**, *10*, 1713–1720. [CrossRef]
25. Li, L.; Qi, G.X.; Fukushima, M.; Wang, B.; Xu, H.; Wang, Y. Insight into the preparation of Fe_3O_4 nanoparticle pillared layered double hydroxides composite via thermal decomposition and reconstruction. *Appl. Clay Sci.* **2017**, *140*, 88–95. [CrossRef]
26. Bernardo, M.P.; Ribeiro, C. Zn-Al-based layered double hydroxides (LDH) active structures for dental restorative materials. *J. Mater. Res. Technol.* **2019**, *8*, 1250–1257. [CrossRef]
27. Elhalil, A.; Elmoubarki, R.; Machrouhi, A.; Sadiq, M.; Abdennouri, M.; Qourzal, S.; Barka, N. Photocatalytic degradation of caffeine by ZnO-ZnAl2O4 nanoparticles derived from LDH structure. *J. Environ. Chem. Eng.* **2017**, *5*, 3719–3726. [CrossRef]
28. Valente, J.S.; Lima, E.; Toledo-Antonio, J.A.; Cortes-Jacome, M.A.; Lartundo-Rojas, L.; Montiel, R.; Prince, J. Comprehending the thermal decomposition and reconstruction process of sol-gel MgAl layered double hydroxides. *J. Phys. Chem. C* **2010**, *114*, 2089–2099. [CrossRef]
29. Venugopal, B.R.; Shivakumara, C.; Rajamathi, M. A composite of layered double hydroxides obtained through random costacking of layers from Mg-Al and Co-Al LDHs by delamination-restacking: Thermal decomposition and reconstruction behavior. *Solid State Sci.* **2007**, *9*, 287–294. [CrossRef]
30. Chagas, L.H.; de Carvalho, G.S.G.; Carmo, W.R.D.; Gil, R.A.S.S.; Chiaro, S.S.X.; Leitao, A.A.; Diniz, R.; de Sena, L.A.; Achete, C.A. MgCoAl and NiCoAl LDHs synthesized by the hydrothermal urea hydrolysis method: Structural characterization and thermal decomposition. *Mater. Res. Bull.* **2015**, *64*, 207–215. [CrossRef]
31. Kim, B.K.; Gwak, G.H.; Okada, T.; Oh, J.M. Effect of particle size and local disorder on specific surface area of layered double hydroxides upon calcination-reconstruction. *J. Solid State Chem.* **2018**, *263*, 60–64. [CrossRef]
32. Seftel, E.M.; Ciocarlan, R.G.; Michielsen, B.; Meynen, V.; Mullens, S.; Cool, P. Insights into phosphate adsorption behavior on structurally modified ZnAl layered double hydroxides. *Appl. Clay Sci.* **2018**, *165*, 234–246. [CrossRef]
33. Lee, S.H.; Tanaka, M.; Takahashi, Y.; Kim, K.W. Enhanced adsorption of arsenate and antimonate by calcined Mg/Al layered double hydroxide: Investigation of comparative adsorption Check for mechanism by surface characterization. *Chemosphere* **2018**, *211*, 903–911. [CrossRef] [PubMed]
34. Kang, J.; Levitskaia, T.G.; Park, S.; Kim, J.; Varga, T.; Um, W. Nanostructured MgFe and CoCr layered double hydroxides for removal and sequestration of iodine anions. *Chem. Eng. J.* **2020**, *380*, 122408. [CrossRef]
35. Santos, R.M.M.; Tronto, J.; Briois, V.; Santilli, C.V. Thermal decomposition and recovery properties of ZnAl-CO_3 layered double hydroxide for anionic dye adsorption: Insight into the aggregative nucleation and growth mechanism of the LDH memory effect. *J. Mater. Chem. A* **2017**, *5*, 9998–10009. [CrossRef]
36. Cao, Y.; Wang, Y.X.; Zhang, X.Y.; Cai, X.G.; Li, Z.H.; Li, G.T. Facile synthesis of 3D Mg-Al layered double oxide microspheres with ultra high adsorption capacity towards methyl orange. *Mater. Lett.* **2019**, *257*, 126695. [CrossRef]
37. Belskaya, O.B.; Leont'eva, N.N.; Gulyaeva, T.I.; Cherepanova, S.V.; Talzi, V.P.; Drozdov, V.A.; Likholobov, V.A. Influence of a doubly charged cation nature on the formation and properties of mixed oxides MAlOx (M = Mg^{2+}, Zn^{2+}, Ni^{2+}) obtained from the layered hydroxide precursors. *Russ. Chem. Bull.* **2013**, *62*, 2349–2361. [CrossRef]

38. Zhao, Y.; Li, J.-G.; Fang, F.; Chu, N.; Ma, H.; Yang, X. Structure and luminescence behaviour of as-synthesized, calcined, and restored MgAlEu-LDH with high crystallinity. *Dalton Trans.* **2012**, *41*, 12175–12184. [CrossRef]
39. Ishikawa, K.; Garskaite, E.; Kareiva, A. Sol-gel synthesis of calcium phosphate-based biomaterials -A review of environmentally benign, simple and effective synthesis routes. *J. Sol-Gel Sci. Technol.* **2020**, *94*, 551–572. [CrossRef]

© 2020 by the authors. Licensee MDPI, Basel, Switzerland. This article is an open access article distributed under the terms and conditions of the Creative Commons Attribution (CC BY) license (http://creativecommons.org/licenses/by/4.0/).

Article

Layered Double Hydroxides for Remediation of Industrial Wastewater from a Galvanic Plant

Anna Maria Cardinale [1,*], **Cristina Carbone** [2], **Sirio Consani** [2], **Marco Fortunato** [1] **and Nadia Parodi** [1]

[1] Dipartimento di Chimica e Chimica Industriale, Università di Genova, Via Dodecaneso 31, 16146 Genova, Italy; fortunato.marco@hotmail.it (M.F.); nadia@chimica.unige.it (N.P.)
[2] Dipartimento di Scienze della Terra, dell'Ambiente e della Vita, Università di Genova, Corso Europa 26, 16132 Genova, Italy; cristina.carbone@unige.it (C.C.); sirio.consani@edu.unige.it (S.C.)
* Correspondence: cardinal@chimica.unige.it; Tel.: +39-10-3536156; Fax: +39-10-3536163

Received: 3 May 2020; Accepted: 28 May 2020; Published: 30 May 2020

Abstract: Owing to their structure, layered double hydroxides (LDHs) are nowadays considered as rising materials in different fields of application. In this work, the results obtained in the usage of two different LDHs to remove, by adsorption, some cationic and anionic pollutants from industrial wastewater are reported. The two compounds MgAl-CO_3 and NiAl-NO_3 have been prepared through a hydrothermal synthesis process and then characterized by means of PXRD, TGA, FESEM, and FTIR spectroscopy. The available wastewater, supplied by a galvanic treatment company, has been analyzed by inductively coupled plasma-optical emission spectrometry (ICP-OES), resulting as being polluted by Fe(III), Cu(II), and Cr(VI). The water treatment with the two LDHs showed that chromate is more efficiently removed by the NiAl LDH through an exchange with the interlayer nitrate. On the contrary, copper and iron cations are removed in higher amounts by the MgAl LDH, probably through a substitution with Mg, even if sorption on the OH^- functional groups, surface complexation, and/or precipitation of small amounts of metal hydroxides on the surface of the MgAl LDH could not be completely excluded. Possible applications of the two combined LDHs are also proposed.

Keywords: layered double hydroxides; wastewater; heavy metals removal

1. Introduction

Layered double hydroxides (LDHs) belong to a family of minerals, the so-called hydrotalcite supergroup, whose crystal structure consists of brucite-type layers, in which a trivalent cation partially substitutes a divalent cation [1,2].

This substitution produces a net positive charge balanced by the entrance of an anionic species in the interlayer, giving as a general formula $M^{2+}_{1-x}M^{3+}_x(A^{z-})_{x/z}(OH)_2 \cdot nH_2O$.

Their capacity to easily exchange the interlayer anions makes LDHs attractive as carriers or scavengers of potential toxic anions [3–6]. Furthermore, their flexible structure, which can be reproduced using several bivalent and trivalent cations, suggests the possibility to use LDHs as a getter of pollutant cations [7,8]. For all of the above, the knowledge of the relationships between metals and LDHs is fundamental to allow the use of these minerals [9].

In a previous study, the relationships between lanthanides metals and a woodwardite (CuAl–SO_4 LDH) structure were investigated [10], with the aim to use these materials for the recovery of the rare earth elements from waste electric and electronic equipment for both georemediation and georecovery exploitation. The aim of this work is the synthesis and characterization of two different LDHs to experimentally investigate their capability to recover pollutants from wastewater. Such wastewater has been taken from a galvanic plant. Its composition is enriched in environmentally hazardous anions and metal cations, which derive from the treatment process. This problem is widespread [11]. Many different treatments have been proposed to solve this contamination issue [12,13]. In this study, the exchange capability of the two compounds is tested directly on this real wastewater, based on the results previously obtained in a laboratory-prepared $Cr_2O_7^{2-}$ solution [14]. The selected compounds are LDHs, as they represent a cheap and effective method to remove pollutants from aqueous solutions. The two compounds investigated are a NiAl-NO_3 and a MgAl-CO_3 LDHs. The choice of these compositions is based on the fact that previous works have shown the chromate uptake capacity by similar LDHs both intercalated with other sorbents [15] and through an ion exchange reaction with the interlayer anion [16,17].

2. Materials and Methods

2.1. Samples Synthesis

The compounds were synthesized via the co-precipitation route (direct method), followed by hydrothermal treatment obtaining nanoscopic crystallites with a partially disordered (turbostratic) structure. The addition of urea to the reactants helps to keep the pH to the desired value and the development of CO_2 allows to obtain a high exchange surface structure [15]. Both the LDHs were with the M^{2+}/M^{3+} ratio = 2.

The NiAl-NO_3 LDH was synthesized following the pathway suggested by [18], starting from Ni(NO_3)$_2 \cdot 6H_2O$ (99.0% purity, supplied by Merck KGaA, Darmstadt, Germany), Al(NO_3)$_3 \cdot 9H_2O$ (98.8% purity, supplied by VWR CHEMICALS, Leuven, Belgium), and urea (99.8mass% purity, supplied by CARLO ERBA, Milan, Italy). The two salts and urea, in a proper stoichiometric amount, were dissolved in 200 mL of deionized water under magnetic stirring, then the solution was transferred in a Teflon vessel autoclave and kept at 100 °C for 24 h. After heating, the system was cooled at room temperature naturally, and the green solid compound obtained was vacuum filtered and washed with water and ethanol. Subsequently, it was dried in a stove at 60 °C for 24 h. To synthesize the MgAl-CO_3 LDH, as suggested by [19], the reagents Al(NO_3)$_3 \cdot 9H_2O$ (98.8% purity, supplied by VWR CHEMICALS, Leuven, Belgium), Mg(NO_3)$_2 \cdot 9H_2O$ (98.9% purity, supplied by VWR CHEMICALS, Leuven, Belgium), and urea (99.8 mass% purity, supplied by CARLO ERBA, Milan, Italy) were used. After the dissolution of the due amount of the reagents, as a function of the Mg/Al desired ratio, the reaction was continued in a Teflon vessel autoclave at 180 °C temperature for one hour. The white compound obtained was separated from the solution by centrifugation at 7000 rpm 10 min^{-1}, repeatedly washed with water, and dried in a stove at 60 °C for 24 h.

2.2. Samples Characterizations

The compounds were characterized by means of the following techniques: thermo gravimetric analysis (TGA), X-ray diffraction analysis on powders (PXRD), field emission scanning electron microscope analysis (FESEM), Fourier transform infrared spectroscopy (FTIR), and inductively coupled plasma optical emission spectroscopy (ICP-OES) chemical analysis.

2.2.1. Thermo Gravimetric Analysis

In order to perform the TGA, about 30–50 mg of the dried powdered LDH was placed in alumina open crucibles; the measurements were carried out by means of a H/LABSYSEVO-1A SETARAM apparatus (Setaram, Caluire, France), at a heating rate of 5 °C min^{-1} under argon flux of 30 mL min^{-1}.

2.2.2. Field Emission Scanning Electron Microscope Analysis

To investigate the morphology and the structure of the synthesized compounds, an FESEM analysis has been performed. The samples were adhered on a conductive resin support, then analyzed by applying an acceleration voltage of 5 kV for 50 s, and a cobalt standard was used for the calibration.

2.2.3. Powder X-ray Diffraction

To determine the crystal structures and to calculate the lattice parameters of the phases, PXRD analysis was carried out by the vertical diffractometer X'Pert MPD (Philips, Almelo, The Netherlands) equipped with a Cu tube (Kα1 wavelength: 1.5406 Å). The samples were grounded in an agate mortar. The patterns were collected between 10° and 100° 2θ with a step of 0.001° and measuring time of 50 s/step. The indexing of the obtained diffraction data was performed by a comparison with the literature or calculated data (the program Powder Cell-version1999 [20]).

2.2.4. FTIR Spectroscopy

To exclude the possible incorporation during the synthesis of undesirable anions or other species in the interlayer (e.g., CO_2 from the atmosphere), FTIR spectroscopy was performed using a Spectrum 65 FT-IR Spectrometer (PerkinElmer, Waltham, MA, USA) equipped with a KBr beamsplitter and a DTGS detector by use of an ATR accessory with a diamond crystal. All spectra were recorded from 4000 to 600 cm^{-1}.

2.2.5. Inductively Coupled Plasma Optical Emission Spectroscopy

The chemical analyses were performed by ICP-OES after the dissolution of the samples in the concentrated nitric acid solution.

2.3. Pollutant Removal from Wastewater

Based on the results reported in the literature about the adsorption of anions and cations [7,11], in this work the adsorption efficiency of different pollutants (cationic and anionic) was tested on a real industrial wastewater sample. The sample available was analyzed by means of ICP-OES in order to determine the chemical composition of the dissolved elements. The results are the following: 127 ppm Cu(II), 460 ppm Fe(III), and 8780 ppm Cr(VI). The pH value of the wastewater was about 3. Owing to the very high metals concentration, the sample was diluted to 1/100 with water, before the adsorption tests. The water pH was corrected at a value of about 5 after dilution, to avoid both the LDH dissolution and the iron hydroxide precipitation. For both the LDHs, a weight of 0.5 g of the compound was added to a volume of 100 mL wastewater, and the mixture was shaken for 24 h. The solid and the liquid phases were separated throughout centrifugation at 7000 rpm 10 min^{-1}, then the solid was repeatedly washed with deionized water. The residual wastewater and the solid phase (after dried and acid dissolution) were analyzed by ICP-OES. After the batch equilibration procedure, the pH value of the residual wastewater did not change significantly, and no precipitation of iron hydroxide was observed.

3. Results and Discussions

The compounds were characterized after synthesis and in Figure 1a,b, their PXRD patterns are reported.

Both diagrams had peculiar features of the presence of the LDH structure, such as the strong basal reflections (003 and 006) around 12° and 24° 2θ, respectively. X-ray data confirm the presence of carbonate and nitrate in the two LDHs. The MgAl-CO_3 LDH has an interplanar distance associated with the main basal reflection of 7.54 Å, which is comparable to those of similar natural and synthetic compounds [1,21], whereas the NiAl-NO_3 LDH has an interplanar distance for the (003) reflection of 7.91 Å, as other similar compounds [22]. Comparing the two diagrams, it is also possible to state that the synthesis of the MgAl-CO_3 LDH yielded a material with a crystallite size significantly higher than the NiAl-NO_3 LDH. In fact, the NiAl-NO_3 LDH has broader reflections. Moreover, the PXRD pattern of the NiAl-NO_3 LDH shows a doublet in the (006) reflection, coupled with a diffuse and broad (003) reflection. This is probably due to the different hydration state of this LDH, which leads to the presence of crystallites with small differences in the interlayer distance. Water content is well known to influence the basal reflection of LDHs, but also the different orientation of nitrate in the interlayer and M^{2+}/M^{3+} ratio could explain this effect [23].

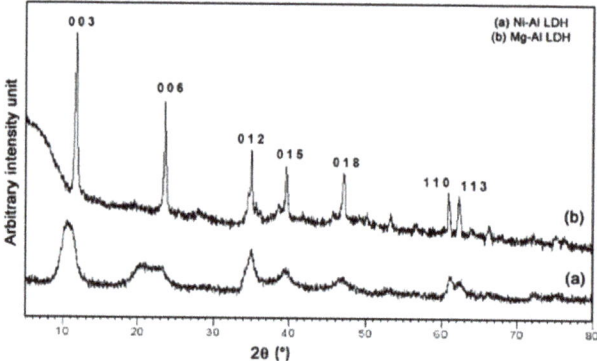

Figure 1. PXRD patterns of (a) NiAl nitrate and (b) MgAl carbonate layered double hydroxides (LDHs). On the graph, the hkl indexes of the main reflections are superimposed.

In Figure 2a,b, the FTIR spectra obtained for the synthesized compounds, the NiAl-NO_3 LDH and MgAl-CO_3 LDH, respectively, are shown, and the presence of the desired functional groups was confirmed.

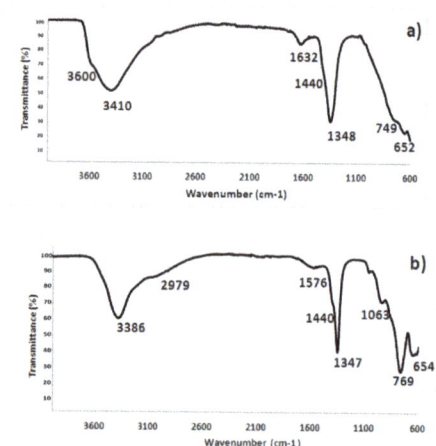

Figure 2. FTIR spectra of (a) the NiAl nitrate LDH, (b) the MgAl carbonate LDH.

Sample (a): A 3600 cm^{-1} OH group stretching, 3410 cm^{-1} hydrogen bond stretching in coordination with the cations, and 1632 cm^{-1} bending of the H_2O bond into the interlayer and at 1348 cm^{-1} can be found in the nitrate group stretching; the two slight bands at 749 and 652 cm^{-1} together with the one at 1440 cm^{-1} partially overlapped with the nitrate bending are related to the undesired interlayer carbonate groups (probably from the atmospheric CO_2). Sample (b): the broad band at 3386 cm^{-1} concerns the stretching of the OH groups in coordination with the cations, at 2979 cm^{-1} there is the large band related to the H-bonding between the water and carbonate anions in the interlayer, the 1576 cm^{-1} band originates from the H_2O bending in the interlayer, and the main carbonate anion adsorption band at 1440 cm^{-1} is overlapped with the 1347 cm^{-1} band related to the nitrate group stretching; ongoing to the lower frequency at 1063 cm^{-1}, there is the signal due to the carbonate vibration and the two bands at 769 and 654 cm^{-1} are related to the interlayer carbonate groups.

From these spectra, it seems that in the MgAl carbonate LDH there are some impurities of the nitrate anions, while the NiAl LDH is affected by a little impurity of carbonate.

The FESEM images are shown in Figure 3a,b, and from the micrographic appearance, the two structures appear similar in morphology, showing the typical hydrotalcite structure, and the thickness of the constituent lamellae was estimated between 10 and 20 nm.

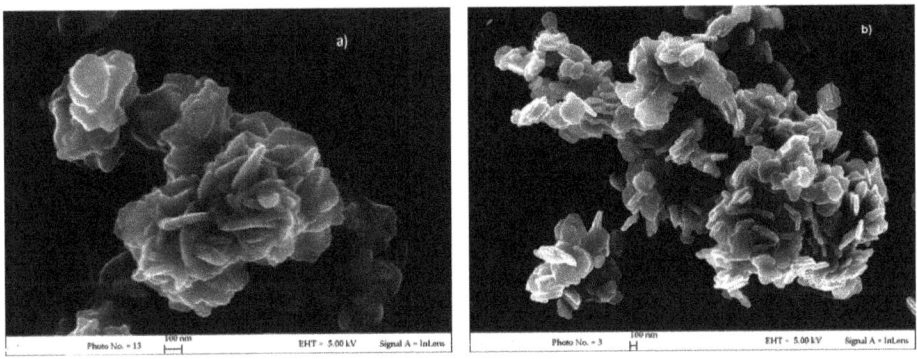

Figure 3. FESEM images of (a) the NiAl-NO$_3$ LDH, (b) the MgAl-CO$_3$ LDH.

For the NiAl-NO$_3$ LDH the specific surface area, calculated by the Brunauer–Emmett–Teller (BET) method, has been measured, resulting in 46.8504 m^2 g^{-1}.

The thermogravimetric analysis, whose results are reported in Figure 4a,b, revealed that the NiAl-based LDH (Figure 4a) loses 2.7 mass% due to humidity at about 150 °C, and has a second mass decreasing at 309.3 °C, involving both the interlayer water and the nitrogen oxide for a whole amount of 24.7 mass%. The MgAl-based LDH (Figure 4b) loses mass in three steps. The first, due to humidity, at 231 °C with a mass decrease of 10 mass%, the second ascribable to the removal of the interlayer water (6.5 mass% loss at 328 °C), and the third of about 16 mass% was due to the CO$_2$ loss, which started at about 400 °C.

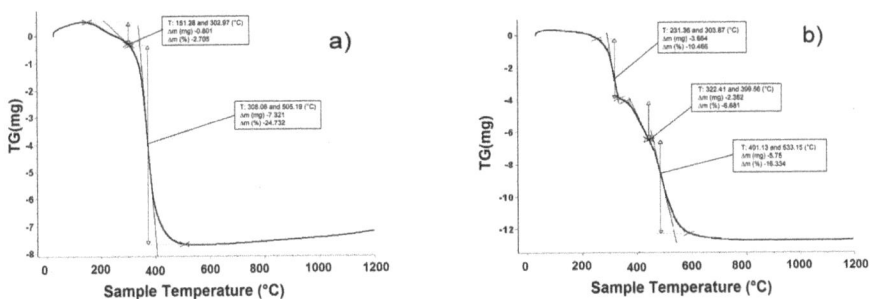

Figure 4. TG thermograms of (a) the NiAl nitrate LDH, (b) the MgAl carbonate LDH.

As the pollutant removal, both the samples have been kept to a batch equilibration with the wastewater, in the experimental conditions previously described. In Tables 1 and 2, the results obtained for the extraction of copper, iron, and chromium, the three pollutants of the industrial wastewater, are reported.

Table 1. Pollutant amount in the diluted wastewater before and after treatment with the NiAl-NO$_3$ LDH.

Pollutant	Concentration [ppm]			Recovery Efficiency
	Diluted Water		LDH	
	before treatment	after treatment	after water treatment	
Cu(II)	1.27	0.57	32	25.2
Fe(III)	4.60	0.51	112	24.3
Cr(VI)	87.8	1.18	6840	77.9

Table 2. Pollutant amount in the diluted wastewater before and after treatment with the MgAl-CO$_3$ LDH.

Pollutant	Concentration [ppm]			Recovery Efficiency
	Diluted Water		LDH	
	before treatment	after treatment	after water treatment	
Cu(II)	1.27	0.01	220.0	173
Fe(III)	4.60	0.02	4700.0	1021
Cr(VI)	87.8	62.0	2370	27

In the tables, the concentration of the three species investigated in the wastewater, before and after the adsorption procedure, and in the solid LDH used to extract the pollutants are shown. The recovery efficiency value expresses the ratio of the element concentration in the LDH to the element concentration in the medium, calculated in each test [24].

The NiAl-NO$_3$ compound demonstrated greater affinity for the CrO$_4^{2-}$ anion than for the Fe(III) and Cu(II) cations. On the contrary, the MgAl-CO$_3$ structure did not adsorb significantly the CrO$_4^{2-}$ anion, while it seemed more useful for the two cations.

The chromium adsorption is related to the different exchange capacity of the two anions (CO$_3^{2-}$ and NO$_3^-$) in the interlayers, where nitrate can be easily substituted by chromate, as confirmed from the PXRD. The comparison between the position of the three main reflections of the compound before and after the chromium extraction from the water are reported in Figure 5. The shift of the main basal reflections of the NiAl LDH toward lower 2θ values (Figure 5a) indicates an enlargement of the interlayers due to the substitution of NO$_3^-$ with CrO$_4^{2-}$. This shift is reflected in a change in the cell parameter c (calculated as $c = 3 \times d_{(003)}$), which changed from 23.85 to 24.99 Å, caused by the swelling of the interlayer due to the NO$_3^-$ with CrO$_4^{2-}$ substitution. The low crystallinity of the samples prevented to meaningfully discuss the cell parameter a, calculated as $a = 2 \times d_{(110)}$.

The values of the cell parameter c of the MgAl-CO$_3$ LDHs before and after the experiment are 22.71 and 22.56 Å, respectively. This fact suggests that little changes took place in the interlayer of these LDHs, as confirmed by Figure 5b. The high affinity of the MgAl-CO$_3$ LDH for the two cations is worthy of further study, but as for copper, it seems to be imputable to an exchange between the copper and magnesium divalent cations due to the similar ionic radius values of the two elements in octahedral coordination (72 and 73 pm, respectively). No significant change in the a parameter before and after the experiment is observed, as its value remains constant at 3.04 Å. Probably, the disordered nature of the cation arrangement in the brucite-like layers prevents the observation of the changes in the a parameter [2]. The Cu-Mg substitution has already been proposed for MgAl LDHs in contact with solutions in which bivalent cations are dissolved [25], even if sorption on the OH$^-$ functional groups, surface complexation, and/or precipitation of small amounts of Me(OH)$_2$ on the surface of the MgAl LDH could not be completely excluded [9]. Regarding iron adsorption, the characterization analysis did not provide precise information on where it might have accumulated It is possible that Fe has been removed via interaction with the functional groups on the surface of the mineral.

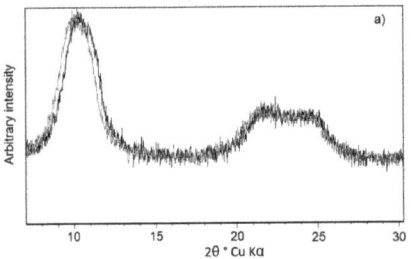
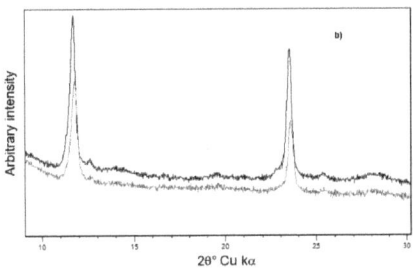

Figure 5. PXRD for (a) the NiAl nitrate LDH before (dark grey) and after (light grey) pollutant removal from the wastewater and (b) the MgAl carbonate LDH before (dark grey) and after (light grey) pollutant removal from the wastewater.

4. Conclusions

The experimental results obtained lead to the following conclusions:

- The LDHs' syntheses are relatively easy to prepare, leading to the required products as shown by the characterization analyses conducted;
- As the tested LDHs:
- The NiAl-NO$_3$ compound demonstrates greater affinity for the CrO$_4^{2-}$ anion than for the Fe(III) and Cu(II) cations;
- The MgAl-CO$_3$ structure does not significantly adsorb the CrO$_4^{2-}$ anion, while it seems more effective for the two cations;
- By comparing the residual concentration of the three elements studied with the Italian legal limit for industrial wastewater, reported in Table 3, with respect to the disposal in superficial water or in the drainage system:

Table 3. Italian legal concentration limit for chromium (VI), iron, and copper for disposal in both superficial water and drainage systems. (All. 5, P. Terza, D. Lgs n.152 del 03-04-06).

Element	Superficial Water/ppm	Drainage System/ppm
Cu(II)	0.1	0.4
Fe(III)	2.0	4.0
Cr(VI)	0.2	0.2

The NiAl-NO$_3$ compound reduces the Cr(VI) content at a value close to the legal limit, but is not sufficiently low. Some improvements of the methods (reaction time, sorbent/wastewater ratio) could help in reaching the goal.

The MgAl-CO$_3$ LDH demonstrates to be very effective in lowering the Cu(II) and Fe(III) concentration to below the legal limit.

For all of the above, the next step will be to test the effectiveness of a combined use of the two metal getters and their transformation into a spinel-like structure after heating for a potential reuse.

Author Contributions: Conceptualization: A.M.C., C.C.; investigation: A.M.C., S.C., M.F., N.P.; methodology: A.M.C., S.C.; writing—original draft preparation: A.M.C., S.C.; writing—review and editing: A.M.C., S.C.; resources: A.M.C., C.C. All authors have read and agreed to the published version of the manuscript.

Funding: This research received no external funding.

Conflicts of Interest: The authors declare no conflict of interest.

References

1. Mills, S.J.; Christy, A.G.; Génin, J.-M.R.; Kameda, T.; Colombo, F. Nomenclature of the hydrotalcite supergroup: Natural layered double hydroxides. *Miner. Mag.* **2012**, *76*, 1289–1336. [CrossRef]
2. Evans, D.G.; Slade, R.C.T. *Layered Double Hydroxides*; Duan, X., Evans, D.G., Eds.; Springer: Berlin/Heidelberg, Germany, 2006; Chapter 1; pp. 1–87. ISBN 978-3-540-28279-2.
3. Wang, Y.; Gao, H. Compositional and structural control on anion sorption capability of layered double hydroxides (LDHs). *J. Colloid Interface Sci.* **2006**, *301*, 19–26. [CrossRef] [PubMed]
4. Morimoto, K.; Anraku, S.; Hoshino, J.; Yoneda, T.; Sato, T. Surface complexation reactions of inorganic anions on hydrotalcite-like compounds. *J. Colloid Interface Sci.* **2012**, *384*, 99–104. [CrossRef]
5. Golban, A.; Lupa, L.; Cocheci, L.; Rodica, P. Synthesis of MgFe Layered Double Hydroxide from Iron-Containing Acidic Residual Solution and Its Adsorption Performance. *Crystals* **2019**, *9*, 514. [CrossRef]
6. Dore, E.; Frau, F.; Cidu, R. Antimonate Removal from Polluted Mining Water by Calcined Layered Double Hydroxides. *Crystals* **2019**, *9*, 410. [CrossRef]
7. Mishra, G.; Dash, B.; Pandey, S. Layered double hydroxides: A brief review from fundamentals to application as evolving biomaterials. *Appl. Clay Sci.* **2018**, *153*, 172–186. [CrossRef]
8. Tamzid Rahman, M.; Kameda, T.; Kumagai, S.; Yoshioka, T. A novel method to delaminate nitrate-intercalated MgAl layered double hydroxides in water and application in heavy metals removal from waste water. *Chemosphere* **2018**, *203*, 281–290. [CrossRef]
9. Liang, X.; Zang, Y.; Xu, Y.; Tan, X.; Hou, W.; Wang, L.; Sun, Y. Sorption of metal cations on layered double hydroxides. *Colloids and Surfaces A: Physicochem. Eng. Asp.* **2013**, *433*, 122–131. [CrossRef]
10. Consani, S.; Balić-Žunić, T.; Cardinale, A.M.; Sgroi, W.; Giuli, G.; Carbone, C. A Novel Synthesis Routine for Woodwardite and Its Affinity towards Light (La, Ce, Nd) and Heavy (Gd and Y) Rare Earth Elements. *Materials* **2018**, *11*, 130. [CrossRef]
11. Zinicovskaia, I.; Safonov, A.V.; Khijniak, T.V. *Heavy Metals and Other Pollutants in the Environment: Biological Aspects*; Zaikov, G.E., Weisfeld, L.I., Lisitsyn, E.M., Bekuzarova, S.A., Eds.; Apple Academic Press: Waretown, NJ, USA, 2017; Chapter 17; pp. 333–360. ISBN 978-1-315-36602-9.
12. Kobya, M.; Erdem, N.; Demirbas, E. Treatment of Cr, Ni and Zn from galvanic rinsing wastewater by electrocoagulation process using iron electrodes. *Desalin. Water Treat.* **2014**, *56*, 1191–1201. [CrossRef]
13. Bednarik, V.; Vondruska, M.; Koutny, M. Stabilization/solidification of galvanic sludges by asphalt emulsions. *J. Hazard. Mat.* **2005**, *B122*, 139–145. [CrossRef] [PubMed]
14. Goswamee, R.L.; Sengupta, P.; Bhattacharyya, K.G.; Dutta, D.K. Adsorption of Cr(VI) in layered double hydroxides. *Appl. Clay Sci.* **1998**, *13*, 21–34. [CrossRef]
15. Chen, S.; Huang, Y.; Han, X.; Wu, Z.; Lai, C.; Wang, J.; Deng, Q.; Zeng, Z.; Deng, S. Simultaneous and efficient removal of Cr(VI) and methyl orange on LDHs decorated porous carbons. *Chem. Eng. J.* **2018**, *352*, 306–315. [CrossRef]
16. Chao, H.-P.; Wang, Y.-C.; Tran, H.N. Removal of hexavalent chromium from groundwater by Mg/Al-layered double hydroxides using characteristics of in-situ synthesis. *Environ. Pollut.* **2018**, *243*, 620–629. [CrossRef]
17. Lei, C.; Zhu, X.; Zhu, B.; Jiang, C.; Le, Y.; Yu, J. Superb adsorption capacity of hierarchical calcined Ni/Mg/Al layered double hydroxides for Congo red and Cr(VI) ions. *J. Hazard. Mat.* **2017**, *321*, 801–811. [CrossRef]
18. Liu, H.; Yu, T.; Su, D.; Tang, Z.; Zhang, J.; Liu, Y.; Yuan, A.; Kong, Q. Ultrathin NiAl layered double hydroxide nanosheets with enhanced supercapacitor performance. *Ceram. Int.* **2017**, *43*, 14395–14400. [CrossRef]
19. Rao, M.M.; Reddy, B.R.; Jayalakshmi, M.; Jaya, V.S.; Sridhar, B. Hydrothermal synthesis of Mg–Al hydrotalcites by urea hydrolysis. *Mat. Res. Bull.* **2005**, *40*, 347–359. [CrossRef]
20. Kraus, W.; Nolze, G. *Powder Cell for Windows*; Federal Institute for Materials Research and Testing: Berlin, Germany, 1999.
21. Somosi, Z.; Muráth, S.; Nagy, P.; Sebők, D.; Szilagyi, I.; Douglas, G. Contaminant removal by efficient separation of in situ formed layered double hydroxide compounds from mine wastewaters. *Environ. Sci. Water Res. Technol.* **2019**, *5*, 2251–2259. [CrossRef]
22. Ravuru, S.S.; Jana, A.; De, S. Synthesis of NiAl- layered double hydroxide with nitrate intercalation: Application in cyanide removal from steel industry effluent. *J. Hazard. Mater.* **2019**, *373*, 791–800. [CrossRef]
23. Marappa, S.; Radha, S.; Kamath, P.V. Nitrate-intercalated layered double hydroxides—Structure model, order, and disorder. *Eur. J. Inorg. Chem.* **2013**, *2013*, 2122–2128. [CrossRef]

24. Jakubiak, M.; Giska, I.; Asztemborska, M.; Bystrzejewska-Piotrowska, G. Bioaccumulation and biosorption of inorganic nanoparticles: Factors affecting the efficiency of nanoparticle mycoextraction by liquid-grown mycelia of Pleurotus eringi and Trametes versicolor. *Mycol. Progress.* **2013**, *13*, 525–532. [CrossRef]
25. González, M.A.; Pavlovic, I.; Barriga, C. Cu(II), Pb(II) and Cd(II) sorption on different layered double hydroxides. A kinetic and thermodynamic study and competing factors. *Chem. Eng. J.* **2015**, *269*, 221–228. [CrossRef]

© 2020 by the authors. Licensee MDPI, Basel, Switzerland. This article is an open access article distributed under the terms and conditions of the Creative Commons Attribution (CC BY) license (http://creativecommons.org/licenses/by/4.0/).

Article

Curcumin Incorporation into Zn$_3$Al Layered Double Hydroxides—Preparation, Characterization and Curcumin Release

Octavian D. Pavel [1], Ariana Șerban [1,2], Rodica Zăvoianu [1,*], Elena Bacalum [1] and Ruxandra Bîrjega [3]

1. Biochemistry and Catalysis, Department of Organic Chemistry, Faculty of Chemistry, University of Bucharest, 030018 Bucharest, Romania; octavian.pavel@chimie.unibuc.ro (O.D.P.); ariana.serban@infim.ro (A.Ș.); elena.bacalum@gmail.com (E.B.)
2. National Institute for Materials Physics, 077125 Măgurele, Romania
3. National Institute for Lasers, Plasma and Radiation Physics-INFLPR, 077125 Măgurele, Romania; ruxandra.birjega@inflpr.ro
* Correspondence: rodica.zavoianu@chimie.unibuc.ro; Tel.: +40-746-171-699

Received: 29 February 2020; Accepted: 25 March 2020; Published: 26 March 2020

Abstract: Curcumin (CR) is a natural antioxidant compound extracted from *Curcuma longa* (turmeric). Until now, researches related to the incorporation of CR into layered double hydroxides (LDHs) were focused only on hybrid structures based on a MgxAl-LDH matrix. Our studies were extended towards the incorporation of CR in another type of LDH-matrix (Zn3Al-LDH) which could have an even more prolific effect on the antioxidant activity due to the presence of Zn. Four CR-modified Zn3Al-LDH solids were synthesized, e.g., PZn3Al-CR(Aq), PZn3Al-CR(Et), RZn3Al-CR(Aq) and RZn3Al-CR(Et) (molar ratio CR/Al = 1/10, where P and R stand for the preparation method (P = precipitation, R = reconstruction), while (Aq) and (Et) indicate the type of CR solution, aqueous or ethanolic, respectively). The samples were characterized by XRD, Attenuated Total Reflectance Fourier Transformed IR (ATR-FTIR) and diffuse reflectance (DR)-UV–Vis techniques and the CR-release was investigated in buffer solutions at different pH values (1, 2, 5, 7 and 8). XRD results indicated a layered structure for PZn3Al-CR(Aq), PZn3Al-CR(Et), RZn3Al-CR(Aq) impurified with ZnO, while RZn3Al-CR(Et) contained ZnO nano-particles as the main crystalline phase. For all samples, CR-release revealed a decreasing tendency towards the pH increase, and higher values were obtained for RZn3Al-CR(Et) and PZn3Al-CR(Et) (e.g., 45% and 25%, respectively at pH 1).

Keywords: layered double hydroxides; reconstruction; curcumin; drug release

1. Introduction

Turmeric (*Curcuma longa*) is a notorious spice, highly esteemed not only by the scientific world but also by gastronomes as it is the primary source of curcumin (CR) or (1E,6E)-1,7-bis(4-hydroxy-3-methoxyphenyl)hepta-1,6-diene-3,5-dione, a renowned natural antioxidant polyphenol that can scavenge free radicals undergoing electron transfer or abstract H-atoms either from the phenolic OH groups or the CH$_2$ group of the β-diketone moiety [1–4]. Depending on the chemical environment, CR, an α,β-unsaturated β-diketone can adopt two different conformations, either the diketonic form or the enolic one (Figure 1) [2].

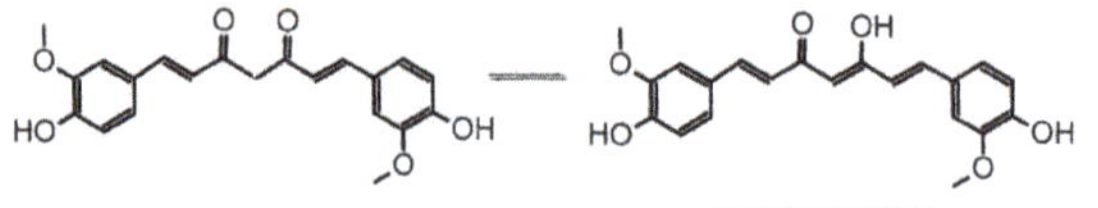

Figure 1. Keto-enolic tautomerization of curcumin.

The antioxidant activity of this natural polyphenol is a controversy for the scientific world. Some consider that the enol tautomer characterized by a better conjugation between the two aromatic rings containing the phenolic OH groups holds the main responsibility for the presence of the antioxidant activity highlighted as an inhibition of superoxide radicals, hydrogen peroxide and nitric oxide radical [3,5], while others acknowledge better the keto form due to its existence in slightly acidic media [6,7]. It was also suggested that the presence of CR accelerates the processes catalyzed by several antioxidant enzymes such as catalase, superoxide dismutase (SOD), glutathione peroxidase (GPx) and heme oxygenase-1 (OH-1) [5]. Since this compound has low solubility in water at neutral pH, its absorption and bioavailability are poor. In addition to that, it has also reduced stability towards oxidation, light, alkalinity, enzymes and heat. Therefore in order to increase its pharmacological effectiveness, the latest researches were focused on improving the bioavailability, chemical and photochemical stability of CR using different methods such as: conjugation with cyclodextrin, β-diglucoside, αS1-Casein, β-Lactoglobulin [8–11], encapsulation in nanoparticles of biocompatible polymers, exosomes, lipid nanoparticles, dextrin nanogels, dendrimers, metal oxide nanoparticles [12–18], complexation with multivalent metal cations [19,20].

Layered double hydroxides are substances belonging to the class of anionic clays, having the general formula $[M^{2+}_{1-x}M^{3+}_x(OH)_2]^{x+}[A^{n-}_{x/n}]^{x-} \cdot mH_2O$ (M^{2+} and M^{3+} are metal cations that can adopt an octahedral arrangement similar to the one adopted by Mg^{2+} in brucite, A^{n-} is a compensation anion, x is a value in the range of 0.2–0.33 and m is the number of water molecules) [21]. The presence of the trivalent cations leads to an excess of positive charge in the brucite-type layer which is compensated by anions A^{n-}, which are located in the interlayer region along with the crystallization water molecules. Even though the hydrotalcite ($Mg_6Al_2(OH)_{16}CO_3 \cdot 4H_2O$) is the most renowned representative of this type of materials, there are also other natural occurring layered double hydroxide (LDH) compounds such as meixnerite ($Mg_6Al_2(OH)_{18} \cdot 4H_2O$), zaccagnaite ($Zn_4Al_2(OH)_{12}[CO_3] \cdot 3H_2O$) and pyroaurite ($Mg_6Fe_2(OH)_{16}CO_3 \cdot 4.5H_2O$). Structurally, these solids consist of positively charged brucite-type layers with balancing anions and water molecules in the interlayer space [21]. Despite the fact that there is a scarce spreading of LDHs in the earth crust, several laboratory methods for their obtaining were developed: co-precipitation at a variable or constant pH, under low or high supersaturation conditions, sol-gel, hydrothermal and mechanochemical synthesis [21–24]. An interesting feature of the LDHs is the so-called "memory effect" which allows the reconstruction of the layered structure when the mixed oxide obtained by thermal decomposition of an LDH precursor at temperatures lower than 550 °C is immersed in an aqueous solution containing the desired compensation anion which can be either inorganic or organic [21,22]. LDH compounds are considered to have low toxicity, good biocompatibility and a buffering action when immersed in aqueous solutions, therefore they were also utilized in medical formulations (such as the antiacid TALCID®), for ibuprofen slow release, in colon-targeted drug-delivery, for anticancer methotrexate delivery [25–28] and as matrices for bioinorganic hybrid materials [29–31]. Due to their structure LDHs compounds have anion exchange properties and are frequently utilized as anion exchangers and adsorbents. The insertion of organic anions in the LDH can be performed by ionic exchange, co-precipitation or reconstruction. However, the process is more difficult due to the poorer aqueous solubility of the organic species [22,27,32]. Recently, Kottegoda and coworkers showed that the inhibitory activity against several microbial species (e.g., *Candida albicans, Candida dubliniensis, Pseudomonas aeruginosa, Escherichia coli* and *Staphylococcus aureus*) can be enhanced by CR encapsulation into an inorganic host, namely a layered double hydroxide (LDH) containing Mg and Al as metal cations [33–35].

Considering this state of the art, the present contribution aims to extend the studies towards the incorporation of CR in another type of LDH-matrix, namely Zn3Al-LDH that is less basic than MgxAl-LDH and could bring its own contribution in increasing the antioxidant activity due to the presence of Zn which is known for its antiseptic properties [36]. The target of this study was to choose the modality for obtaining the solid with the best capacity to incorporate CR during synthesis and the best ability to release it in vitro under controlled buffer conditions. Therefore, for the synthesis of CR-containing Zn3Al-LDH, two different methods were applied: (*i*) direct co-precipitation (P) and (*ii*) reconstruction (R) of the LDH in the presence of CR, which was added either as an aqueous alkaline solution (Aq) or as an ethanolic solution (Et). Depending on the applied preparation protocol, the names of the synthesized solids were abbreviated as PZn3Al-CR(Aq), PZn3Al-CR(Et), RZn3Al-CR(Aq) and RZn3Al-CR(Et). The structural characterization of the samples was performed using X-ray diffraction (XRD), attenuated total reflectance Fourier transformed infrared spectroscopy (ATR-FTIR) and diffuse reflectance UV–Vis spectroscopy (DR-UV–Vis). The release of CR from the solids was investigated in buffer solutions at different pH values (1, 2, 5, 7 and 8) and the spent solid samples recovered after 24 h were also characterized by ATR-FTIR and DR-UV–Vis. Based on the obtained results, the sample allowing the highest CR-release was selected for further antimicrobial activity tests which are going to be the subject of a future publication.

2. Materials and Methods

Curcumin (CR) from Sigma-Aldrich, $Zn(NO_3)_2 \cdot 2H_2O$, $Al(NO_3)_3 \cdot 9H_2O$, NaOH and Na_2CO_3 from Merck were utilized as raw materials for the synthesis of the CR-functionalized LDHs. Absolute ethanol from Fluka and deionized water were used as solvents. Certified SUPELCO buffer solutions of pH 1 (glycine, sodium chloride, hydrochloric acid), pH 2 (citric acid, sodium hydroxide, hydrochloric acid), pH 5 (citric acid, sodium hydroxide) and pH 7 (potassium dihydrogen phosphate/di-sodium hydrogen phosphate) were purchased from Merck, while the certified Fischer Chemical buffer solution of pH 8 (potassium dihydrogen phosphate/sodium hydroxide) was purchased from Fischer Scientific.

A pristine Zn3Al-LDH with interlayer carbonate anion was prepared by co-precipitation at pH 9 using 160 mL of a metal nitrates $Zn(NO_3)_2 \cdot 6H_2O$, $Al(NO_3)_3 \cdot 9H_2O$ aqueous solution (1.5 M, molar ratio Zn/Al = 3/1) and 160 mL of an aqueous solution containing Na_2CO_3 and NaOH 160 mL (1 M concentration of Na_2CO_3, 2.5 M concentration of NaOH) for pH adjustment. The resulting gel was aged 18 h at 50 °C. The solid recovered by filtration was washed with deionized water until the conductivity of the wastewater was lower than 100 µS/cm and dried at 90 °C for 24 h and the reference material Zn3Al-LDH was finally obtained.

The preparation of CR-containing Zn3Al-LDH was performed using always an amount of CR corresponding to a molar ratio CR/Al = 1/10. For the synthesis by co-precipitation at pH 9 the above-mentioned amount of metal nitrates in aqueous solution were utilized and the pH was adjusted with an aqueous solution of NaOH (2.5 M). Two samples, e.g., PZn3Al-CR(Aq) and PZn3Al-CR(Et), were obtained following this procedure since CR was added either as an aqueous alkaline solution (Aq) or as an ethanolic solution (Et). Then, 100 mL of deionized water were poured in the reactor before starting the precipitation of PZn3Al-CR(Aq) by concomitantly adding the metal nitrates solution and the alkaline solution containing CR under vigorous stirring (350 rot/min). For the obtaining of PZn3Al-CR(Et), 100 mL of CR ethanolic solution were first added in the reactor and then the precipitation took place by concomitantly adding the solutions containing the metal nitrates and the NaOH under similar conditions of stirring. The flowchart for these preparations is presented in Figure 2. The aging of the precipitates was performed under an inert atmosphere (He flow 10 mL/min).

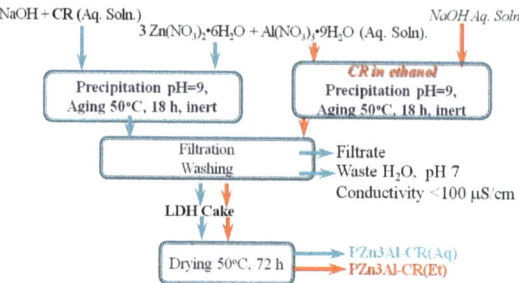

Figure 2. Flowchart for the preparation of curcumin (CR)-containing Zn3Al-layered double hydroxide (LDH) by co-precipitation.

For the preparations performed by reconstruction a mixed oxide called CZn3Al obtained by the thermal decomposition of the pristine Zn3Al-LDH at 460 °C during 18 h was utilized as raw material. The reconstructions were performed in brown glass vessels at 25 °C by contacting CZn$_3$Al powder with a CR-containing solution under magnetic stirring and inert atmosphere (He 1atm) during 24 h. Two CR-containing solutions were prepared, an alkaline aqueous solution containing CR and NaOH at a concentration of 4×10^{-3} M, and an ethanolic solution containing 4×10^{-3} M CR. Depending on the type of CR-solution utilized for reconstruction, two solid samples were obtained, e.g., RZn3Al-CR(Aq) and RZn3Al-CR(Et), respectively. The washing of the recovered solids was performed with deionized water for RZn3Al-CR(Aq), and with ethanol for RZn3Al-CR(Et). The flowchart for these preparations is presented in Figure 3.

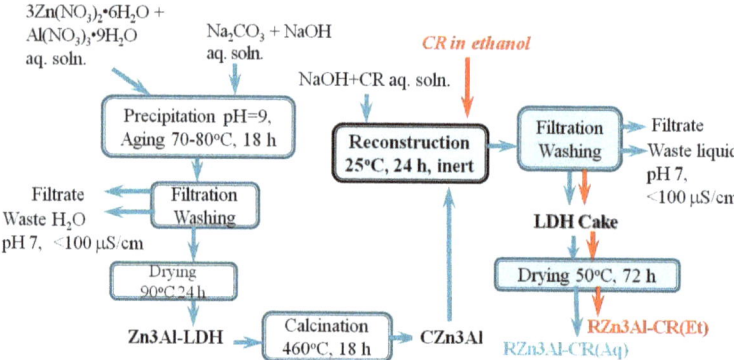

Figure 3. Flowchart for the preparation of CR-containing Zn3Al-LDH by reconstruction.

The content of Zn and Al in the solids was determined by atomic absorption spectrometry (AAS) using a Thermoelemental Solar AAS apparatus, while the content of CR in the solid samples was calculated based on the determination of total organic carbon (TOC) content using HiPerTOC–Thermo carbon analyzer according to a previously described protocol [37]. The value of TOC was calculated as the difference between total carbon (TC content obtained by UV-persulfate oxidation of the samples) and total inorganic carbon (TIC content obtained by mineralization of the samples with HNO_3 to convert the bicarbonate and carbonate ions to CO_2).

The XRD patterns of the samples were recorded on PANalytical MPD system using Ni-filtered CuK$_\alpha$ radiation ($\lambda = 1.5418$ Å), with a scan step of 0.02° and a counting time of 20 s per step, for 2θ ranging between 5 and 70°. The average crystallite size (D) of the different phases in the samples was determined using the Scherrer formula applied to particular reflections/crystallographic directions.

The characterization of the samples by infrared spectroscopy was performed using a JASCO 4700 FT-IR spectrophotometer equipped with ATR PRO ONE Single-reflection ATR accessory and monolithic diamond crystal on the 4000–400 cm^{-1} domain at 128 scans and a resolution of 4 cm^{-1}.

Shimadzu 3600 UV–Vis NIR spectrometer equipped with an integration sphere was utilized for recording the DR-UV–Vis spectra of the solids in the range of 200–800 nm, using BaSO$_4$ as white reference.

In vitro CR release studies were performed in dark brown bottles where 0.5 g of solid sample were contacted with 50 mL of the adequate buffer solution (pH 1, pH 2, pH 5, pH 7 and pH 8) at 25 °C during 24 h under mild stirring (100 rot/min). The amount of released CR was determined by UV–Vis spectrometry using a JASCO V650 UV–Vis double-beam spectrophotometer with a photomultiplier tube detector. Liquid samples were withdrawn from the bottles hourly in the first four hours and finally after 24 h and their absorption spectra were recorded in the range 350–550 nm against the corresponding buffer solution as blank. The concentration of CR in the solution was calculated with Equations (1)–(3):

$$C_{CR\,solution} = A_{423nm}/8153.5 \text{ [mol/L]} = A_{423nm}/22.13 \text{ [g/L]} \tag{1}$$

$$m_{CR\text{ in 50 mL solution}} = C_{CR\,solution}/20 \text{ [g]} \tag{2}$$

$$\%CR_{released} = m_{CR\text{ in 50 mL solution}} \times 100/(m_{solid} \times C_{CR\text{ in the solid (see Table 1)}}) \tag{3}$$

Table 1. Chemical composition of the solids.

Sample	Zn (wt. %)	Al (wt. %)	CO$_3$ (wt. %) [1]	CR (wt. %) [2]	H$_2$O (wt. %) [3]	Molar Ratio Zn/Al	Molar Ratio CR/Al
Zn3Al-LDH	46.2	6.4	7.1	0	8.5	2.97	0
PZn3Al-CR(Aq)	49.9	7.2	2.8	12.1	8.9	2.86	1/8.5
PZn3Al-CR(Et)	48.1	6.8	1.7	9.3	8.8	2.90	1/10
RZn3Al-CR(Aq)	51.1	7.2	1.2	10.0	5.6	2.91	1/9.9
RZn3Al-CR(Et)	60.3	8.4	0.2	11.5	2.9	2.97	1/9.9

[1] Calculated from TIC (wt. %) values CO$_3$ (wt. %) = TIC/0.2; [2] Calculated from TOC (wt. %) values: CR (wt. %) = TOC (wt. %)/0.684; [3] H$_2$O calculated considering the loss of weight in the temperature range 105-200 °C [21].

3. Results

3.1. Characterization of the Curcumin Containing Zn3Al-LDHs

3.1.1. Chemical Composition

The results obtained by AAS for the determination of Zn and Al content and the content of carbonate and CR in the samples calculated from the determination of the carbon content (see Table 1) showed that the molar ratios Zn/Al and CR/Al were very close to 3/1 and 1/10, respectively (calculated from the amounts introduced in the synthesis mixture) for all the samples besides PZn3Al-CR(Aq) which showed lower values. The higher concentrations of Zn and Al in RZn3Al-CR(Et) are explained by the lower value of the H$_2$O concentration in this sample.

3.1.2. XRD Characterization of CR Functionalized Zn3Al-LDH Samples

The XRD patterns of the CR-loaded powders prepared by both direct precipitation and reconstruction are presented in Figure 4. The XRD patterns of curcumin-loaded powders are compared with those of the curcumin free powders obtained either by precipitation (Zn3Al-LDH) or by reconstruction in water (RZn3Al-LDH). The structural data are gathered in Table 2. The XRD patterns of the powders prepared by coprecipitation reveal that the precipitation in the presence of curcumin generates the formation of a zincite-phase (ZnO, ICDD card no. 36-1451) as by-product alongside with the layered structure. In the pristine Zn3Al-LDH reference sample, the LDH is the dominant phase and it is similar to the carbonate-intercalated Zn,Al-LDH having a Zn/Al molar ratio of 3 standard, $(Zn_6Al_2(OH)_{16}CO_3 \cdot 4H_2O$, ICDD card no. 38-0486). Small reflections assignable to hydrozincite, $Zn_5(CO_3)_2(OH)_2$, as a minor phase impurity are also observable ($Zn_5(CO_3)_2(OH)_2$, ICDD card no. 19-1458) (marked by * in Figure 4a). The lattice parameters are smaller for the PZnAl-CR(Aq) sample, denoting a lower Zn/Al molar ratio due to the formation of the ZnO phase. For the PZn3Al-CR(Et) solid, an extra layered phase with having a larger interlayer space has appeared, thus indicating the intercalation of larger-sized anions. Moreover, the small D_{003} value obtained for this extra-phase denotes a degree of crystalline disorder along the c-axis, the axis on which the brucite-like layers are stacked.

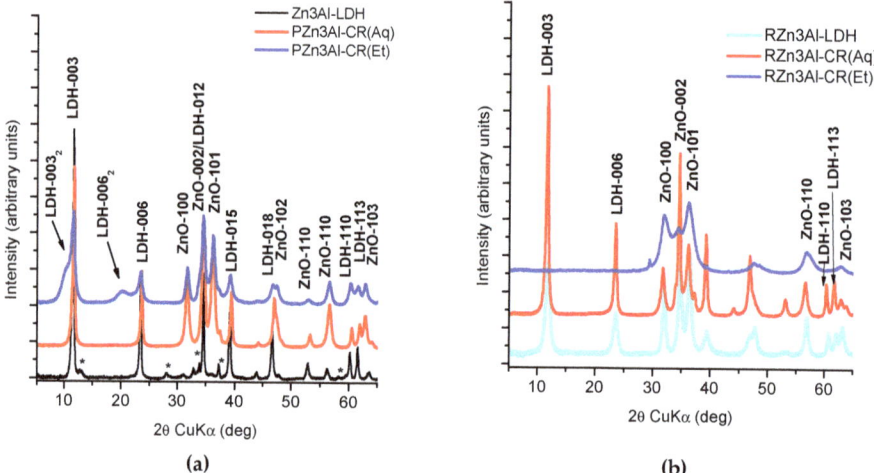

Figure 4. XRD patterns: (a) precipitated samples Zn3Al-LDH, PZn3Al-CR(Aq), PZn3Al-CR(Et); (b) reconstructed samples RZn3Al-LDH, RZn3Al-CR(Aq), RZn3Al-CR(Et).

Table 2. Structural data of the samples obtained from XRD analysis.

Samples	LDH Phase					ZnO Phase				
	a (Å)	c (Å)	I_{003}/I_{110}	D_{110} (nm)	D_{003} (nm)	a (Å)	c (Å)	Vol (Å³)	D (nm)	ZnO (%)
Zn3Al-LDH	3.075(2)	22.9026	7.66	35.6	23.4					
RZn3Al-LDH [1]	3.07(2)	22.84(7)	5.10	16.6	8.9	3.243(2)	5.183(6)	47.21	13.2	42
PZn3Al-CR(Aq)	3.061(5)	22.56(4)	7.71	24.4	20.2	3.251(6)	5.21(1)	47.69	12.0	45
PZn3Al-CR(Et)	3.071(6)	22.99(7) / 25.476(11)	3.37 / 1.34	21.9	11.9 / 3.9	3.262(4)	5.20(1)	47.92	11.6	22
RZn3Al-CR(Aq)	3.071(4)	22.62(5)	5.88	26.7	16.5	3.263(9)	5.21(3)	48.04	11.3	41
RZn3Al-CR(Et)						3.234(2)	5.238(3)	47.44	5.3	100

[1] RZn3Al-LDH is the structure obtained after the rehydration of CZn3Al in water for 24 h at 25 °C.

The XRD patterns of the reconstructed samples (Figure 4b) show the partial reconstruction of the layered structure for the sample exposed to an aqueous solution. According to Kooli et al. [38], the hydration of the Zn(Al)O mixed-oxides leads to the formation of LDH with an insignificant amount of a zincite phase only for the calcined sample with a molar ratio Zn/Al=2 while, for higher Zn/Al ratios, residual ZnO is always present. The XRD pattern of the powder reconstructed in a CR ethanolic solution displays only the reflections of a ZnO-phase. The peaks are extremely broad, typically for a ZnO-phase calcined under mild conditions (400 °C–500 °C) [39]. The amount of ZnO-phase reported to the layered LDH phases in PZn3Al-CR(Aq), PZn3Al-CR(Et) and RZn3Al-CR(Aq) fresh samples was estimated by considering the integrated intensities of the main single reflections of the ZnO-phase in RZn3Al-CR(Et) as reference. The data are included in the last column of Table 2 and disclosed values between 22 and 45% from the totality of crystalline products. However, it should also be acknowledged that the procedures used for the preparation of these powders, namely precipitation, thermal treatment and rehydration generate also amorphous oxide or hydroxides phases undetectable by XRD [38].

The reflections derived from crystalline CR were not detectable as a separate phase in the diffraction patterns of any of the CR-loaded samples. This fact may be a consequence of its dispersion as amorphous nano-particles in the inorganic matrix.

3.1.3. ATR-FTIR Characterization

The ATR-FTIR spectra of the fresh CR-containing samples are displayed in Figure 5a,b. The spectrum of the reference sample (Zn3Al-LDH) (Figure 5a) presents all the absorption bands specific to carbonate intercalated Zn,Al-LDH at 3428 cm^{-1} (υ-OH), 2981 cm^{-1} (interaction of carbonate and H$_2$O in the interlayer through hydrogen bonds), 1630 cm^{-1} and 772 cm^{-1} (deformation vibrations of interlayer H$_2$O), 1363 cm^{-1} (deformation vibration of carbonate anion), 770 cm^{-1} (Al-OH out-of-plane) 617 cm^{-1} (deformation of Zn-OH bond), 551 cm^{-1} and 427 cm^{-1} (vibrations in Al-O-Al and Zn-O-Zn condensed groups) [40,41].

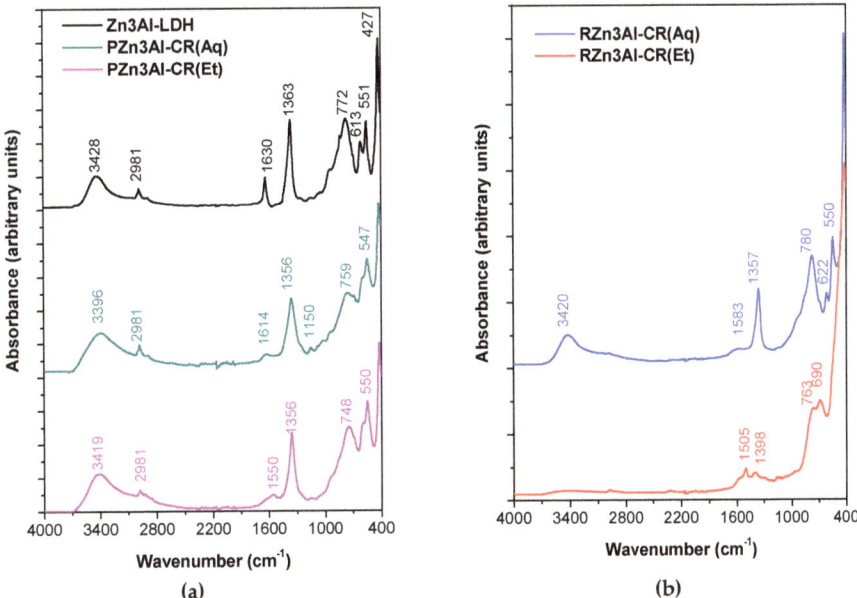

Figure 5. Normalized attenuated total reflectance (ATR)-FTIR spectra: (a) precipitated samples Zn3Al-LDH, PZn3Al-CR(Aq), PZn3Al-CR(Et); (b) reconstructed samples RZn3Al-CR(Aq), RZn3Al-CR(Et).

The spectra of CR-containing samples were similar to that of Zn3Al-LDH, except the one attributed to RZn3Al-CR(Et) (Figure 5b). In addition, the bands characteristic for neat curcumin could not be delimited from those of the LDH. However, following CR-incorporation by precipitation, the bands of the parent LDH present in the region 4000–2800 cm^{-1} have increased their relative intensity due to the overlapping of the bands attributed to CR (see Figure 6) with those of Zn3Al-LDH (Figure 5a). There is also a noticeable red shifting of the bands at 3428 cm^{-1}, 1363 cm^{-1} and 772 cm^{-1} to 3396 cm^{-1}

for PZn3Al-CR(Aq), 3419 cm^{-1} for PZn3Al-CR(Et), 1356 cm^{-1}, 759 cm^{-1} for PZn3Al-CR(Aq) and 748 cm^{-1} for PZn3Al-CR(Et), respectively. The presence of the band at 1356 cm^{-1} indicates the contamination of these samples with carbonate most probably caused by the carbonation of NaOH during the manipulation before its utilization in the synthesis. This assumption is sustained by the results obtained in the analysis of carbon content presented in Table 1. In addition to that, the more pronounced asymmetry, the shifting of the bands in the region 3600–3300 cm^{-1} as well as the significant attenuation of the band corresponding to H_2O deformation vibrations at 1630 cm^{-1} compared to the reference sample Zn3Al-LDH, emphasizes the contribution of CR interactions with the inorganic matrix through hydrogen bonds. The higher intensity of the band around 550 cm^{-1} compared to the one in the region 780–748 cm^{-1} indicates an increased amount of Zn-O-Zn condensed groups in both CR-functionalized samples obtained by co-precipitation and reconstruction with CR-aqueous solution and it may be correlated with the results obtained from XRD characterization. In the spectrum of PZn3Al-CR(Aq) the band corresponding to water deformation vibrations is red shifted to 1614 cm^{-1}, indicating a perturbation in the interlayer region as a consequence of CR-incorporation, while in the spectrum of PZn3Al-CR(Et) the same band is missing. Though, a novel absorption band appears at 1550 cm^{-1} in the spectrum of PZn3Al-CR(Et) indicating the formation of a distorted Zn(II)-CR complex [42]. In the spectrum of RZn3Al-CR(Aq), the band appearing in the hydroxyl vibrations region has a lower relative intensity compared to the one of the reference material indicating a poor reconstruction due to the remaining of a segregate phase of ZnO whose presence was also confirmed by XRD. In addition to that, it was also noticed the absence of the band attributed to water deformation vibrations and the presence of a new band corresponding to Zn(II)-CR complex at 1583 cm^{-1} [42]. The spectrum of RZn3Al-CR(Et) shows five well defined absorption bands at 1505, 1398, 763, 680 and 427 cm^{-1}. Among these bands, the one at 1505 cm^{-1} is correlated to the most intense band of curcumin (Figure 6), the band at 1398 cm^{-1} could be assimilated to a red shift of the curcumin band at 1427 cm^{-1} while the band at 680 cm^{-1} could be also due to a red shift of the curcumin band at 808 cm^{-1}. The bands at 763 and 427 cm^{-1} are related to M-O vibration modes [40,41]. The absence of the band at ca. 3400 cm^{-1} in this spectrum shows the absence of the reconstruction effect.

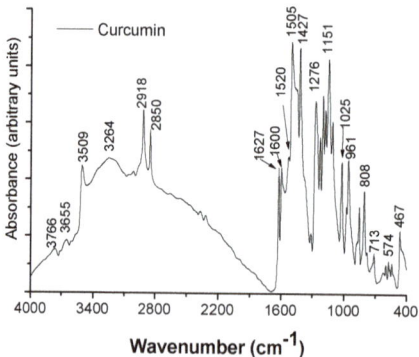

Figure 6. ATR-FTIR spectrum of the neat curcumin (CR) utilized in this study.

3.1.4. DR-UV–Vis Characterization

The DR-UV–Vis spectra of CR-containing samples along with the spectrum of CR are displayed in Figure 7a,b.

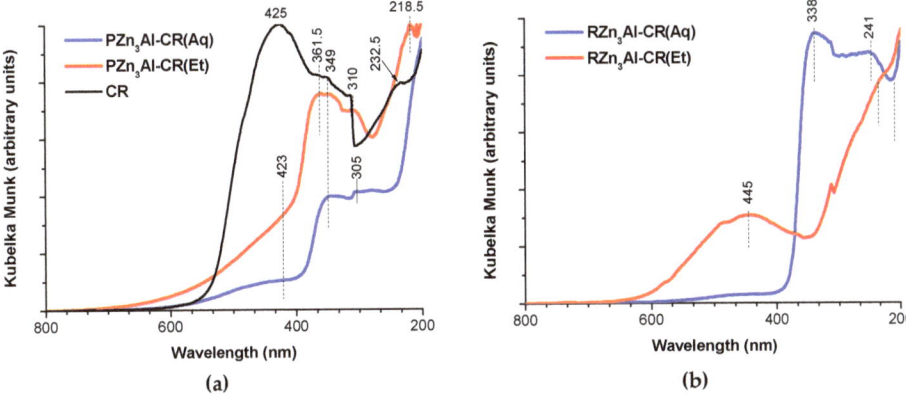

Figure 7. Diffuse reflectance (DR)-UV–Vis spectra: (**a**) neat curcumin (CR) and precipitated samples, PZn3Al-CR(Aq), PZn3Al-CR(Et); (**b**) reconstructed samples RZn3Al-CR(Aq), RZn3Al-CR(Et).

The spectrum of the neat curcumin powder has the highest intensity absorption band in the visible region at 425 nm. This band is shifted to 455 nm and is clearly evidenced only in the spectrum of the sample RZn3Al-CR(Et) which shows also the second absorption band specific to CR at 364 nm. Meanwhile, in the spectra of the precipitated samples PZn3Al-CR(Aq) and PZn3Al-CR(Et) the maximum absorption in the visible domain appears at 361.5 and 349 nm, respectively and only an inflection of the absorption curve is noticed at 423 nm. The absorption maxima observed in these two spectra indicate that during the preparation of the samples curcumin was partially decomposed to a mixture of feruloyl methane and ferulic acid whose most intense absorption peaks are at 340 nm, and 305 nm respectively [43]. In the spectrum of RZn3Al-CR(Aq), only the characteristic absorption bands for feruloyl methane at 340 nm and 241 nm were noticed.

3.2. Curcumin Release Studies

The results of the CR-release studies performed by contacting the synthesized solids with different buffer solutions (pH 1, pH 2, pH 5, pH 7 and pH 8) at 25 °C during 24 h under mild stirring are presented in Figure 8a–d. The release of curcumin from pure curcumin powder into the same buffer solutions was determined using an amount of curcumin powder equal to the average amount of curcumin incorporated in 0.5 g of the LDH samples (e.g., 56.25 mg) which was immersed in 50 mL of buffer solution. The results obtained under operating conditions (temperature, stirring and duration) similar to those employed for CR-containing LDH samples are presented in Figure 9.

The results displayed in Figure 8 indicate that for all samples the release of curcumin was favored at lower pH values and was almost insignificant at pH values higher than 5. It is also noticeable that CR-loaded samples prepared with ethanolic solutions were able to release higher amounts of CR than the samples prepared with aqueous solutions. At each pH value, the amount of curcumin released in the buffer solutions varied in the order: RZn3Al-CR(Et)>PZn3Al-CR(Et)>PZn3Al-CR(Aq)>RZn3Al-CR(Aq). At pH 1, CR was released faster from RZn3Al-CR(Et) since half of the total amount of released curcumin was reached after 2 h, while for the other solids only about one third of the total amount was reached in that period. Under similar conditions, the results plotted in Figure 9 show that the pH decrease led also to an increased release of curcumin from pure curcumin powder into the buffer solutions, but the highest value obtained was lower than 0.8%. When the release tests with pure curcumin powder were performed during 4 h at 37 °C (the physiological temperature) the amount of released CR varied in the order: 1.0 ± 0.1% at pH 1, 0.8 ± 0.1% at pH 2, 0.5 ± 0.1% at pH 5, 0.4 ± 0.1% at pH 7 and 0.3 ± 0.1% at pH 8. The effect of the temperature on the CR-release after 4 h from the sample RZn3Al-CR(Et) was also not significant as it may be seen from the results plotted in Figure 10.

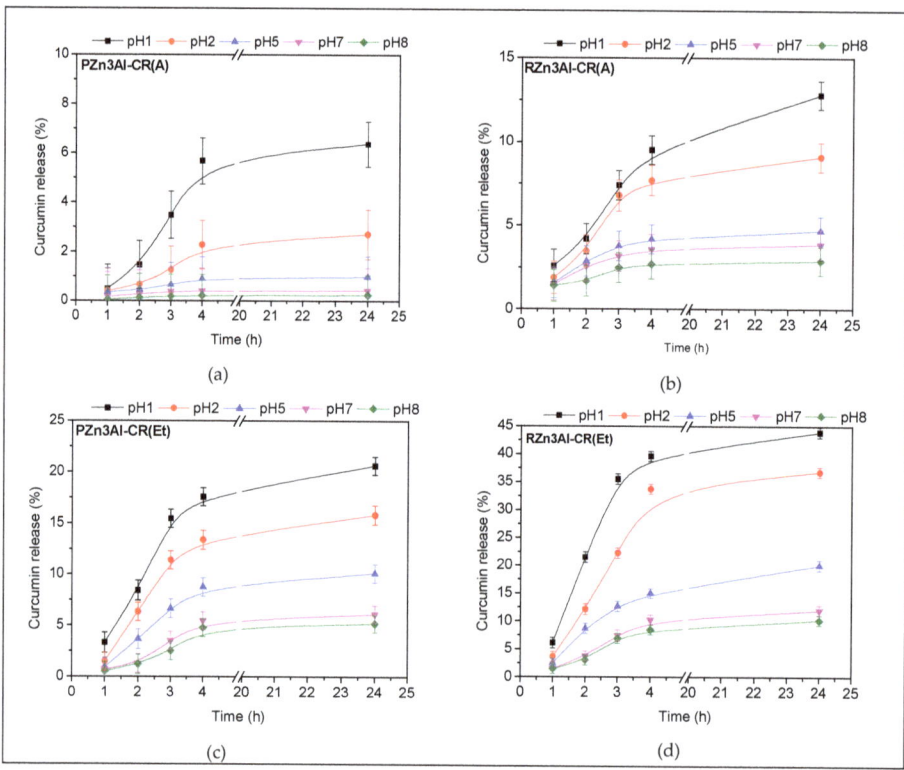

Figure 8. Curcumin release from the solid samples during 24 h in different pH buffers: (**a**) PZn3Al-CR(Aq); (**b**) RZn3Al-CR(Aq); (**c**) PZn3Al-CR(Et); (**d**) RZn3Al-CR(Et).

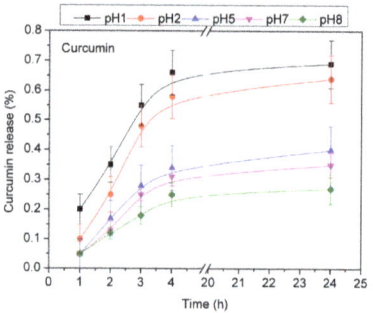

Figure 9. Curcumin release from pure curcumin powder.

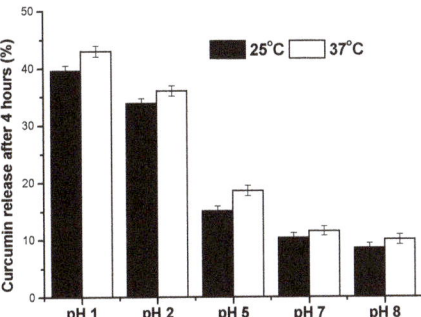

Figure 10. Curcumin release from RZn3Al-CR(Et) in different buffer solutions at 25 and 37 °C.

The solid samples recovered after each CR-release test were analyzed by DR-UV–Vis spectroscopy and the spectra are presented in Figure 11a–d.

The alteration of the spectra of the samples after they were contacted with different buffered solutions indicates how the chemical composition of each buffer affected the CR-functionalized solids. After the tests performed at pH 7 and pH 8 in buffered solutions containing potassium dihydrogen phosphate/di-sodium hydrogen phosphate, and potassium dihydrogen phosphate/sodium hydroxide, respectively, none of the specific absorption bands observed in the spectra of the fresh samples could be noticed. This fact suggests the occurrence of chemical reactions between the components of the buffer and the solid leading to the decomposition of the chemical species responsible for the absorption bands noticed in the DR-UV–Vis spectra of the fresh samples. After the tests performed in acid buffers at pH 1 and 2, the main bands specific to the fresh solids were preserved but their relative intensity compared to the spectra of the fresh solids decreased, suggesting that a partial dissolution of the inorganic host took place. In addition to that, for the samples PZn3Al-CR(Aq), PZn3Al-CR(Et) and RZn3Al-CR(Aq) their position was shifted to lower wavelengths. The decrease was more intense at pH 1 when the highest amount of curcumin was released for each sample. The contact of the CR-loaded solids with the pH 5 buffer solution containing citric acid and sodium hydroxide affected all the samples. In the spectra of the samples prepared with CR-ethanolic solution (Figure 11b,d), after the contact with pH 5 buffer, the absorption maximum characteristic to bicyclopentadione at 232 nm was noticed [43]. This maximum was also noticeable as a shoulder in the spectrum of PZn3Al-CR(Aq) (Figure 11a), but it was absent in the spectrum of RZn3Al-CR(Aq) (Figure 11c).

The interactions between the CR-containing samples and different buffer solutions were also evidenced in the ATR-FTIR spectra of the solids recovered after the CR-release tests performed with each buffer solution which are presented in Figure 12.

The spectra of all the samples recovered from the alkaline buffer solutions (pH 7 and 8) have the main absorption bands around 1000 cm^{-1} (marked with * in Figure 12) and they were attributed to the phosphate anion which was intensely adsorbed on the solids. In this case, the spectra of the two samples prepared by co-precipitation (Figure 12a,b) presented the same absorption maxima, whereas in the spectra of the samples RZn3Al-CR(Aq) and RZn3Al-CR(Et) recovered from pH 8 buffer solution, the band corresponding to water bending at 1620 cm^{-1} was absent. The spectra of the samples recovered from acid buffer solutions (pH 1, pH 2 and pH 5, respectively) were mostly altered in the mid-infrared region 1600–1300 cm^{-1} (delimited by a dashed rectangle in Figure 10) where the bands specific to citrate occur ($v_{as(COO-)}$ at 1612 cm^{-1}; $v_{s(COO-)}$ at 1396 cm^{-1}, $\delta_{(CH2)}$ at 1365 cm^{-1} [44]). For the precipitated samples (Figure 12a,b) the relative intensity of the band around 1600 cm^{-1} was sensibly higher than in the spectra of the fresh samples, while for the reconstructed ones, the relative intensity of all the bands in the domain 1600–1300 cm^{-1} was definitely increased compared to the spectra of the fresh samples (Figure 12c,d) indicating that citrate was better adsorbed on the surface of the reconstructed solids.

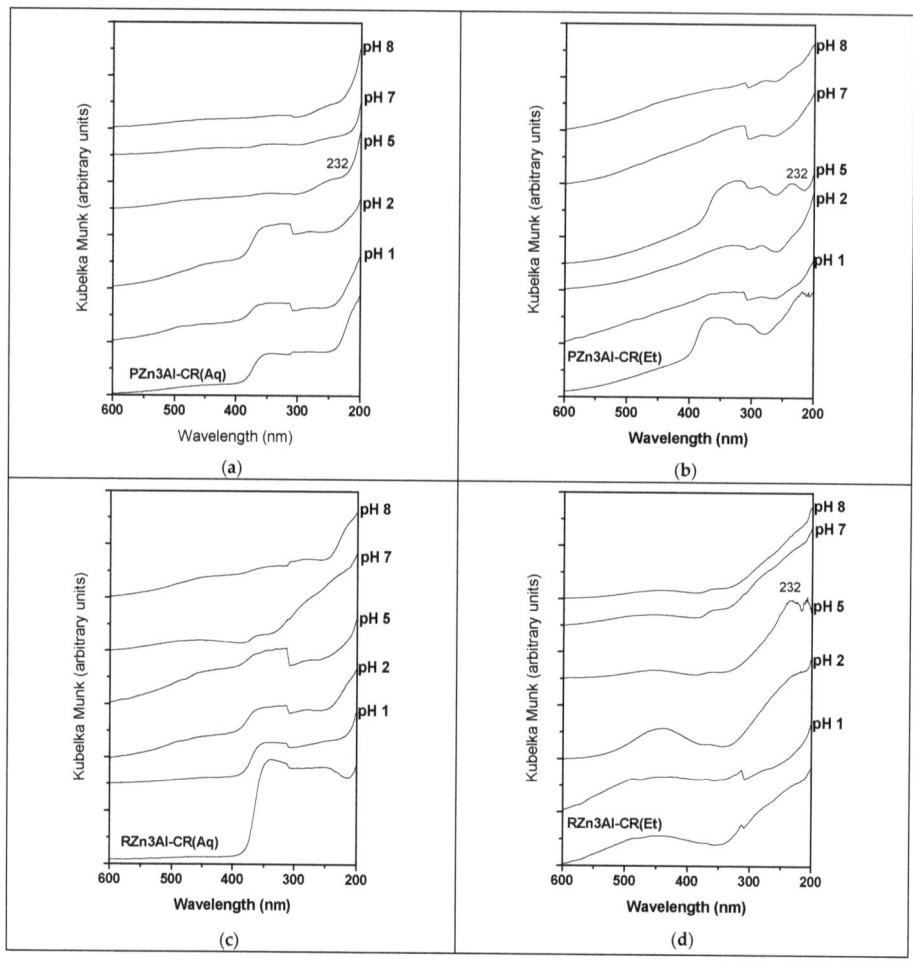

Figure 11. DR-UV–Vis spectra of the CR-containing samples after the CR-release tests at different pH values: (**a**) PZn3Al-CR(Aq); (**b**) RZn3Al-CR(Aq); (**c**) PZn3Al-CR(Et); (**d**) RZn3Al-CR(Et).

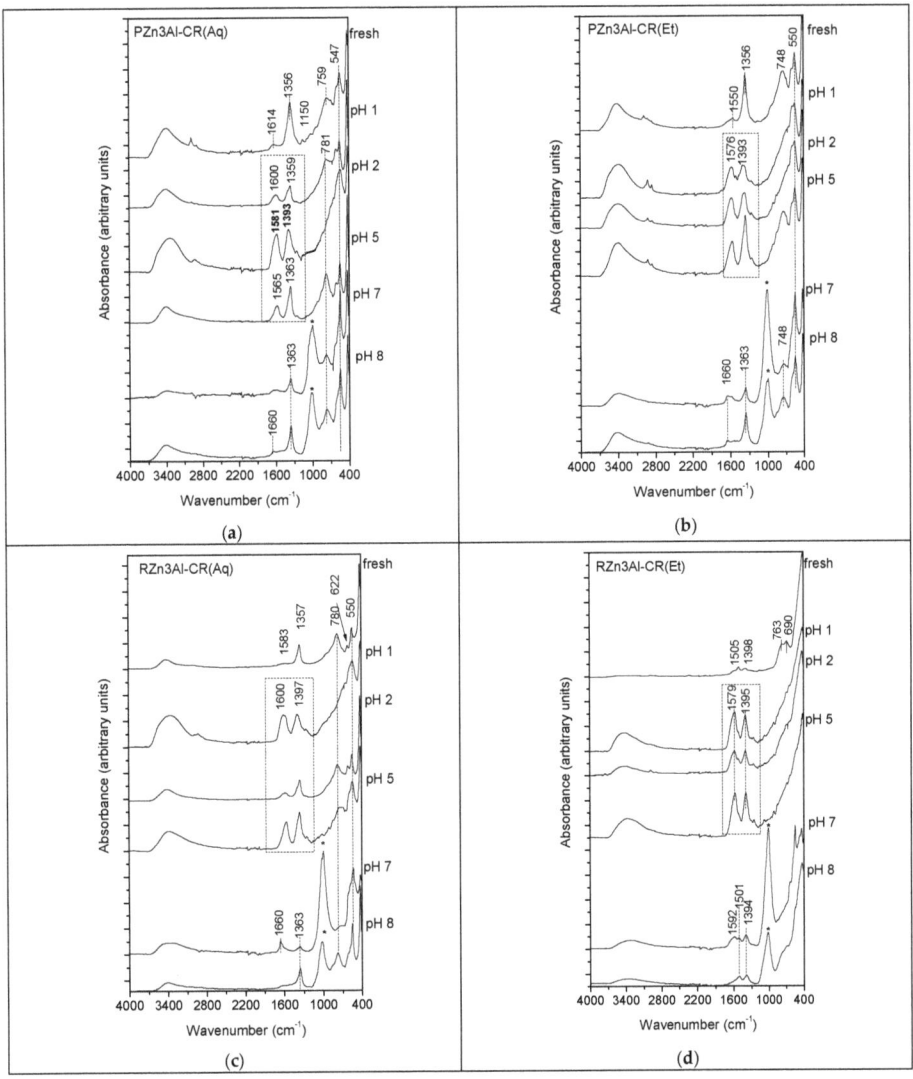

Figure 12. ATR-FTIR spectra of the CR-containing samples after the CR-release tests at different pH values: (**a**) PZn3Al-CR(Aq); (**b**) RZn3Al-CR(Aq); (**c**) PZn3Al-CR(Et); (**d**) RZn3Al-CR(Et).

4. Discussion

The investigation of the collected data has revealed a contradiction between the results of the chemical analysis concerning the curcumin content in the samples and those of the other characterizations which did not clearly indicate the presence of CR in the solids except the one obtained by reconstruction with CR-ethanolic solution (RZn3Al-CR(Et)). A possible explanation for this inconsistency is revolving around the CR-loading present in the solids, which was calculated based on the determination of the total organic carbon content, including also the carbon from the compounds obtained by the degradation of curcumin during the syntheses performed at basic pH. Several studies concerning the stability of curcumin in the alkaline medium have suggested that the degradation products are ferulic acid, feruloyl methane, vanillin and trans-6-(4'-hydroxy-3'-methoxyphenyl)-2,4-dioxo-5-hexenal [2,7,8,43,44]. The results of our DR-UV–Vis analysis (Figure 7) confirmed the presence of feruloyl methane and ferulic acid in the samples obtained by co-precipitation PZn3Al-CR(Aq), PZn3Al-CR(Et), while in the sample RZn3Al-CR(Aq) only the presence of feruloyl methane was evidenced. This fact implies that when curcumin is contacted with an aqueous solution containing NaOH, the main degradation product resulted is feruloyl methane and that ferulic acid is obtained most probably during the ageing of the precipitates. Our findings are in agreement with those of Gordon and Schneider [43], which showed that vanillin and ferulic acid were not the major degradation products of curcumin.

From the results of DR-UV–Vis, it could also be inferred that the degradation of curcumin was less pronounced when the co-precipitation was performed with CR-ethanolic solution since the spectrum of PZn3Al-CR(Et) showed a more intense absorption in the region 400–600 nm compared to the spectrum of PZn3Al-CR(Aq). This fact could be due to the formation of a CR-Zn(II) complex which according to literature data has the maximum absorption in the visible region at 455 nm [42]. The presence of such a complex was also suggested by the ATR-FTIR spectrum of PZn3Al-CR(Et) (Figure 5a) displaying an absorption maximum at 1550 cm^{-1}. The red shifting of the absorption maximum specific to CR-Zn(II) complex from 1583 cm^{-1} [42] to 1550 cm^{-1}, could be related to its distortion under the influence of the LDH matrix and/or to its participation as secondary layered phase (whose presence was indicated by XRD analysis) in the LDH structure. The ATR-FTIR spectrum of the sample RZn3Al-CR(Aq) (Figure 5b) showed a sensibly weaker absorption band specific to CR-Zn(II) confirming that the stabilization of curcumin by complexation with Zn(II) was much lower in this case, most probably due to its degradation during the dissolution in the aqueous alkaline solution. The degradation of curcumin during the co-precipitation and reconstruction with alkaline aqueous CR-solution could also explain why the XRD patterns of the samples PZn3Al-CR(Aq) and RZn3Al-CR(Aq) did not show the specific diffraction lines of CR. The lower degradation of CR during the preparation of the PZn3Al-CR(Et) solid accompanied by the formation of the CR-Zn(II) complex evidenced by ATR-FTIR could be responsible for the obtaining of the extra layered phase with slightly larger interlayer space and a degree of crystalline disorder along the c-axis revealed by the XRD analysis (Figure 4a). Even if both ATR-FTIR and DR-UV–Vis spectra of the sample RZn3Al-CR(Et) obtained by reconstruction with CR-ethanolic solution indicated the presence of curcumin, its presence was not evidenced as a single phase in the XRD pattern most probably because nano-particles of curcumin were dispersed on the surface of this solid which contained also nano-sized particles of ZnO (see Table 2). Considering the results of the characterization studies it may be concluded that curcumin was incorporated without degradation only in RZn3Al-CR(Et), while in the rest of the solids, CR-Zn(II) complex and different degradation products of curcumin were incorporated in various extents. The amount of curcumin released from the synthesized solids was higher for those prepared with CR-ethanolic solutions (e.g., RZn3Al-CR(Et) and PZn3Al-CR(Et)) (Figure 8) and was well correlated to the content of the stabilized curcumin in the samples.

The release of curcumin from the CR-loaded solids was significantly influenced by the pH of the buffer solution utilized in the "in vitro" release studies. The buffers with acid pH (1, 2 and 5) allowed a better release of curcumin than the neutral to basic pH buffers (pH 7 and 8) which have a significant degrading effect on curcumin as it was indicated by Wang et al. [45]. This fact suggests that curcumin will be better released in the stomach where the pH can vary in the range 1.5–6.5 (e.g., pH 1.5–4.0 in the lower portion of the stomach and pH 4.0–6.5 in the upper portion of the stomach where predigestion takes place), than in the duodenum where the pH changes from 7.0 to 8.5 [46,47]. Considering the results of the CR-release tests (Figure 8) it may be inferred that CR-release will be lower in the upper portion of the stomach where the pH is less acidic (release tests at pH 5) and will enhance gradually as the solid will reach the lower portion of the stomach (release tests at pH 2 and pH 1), and it will be negligible when the solid reaches the duodenum (release tests at pH 7 and pH 8).

5. Conclusions

Our studies showed that the incorporation of curcumin into a Zn3Al-LDH structure can be achieved by co-precipitation only if the curcumin is utilized as an ethanolic solution. However, since the co-precipitation is performed with NaOH for pH adjustment, the partial degradation of curcumin could not be avoided. Hence the curcumin release capacity of this sample (PZn3Al-Cr(ET)) under acid pH conditions was by 20% lower than that of the sample prepared by reconstruction RZn3Al-CR(Et) even if the CR-concentration in the fresh samples was not significantly different (9.3 and 11.5 wt. %, respectively). The utilization of alkaline CR-aqueous solution either in the co-precipitation or in the reconstruction method leads to the obtaining of a Zn3Al-LDH structure impurified by ZnO phase where amorphous curcumin and mostly its degradation products were spread flatly on the surface of the inorganic matrix and at the edges of the layers. Taking into account both the results of the characterizations and those of CR-release tests, the preferable method to obtain CR-loaded Zn3Al solids from LDH precursors is the so-called "reconstruction" (since it does not restore the layered LDH structure) with CR-ethanolic solution which does not allow curcumin degradation and ensures also the highest curcumin release from the solid. Therefore, the sample RZn3Al-CR(Et) was the one selected for future antimicrobial activity tests.

Author Contributions: Conceptualization, R.Z.; formal analysis, E.B. and R.B.; investigation, O.D.P., A.Ș. and R.Z.; methodology, O.D.P. and R.Z.; project administration, O.D.P.; visualization, O.D.P.; writing—original draft, A.Ș. and R.Z.; writing—review and editing, R.Z. All authors have read and agreed to the published version of the manuscript.

Funding: This research was funded by the Romanian Ministry of Research and Innovation, CCCDI–UEFISCDI, grant number PN-III-P1-1.2-PCCDI-2017-0387/80PCCDI, within PNCDI III program.

Conflicts of Interest: The authors declare no conflict of interest.

References

1. Aggarwal, S.; Ichikawa, H.; Takada, Y.; Sandur, S.K.; Shishodia, S.; Aggarwal, B.B. Curcumin (diferuloylmethane) down-regulates expression of cell proliferation and antiapoptotic and metastatic gene products through suppression of IkappaBalpha kinase and Akt activation. *Mol. Pharmacol.* **2006**, *69*, 195–206. [CrossRef] [PubMed]
2. Heger, M.; van Golen, R.F.; Broekgaarden, M.; Michel, M.C. The molecular basis for the pharmacokinetics and pharmacodynamics of curcumin and its metabolites in relation to cancer. *Pharmacol. Rev.* **2014**, *66*, 222–307. [CrossRef] [PubMed]
3. Pulido-Moran, M.; Moreno-Fernandez, J.; Ramirez-Tortosa, C.; Ramirez-Tortosa, M.C. Curcumin and Health. *Molecules* **2016**, *21*, 264. [CrossRef] [PubMed]
4. Priyadarsini, K.I.; Maity, D.K.; Naik, G.H.; Kumar, M.S.; Unnikrishnan, M.K.; Satav, J.G.; Mohan, H. Role of phenolic O-H and methylene hydrogen on the free radical reactions and antioxidant activity of curcumin. *Free Radic. Biol. Med.* **2003**, *35*, 475–484. [CrossRef]
5. Kolev, T.M.; Velcheva, E.A.; Stamboliyska, B.A.; Spiteller, M. DFT and Experimental Studies of the Structure and Vibrational Spectra of Curcumin. *Int. J. Quantum Chem.* **2005**, *102*, 1069–1079. [CrossRef]
6. Jovanovic, S.V.; Steenken, S.; Boone, C.W.; Simic, M.G. H-Atom Transfer is a Preferred Antioxidant Mechanism of Curcumin. *J. Am. Chem. Soc.* **1999**, *121*, 9677–9681. [CrossRef]
7. Jagannathan, R.; Abraham, P.M.; Poddar, P. Temperature-Dependent Spectroscopic Evidences of Curcumin in Aqueous Medium: A Mechanistic Study of Its Solubility and Stability. *J. Phys. Chem. B* **2012**, *116*, 14533–14540. [CrossRef]
8. Tønnesen, H.H.; Másson, M.; Loftsson, T. Studies of curcumin and curcuminoids. XXVII. Cyclodextrin complexation: Solubility, chemical and photochemical stability. *Int. J. Pharm.* **2002**, *244*, 127–135. [CrossRef]
9. Parvathy, K.S.; Negi, P.S.; Srinivas, P. Antioxidant, antimutagenic and antibacterial activities of curcumin-β-diglucoside. *Food Chem.* **2009**, *115*, 265–271. [CrossRef]
10. Sneharani, A.H.; Singh, S.A.; Appu Rao, A.G. Interaction of αS1-Casein with Curcumin and Its Biological Implications. *J. Agric. Food Chem.* **2009**, *57*, 10386–10391. [CrossRef]

11. Sneharani, A.H.; Karakkat, J.V.; Singh, S.A.; Appu Rao, A.G. Interaction of Curcumin with β-Lactoglobulin—Stability, Spectroscopic Analysis, and Molecular Modeling of the Complex. *J. Agric. Food Chem.* **2010**, *58*, 11130–11139. [CrossRef] [PubMed]
12. Bisht, S.; Feldmann, G.; Soni, S.; Ravi, R.; Karikar, C.; Maitra, A.M.; Maitra, A.N. Polymeric nanoparticle encapsulated curcumin ("nanocurcumin"): A novel strategy for human cancer therapy. *J. Nanobiotechnol.* **2007**, *5*, 3. [CrossRef] [PubMed]
13. Tiwari, S.K.; Agarwal, S.; Seth, B.; Yadav, A.; Nair, S.; Bhatnagar, P.; Karmakar, M.; Kumari, M.; Chauhan, L.K.; Patel, D.K.; et al. Curcumin-loaded nanoparticles potently induce adult neurogenesis and reverse cognitive deficits in Alzheimer's disease model via canonical Wnt/b-catenin pathway. *ACS Nano* **2014**, *8*, 76–103. [CrossRef] [PubMed]
14. Sun, D.; Zhuang, X.; Xiang, X.; Liu, Y.; Zhang, S.; Liu, C.; Barnes, S.; Grizzle, W.; Miller, D.; Zhang, H.G. A novel nanoparticle drug delivery system: The anti-inflammatory activity of curcumin is enhanced when encapsulated in exosomes. *Mol. Ther.* **2010**, *18*, 1606–1614. [CrossRef]
15. Wang, W.; Zhu, R.; Xie, Q.; Li, A.; Xiao, Y.; Li, K.; Liu, H.; Cui, D.; Chen, Y.; Wang, S. Enhanced bioavailability and efficiency of curcumin for the treatment of asthma by its formulation in solid lipid nanoparticles. *Int. J. NanoMed.* **2012**, *7*, 3667–3677. [CrossRef]
16. Gonçalves, C.; Pereira, P.; Schellenberg, P.; Coutinho, P.J.; Gama, F.M. Self-assembled dextrin nanogel as curcumin delivery system. *J. Biomater. Nanobiotechnol.* **2012**, *3*, 178–184. [CrossRef]
17. Falconieri, M.C.; Adamo, M.; Monasterolo, C.; Bergonzi, M.C.; Coronnello, M.; Bilia, A.R. New dendrimer based nanoparticles enhance curcumin solubility. *Planta Med.* **2017**, *83*, 420–425. [CrossRef]
18. Shome, S.; Talukdar, A.D.; Choudhury, M.D.; Bhattacharya, M.K.; Upadhyaya, H. Curcumin as potential therapeutic natural product: A nanobiotechnological perspective. *J. Pharm. Pharmacol.* **2016**, *68*, 1481–1500. [CrossRef]
19. Zebib, B.; Mouloungui, Z.; Noirot, V. Stabilization of Curcumin by Complexation with Divalent Cations in Glycerol/Water System. *Bioinorg. Chem. Appl.* **2010**, *2010*, 292760. [CrossRef]
20. Jiang, T.; Wang, L.; Zhang, S.; Sun, P.C.; Ding, C.F.; Chu, Y.Q.; Zhou, P. Interaction of curcumin with Al (III) and its complex structures based on experiments and theoretical calculations. *J. Mol. Struct.* **2011**, *1004*, 163–173. [CrossRef]
21. Cavani, F.; Trifirò, F.; Vaccari, A. Hydrotalcite-type anionic clays: Preparation, properties and applications. *Catal. Today* **1991**, *11*, 173–301. [CrossRef]
22. Trifiro, F.; Vaccari, A. Hydrotalcite-like Anionic Clays (Layered Double Hydroxides). In *Comprehensive Supramolecular Chemistry*; Atwood, J.L., Davies, J.E.D., MacNicol, D.D., Vögtle, F., Eds.; Pergamon: Oxford, UK, 1996; Volume 7, pp. 251–291.
23. Isupov, V.P.; Chupakhina, L.E.; Mitrofanova, R.P. Mechanochemical synthesis of double hydroxides. *J. Mater. Synth. Process.* **2000**, *8*, 251–253. [CrossRef]
24. Richetta, M.; Medaglia, P.G.; Mattoccia, A.; Varone, A.; Pizzoferrato, R. Layered Double Hydroxides: Tailoring Interlamellar Nanospace for a Vast Field of Applications. *J. Mater. Sci. Eng.* **2017**, *6*, 360. [CrossRef]
25. Choy, J.H.; Jung, J.S.; Oh, J.M.; Park, M.; Jeong, J.; Kang, Y.K.; Han, O.J. Layered double hydroxide as an efficient drug reservoir for folate derivatives. *Biomaterials* **2004**, *25*, 3059–3064. [CrossRef] [PubMed]
26. Gordijo, C.R.; Barbosa, C.A.S.; Ferreira, A.M.D.C.; Constantino, V.R.L.; Silva, D.D. Immobilization of ibuprofen and copper-ibuprofen drugs on layered double hydroxides. *J. Pharm. Sci.* **2005**, *94*, 1135–1148. [CrossRef] [PubMed]
27. Costantino, U.; Ambrogi, V.; Nocchetti, M.; Perioli, L. Hydrotalcite-like compounds: Versatile layered hosts of molecular anions with biological activity. *Microporous Mesoporous Mater.* **2008**, *107*, 149–160. [CrossRef]
28. Oh, J.-M.; Park, M.; Kim, S.-T.; Jung, J.-Y.; Kang, Y.-G.; Choy, J.-H. Efficient delivery of anticancer drug MTX through MTX-LDH nanohybrid system. *J. Phys. Chem. Solids* **2006**, *67*, 1024–1027. [CrossRef]
29. Bi, X.; Zhang, H.; Dou, L. Layered Double Hydroxide-Based Nanocarriers for Drug Delivery. *Pharmaceutics* **2014**, *6*, 298–332. [CrossRef]
30. Muráth, S.; Szerlauth, A.; Sebők, D.; Szilágyi, I. Layered Double Hydroxide Nanoparticles to Overcome the Hydrophobicity of Ellagic Acid: An Antioxidant Hybrid Material. *Antioxidants* **2020**, *9*, 153. [CrossRef]
31. Muráth, S.; Alsharif, N.B.; Sáringer, S.; Katana, B.; Somosi, Z.; Szilágyi, I. Antioxidant Materials Based on 2D Nanostructures: A Review on Recent Progresses. *Crystals* **2020**, *10*, 148. [CrossRef]

32. Wang, Q.; O'Hare, D. Recent Advances in the Synthesis and Application of Layered Double Hydroxide (LDH) Nanosheets. *Chem. Rev.* **2012**, *112*, 4124–4155. [CrossRef]
33. Supun Samindra, K.M.; Kottegoda, N. Encapsulation of curcumin into layered double hydroxides. *Nanotechnol. Rev.* **2014**, *3*, 579–589. [CrossRef]
34. Megalathan, A.; Kumarage, S.; Dilhari, A.; Weerasekera, M.M.; Samarasinghe, S.; Kottegoda, N. Natural curcuminoids encapsulated in layered double hydroxides: A novel antimicrobial nanohybrid. *Chem. Cent. J.* **2016**, *10*, 35. [CrossRef]
35. Gayani, B.; Dilhari, A.; Wijesinghe, G.K.; Kumarage, S.; Abayaweera, G.; Samarakoon, S.R.; Perera, I.C.; Kottegoda, N.; Weerasekera, M.M. Effect of natural curcuminoids-intercalated layered double hydroxide nanohybrid against Staphylococcus aureus, Pseudomonas aeruginosa, and Enterococcus faecalis: A bactericidal, antibiofilm, and mechanistic study. *MicrobiologyOpen* **2019**, *8*, e723. [CrossRef] [PubMed]
36. Król, A.; Pomastowski, P.; Rafińska, K.; Railean-Plugaru, V.; Buszewski, B. Zinc oxide nanoparticles: Synthesis, antiseptic activity and toxicity mechanism. *Adv. Colloid Interface Sci.* **2017**, *249*, 37–52. [CrossRef]
37. Zăvoianu, R.; Pavel, O.D.; Cruceanu, A.; Florea, M.; Bîrjega, R. Functional layered double hydroxides and their catalytic activity for 1,4-addition of n-octanol to 2-propenonitrile. *Appl. Clay Sci.* **2017**, *146*, 411–422. [CrossRef]
38. Kooli, E.; Depège, C.; Ennaqadi, A.; De Roy, A.; Besse, J.E. Rehydration of Zn-Al layered double hydroxides. *Clays Clay Miner.* **1997**, *45*, 92–98. [CrossRef]
39. Elhalil, A.; Farnane, M.; Machrouhi, A.; Mahjoubi, F.Z.; Elmoubarki, R.; Tounsadi, H.; Abdennouri, M.; Barka, N. Effects of molar ratio and calcination temperature on the adsorption performance of Zn/Al layered double hydroxide nanoparticles in the removal of pharmaceutical pollutants. *J. Sci. Adv. Mater. Devices* **2018**, *3*, 188–195. [CrossRef]
40. Karami, Z.; Aghazadeh, M.; Jouyandeh, M.; Zarrintaj, P.; Vahabi, H.; Ganjalia, M.R.; Torre, L.; Puglia, D.; Sae, M.R. Epoxy/Zn-Al-CO3 LDH nanocomposites: Curability assessment. *Prog. Org. Coat.* **2020**, *138*, 105355. [CrossRef]
41. George, G.; Saravanakumar, M.P. Facile synthesis of carbon-coated layered double hydroxide and its comparative characterisation with Zn–Al LDH: Application on crystal violet and malachite green dye adsorption—Isotherm, kinetics and Box-Behnken design. *Environ. Sci. Pollut. Res.* **2018**, *25*, 30236–30254. [CrossRef]
42. Zhao, X.Z.; Jiang, T.; Wang, L.; Yang, H.; Zhang, S.; Zhou, P. Interaction of curcumin with Zn (II) and Cu (II) ions based on experiment and theoretical calculation. *J. Mol. Struct.* **2010**, *984*, 316–325. [CrossRef]
43. Gordon, O.N.; Schneider, C. Vanillin and ferulic acid: Not the major degradation products of curcumin. *Trends Mol. Med.* **2012**, *18*, 361–363. [CrossRef] [PubMed]
44. Silva, A.M.N.; Kong, X.; Parkin, M.C.; Cammack, R.; Hider, R.C. Iron (III) citrate speciation in aqueous solution. *Dalton Trans.* **2009**, *40*, 8616–8625. [CrossRef] [PubMed]
45. Wang, Y.J.; Pan, M.H.; Cheng, A.L.; Lin, L.I.; Ho, Y.S.; Hsieh, C.Y.; Lin, J.K. Stability of curcumin in buffer solutions and characterization of its degradation products. *J. Pharm. BioMed. Anal.* **1997**, *15*, 1867–1876. [CrossRef]
46. Allegany Nutrition. The Enzyme Specialists. Available online: https://www.alleganynutrition.com/supporting-pages/the-human-digestive-tract-ph-range-diagram/ (accessed on 1 February 2020).
47. Beasley, D.E.; Koltz, A.M.; Lambert, J.E.; Fierer, N.; Dunn, R.R. The evolution of stomach acidity and its relevance to the Human Microbiome. *PLoS ONE* **2015**, *10*, e0134116. [CrossRef] [PubMed]

© 2020 by the authors. Licensee MDPI, Basel, Switzerland. This article is an open access article distributed under the terms and conditions of the Creative Commons Attribution (CC BY) license (http://creativecommons.org/licenses/by/4.0/).

Review

Effect of LDHs and Other Clays on Polymer Composite in Adsorptive Removal of Contaminants: A Review

Maleshoane Mohapi [1], Jeremia Shale Sefadi [2,*], Mokgaotsa Jonas Mochane [1,*], Sifiso Innocent Magagula [1] and Kgomotso Lebelo [1]

1. Department of Life Sciences, Central University of Technology, Free State, Private Bag X20539, Bloemfontein 9300, South Africa; shoanymohapi@gmail.com (M.M.); sifisom61@gmail.com (S.I.M.); KLebelo@cut.ac.za (K.L.)
2. Department of Physical and Earth Sciences, Sol Plaatje University, Kimberley 8301, South Africa
* Correspondence: jeremia.sefadi@spu.ac.za (J.S.S.); mochane.jonas@gmail.com or mmochane@cut.ac.za (M.J.M.)

Received: 15 August 2020; Accepted: 15 October 2020; Published: 22 October 2020

Abstract: Recently, the development of a unique class of layered silicate nanomaterials has attracted considerable interest for treatment of wastewater. Clean water is an essential commodity for healthier life, agriculture and a safe environment at large. Layered double hydroxides (LDHs) and other clay hybrids are emerging as potential nanostructured adsorbents for water purification. These LDH hybrids are referred to as hydrotalcite-based materials or anionic clays and promising multifunctional two-dimensional (2D) nanomaterials. They are used in many applications including photocatalysis, energy storage, nanocomposites, adsorption, diffusion and water purification. The adsorption and diffusion capacities of various toxic contaminants heavy metal ions and dyes on different unmodified and modified LDH-samples are discussed comparatively with other types of nanoclays acting as adsorbents. This review focuses on the preparation methods, comparison of adsorption and diffusion capacities of LDH-hybrids and other nanoclay materials for the treatment of various contaminants such as heavy metal ions and dyes.

Keywords: layered double hydroxides (LDHs); other nanoclays; organically modified LDH; water purification; adsorption; adsorption interaction; diffusion

1. Introduction

A reliable, affordable, sustainable and easily accessible clean water supply chain for many societies in the entire world is an essential component for healthier life and safe environment. However, due to limited economical resources or lack of infrastructure, millions of poor and vulnerable people including children die annually from diseases caused by an inadequate water supply, poor water quality, sanitation and hygiene. Recently, many countries and communities experienced the global challenge/phenomenon known as "Coronavirus (COVID-19) or COV2 infections", which required a frequent washing of hands with clean water and soap or hand sanitizer to avoid or curb the spread (flatten the curve). These key risk aspects or factors adversely impact on food security, livelihood diversities and learning opportunities for poor and most susceptible households across the world. According to the World Health Organization (WHO), almost 1.7 million people lost their lives because of water pollution, and four billion cases of diverse health issues were reported every year due to water borne diseases [1]. Table 1 represents various types of water contaminants, their sources and negative effects. To improve access to quality and safe drinking water, sanitation, and hygiene (WASH), there must be value-added infrastructure investment in dealing with and managing the freshwater ecosystems and sanitation facilities on a local level in many developing countries. The improved WASH is thus fundamental to poverty reduction, promotion of equality, and support for socioeconomic development under the sustainable development goals (SDGs) [2,3]. The most essential requirements for clean water supply chain is a proper material with high degree of separation capacity, low cost, porosity, and reusability [4–7]. Nanotechnology presents a set of opportunities to develop nanomaterials for effective water purification systems. Optimization of the properties like hydrophilicity, hydrophobicity, porosity, mechanical strength and dispersibility [8–10] is the best option to treat wastewater. Due to their high surface area, high chemical reactivity, adsorption capabilities, excellent mechanical strength and cost-effectiveness, nanomaterials have a huge potential to effectively purify water in numerous ways [8,10–12] by removing various contaminants. This can be done by using different purifiers with different pore sizes such as: microfiltration (MF), ultrafiltration (UF), nanofiltration (NF) and reverse osmosis (RO) (Figure 1). However, the main stumbling block associated with addition of 2D nanomaterials is the aggregation or agglomeration that restricts their effective use in many industrial applications. This daunting aggregation or agglomeration challenge of nanomaterials can be minimized by (i) transforming 2D nanomaterials into nanocomposites and (ii) surface modification of 2D nanomaterials, owing to their excellent interfacial interaction between the surface of 2D nanomaterials and polymer matrices. Surface modification of nanomaterials (SMNs), compared to unmodified nanomaterials, has attracted a considerable interest in science communities.

Table 1. List of different water pollutants with their sources and adverse effects.

Water Pollutants	Sources of Pollutants	Effects of Pollutants	References.
Pathogens	Viruses and bacteria	Cause water borne diseases which can affect anyone. Those at high risk are infants, younger children, the elderly and patients with underlying illnesses (diabetes, chronic diseases of heart disease and kidney).	[14]
Agricultural Pollutants	Agricultural chemicals	Directly affect the freshwater resources and can cause health-related problems contributing to blue baby syndrome leading to the death in infants.	[15]
Sediments and suspended solids	Land cultivation, demolition, mining operations	Affect water quality and bring about toxicity on fish life and involve reduced oxygen transfer at the gills, reduced ability to clear sediment from the gills and diminished bloodstream.	[16]
Inorganic pollutants	Metals compounds, trace elements, inorganic salts, heavy metals, mineral acids	Cause several human health–related problems on the flora and fauna of the Earth system such as abnormal growth, high risk of cancer, diabetes and obesity.	[17]
Organic pollutants	Detergents, insecticides, herbicides	They are resistant to degradation and tend to bioaccumulate within the food chain. Cause various negative health issues including cancer, immune system suppression, decrements in cognitive and neurobehavioral function, and at least some of them increase the risk of chronic diseases, such as hypertension, cardiovascular disease and diabetes.	[18]
Industrial pollutants	Municipal pollutant water	Cause air, water and land pollution leading to many environmental problems, illnesses and loss of life.	[19]
Radioactive pollutants	Different Isotopes	Exposure to high levels of radiation causes acute health problems like bones, teeth, skin burns and cancer as well as cardiovascular disease.	[20]
Nutrients pollutants	Plant debris, fertilizer.	Cause serious environmental and human health issues which influence the socio-economic issues. Causes algae to grow and expand higher than ecosystems can handle.	[21]
Macroscopic pollutants	Marine debris	Macroscopic pollutants are non-biodegradable materials which cause garbage wastes and plastic pollution.	[22]
Sewage and contaminated water	Domestic wastewater	Causes the quality of the water to worsen, water borne diseases and affects aquatic ecosystems.	[23]

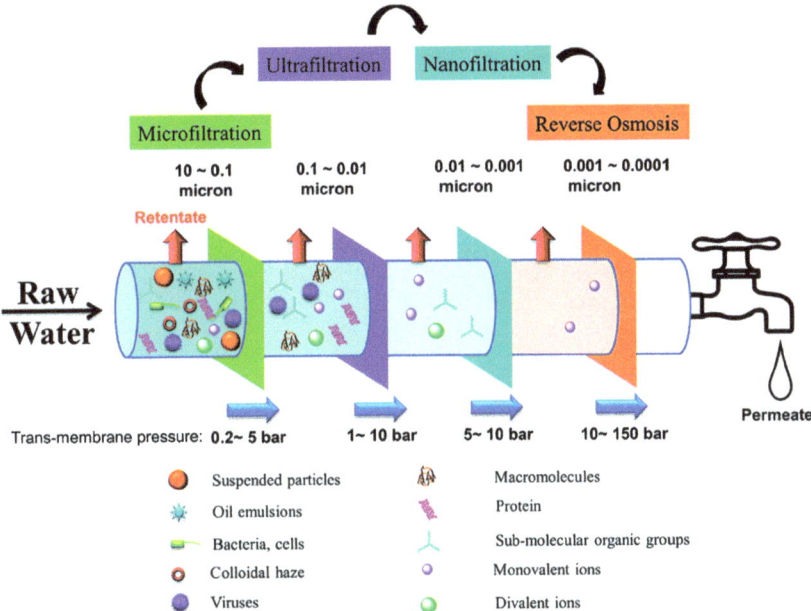

Figure 1. Trans-membrane pressure processes for water treatment technologies with different pore sizes. Reproduced with permission from Reference [13]. Copyrights 2018, Elsevier Science Ltd.

Nanocomposites are multi-phasic materials, in which at least one of the phases shows dimensions in the nano range of 10–100 nm [24,25]. Currently, these materials have emerged as alternatives to overcome deficiencies of different engineering materials and are said to be the 21st century materials, due to their design uniqueness and property combinations which are different from conventional composites. Nanocomposite materials can be classified according to their primary phase (matrix) and secondary phase (reinforcing filler) [26,27]. Among different nanocomposites, polymer-based nanocomposites (PNCs) have become a noticeable field of current research interest and innovation development. PNCs have a lot of advantageous multifunctional properties such as film forming ability, dimensional variability, and activated functionalities [8,28]. Generally, the properties of PNCs are strongly related to the type of polymer matrix and the extent of dispersion of nanomaterials incorporated into the polymer matrix, as well as interfacial interactions between the polymer and nanomaterials [29–31]. The improved interfacial interactions of the nanomaterials with pure polymer change the overall morphology leading to synergistic effects in the nanocomposite properties. The accomplished properties are much better than the individual constituents. Finally, the properties of the PNCs are directly dependent on the volume fraction of nanomaterials, aspect ratio, alignment in matrix and other geometrical factors [32,33]. The main challenges in the development of superior PNCs are (i) the selection of appropriate nanomaterials that possess specific interfacial interaction, (ii) compatibility of nanomaterials with polymer matrix and (iii) suitable processing method to evenly disperse and dispense these nanoparticles within a polymer matrix. The impact of polymer nanocomposites (PNCs) in wastewater treatment can be recognized by an uninterrupted rise in publications over the past ten years. This is contrary to the less extensively instigated environmental impacts of nanomaterials in polymer nanocomposites in water purification (Figure 2).

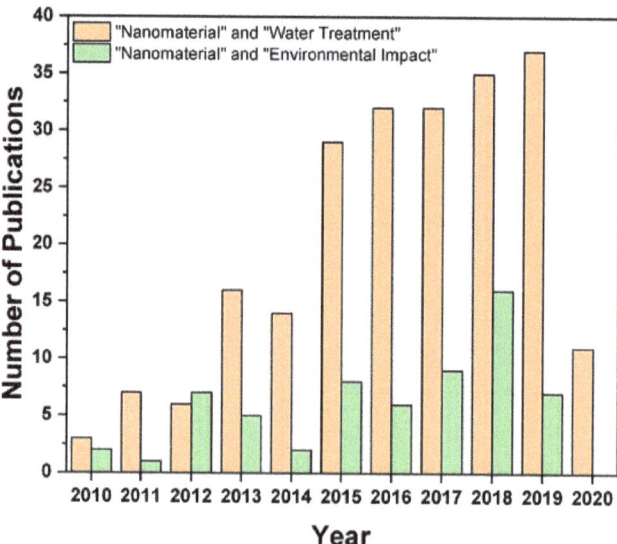

Figure 2. Number of publications in polymer nanocomposites (PNCs) and water treatment. Reproduced with permission from Reference [34]. Open Access 2020, MDPI Water.

In this review, we focus on the comparison of 2D nanomaterials such as layered double hydroxides (LDHs) and other nanoclays in polymer-based nanocomposites (PNCs), their preparation methods and multifunctional properties and their use for water purification. The primary goal is to highlight an optimal efficiency of LDHs and other nanoclays adsorption capacities and recent progress of nanomaterials in decontamination ability of various pollutants with selectivity including practical and potential applications.

2. Comparison of Other Nanoclays and LDHs Crystal Structures

2.1. Other Nanoclays Crystal Structure

Nanoclays (NCs) are a broad class of naturally occurring inorganic minerals optimized for use in polymer-clay nanocomposites for water purification and environmental protection. NCs are versatile and two-dimensional (2D) building blocks for multifunctional material systems with several property enhancements targeted for many applications. Based on their chemical composition and particle morphology, clay minerals are categorized into many classes such as smectite, chlorite, kaolinite, illite and halloysite. Nanoclays have been studied and developed for various applications [29,31,35] and are abundantly available, very cheap and low environmental impact. Clay minerals are members of the phyllosilicate or sheet clay silicates consisting of hydrated alumina–silicates and can be used as natural nanomaterials or nano-absorbent since the dawn of nanotechnology [36]. Nanoclays are nanoparticles of layered mineral silicates with layered structural units that can lead to the formation complex/multifaceted clay crystallites by stacking these layers [37]. The basic building blocks of clay minerals are tetrahedral silicates and octahedral hydroxide sheets [38]. Octahedral sheets consist of aluminum or magnesium in a six-fold coordination with oxygen from a tetrahedral sheet and with hydroxyl (Figure 3). Tetrahedral sheets consist of silicon–oxygen tetrahedra concomitant to neighboring tetrahedral sharing three corners, while the fourth corner of each tetrahedron sheet is connected to an adjacent octahedral sheet via a covalent bond.

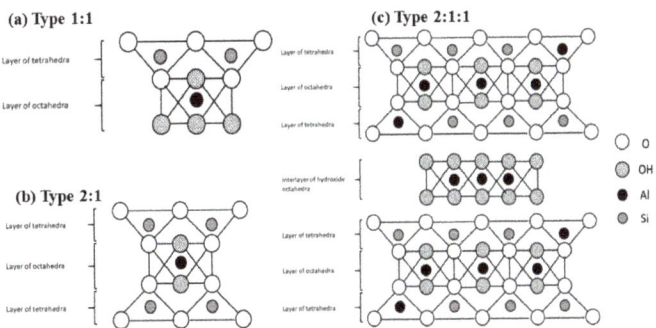

Figure 3. The layer phyllosilicate structures: (**a**) Type 1:1, (**b**) Type 2:1, and (**c**) Type 2:1:1. Reproduced with permission from Reference [39]. Open Access 2019, MDPI Animals.

The arrangements of these sheets influence a number of contributing factors in clay silicates. Based on their mineralogical composition, there are nearly thirty different types of nanoclays used in various applications [40,41]. Table 2 depicts three major types of phyllosilicates which are distinguished as 1:1 layer type (T-O), 2:1 layer type (T-O-T) and 2:1:1 layer type (T-O-T:O) common in nanoclay materials. In 1:1 lattice structures (T-O), each tetrahedral is connected to one octahedral sheet, while in 2:1 lattice structures (T-O-T), each octahedral sheet is connected to two tetrahedral sheets, one sheet on each side. Lastly, in 2:1:1 lattice structures (T-O-T:O), each octahedral sheet is adjacent to another octahedral sheet and connected to two tetrahedral sheets [31,42–44].

Table 2. Classification of clay minerals and their characteristics.

Clay Minerals Group	Layer Type Ratio	Characteristics	References.
Rectorite, Kaolinite, Halloysite, Chyrsotile.	1:1 (T-O) Dioctahedral	Non-expansive, no layer charge and very little isomorphic substitution.	[41,42]
Smectite Vermiculite	2:1 (T-O-T) dioctahedral Trioctahedral	Highly expansive, low layer charge moderately expansive, Intermediate layer charge. extensive isomorphic substitution.	[43]
Pyrophylite talc, mica, brittle mica.	2:1 (T-O-T) dioctahedral Trioctahedral	Non-expansive, high layer charge, extensive isomorphic substitution.	[43]
Chlorite	2:1:1 (T-O-T-O) dioctahedral Trioctahedral Di, Trioctahedral	Non-expansive, high layer charge, extensive isomorphic substitution.	[31,44]

Halloysite nanoclay is an aluminosilicate nanotube naturally occurring clay material with the average dimensions of 15 nm × 1000 nm [45]. This halloysite nanoclay has (1:1-layer type) and the hollow tube structure is primarily utilized in medical applications, food packaging industry and rheology modification [46]. The most commonly used nanoclay in materials applications is plate-like montmorillonite (MMT) material. This MMT has approximately 1 nm of aluminosilicate layers which are surface coated with metal cations in a multilayer stacks of ~10 µm. Depending on surface modification of the clay layers, MMT can be dispersed in a polymer matrix to form polymer-clay nanocomposites with applications such as, flame-resistance, solidifying agents, water purification and gas permeability modification. MMT clay layers with 2:1 layered silicates of T-O have high cation exchange capacity (CEC) on the siloxane surface that can interact well with different substances like organic or biological molecules [47,48]. The MMT nanoclay stacks have attracted a lot of interest because of outsized surface area, swelling behavior and high cation exchange capacity [49,50]. Unlike MMTs, halloysite materials are easily dispersed in many polymers showing no exfoliation due to scarcity of OH groups on their surfaces. In addition, these tube-like nanoclays are excellent nanomaterials for numerous chemical molecules [51]. Therefore, the modified clays are used as effective reinforcing phase for polymers to improve their mechanical and thermal properties. Nanoclays acting as carriers continuously and constantly released some active molecules such as flame-retardants, antioxidants, anticorrosion and antimicrobial agents [52,53]. In recent years, the research and development of novel polymer/nanoclay composites for water purification has attracted a lot of attention in the field of material chemistry [54]. Rigid nanoclay like layered double hydroxides (LDHs) must be used as an effective reinforcing filler to polymer structures and impede the polymer chains free movement adjacent to the filler [29–31,54].

2.2. Layered Double Hydroxides (LDHs) Crystal Structure

Layered double hydroxides (LDHs) also known as hydrotalcite (HT)-like materials are a class of synthetic two-dimensional (2D) nanostructured anionic clays with a highly tunable brucite [Mg(OH)$_2$]-like layered crystal structure (Figure 4). These inorganic materials contain layers of positively charged metal hydroxides with multivalent anions for neutrality. The LDHs are generally represented by formula

$$[M^{2+}_{1-x} M^{3+}_{x} (OH)_2]^{X+} [A^{n-}]_{X/n} \, mH_2O \tag{1}$$

In this formula, M^{2+} and M^{3+} represent the divalent and trivalent layer cations, respectively. A^{n-} is the exchangeable anion such as OH^-, F^-, NO_3^-, Cl^-, CO_3^{2-} and/or SO_4^{2-}. Reasonably stable LDH phases are often observed only when the value of x varies in the range 0.22–0.33 resulting in M^{2+}/M^{3+} molar ratios of 2:1 to 4:1 [55–59]. If x is more than 0.33, then an increased number of neighboring M^{3+} containing octahedra leads to the formation of $M(OH)_3$. If x is less than 0.2, then an increased number of neighboring M^{2+} containing octahedra in the brucite-like sheets resulted in the precipitation of $M(OH)_2$. However, these limits of the value of x must be regarded as the maximum interval, which can be narrower depending on the composition of the LDH.

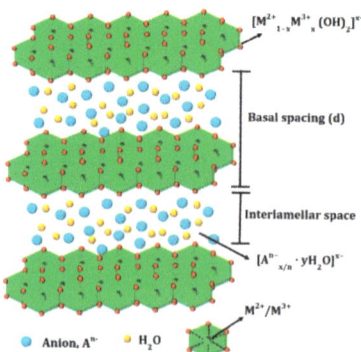

Figure 4. Structure of layered double hydroxide (LDH). Reproduced with permission from Reference [60]. Copyrights 2018, Elsevier Science Ltd.

As result, a large class of isostructural materials, which can be well-thought-out complementary to aluminosilicate clays, with useful physical and chemical properties can be achieved. This can be carried out by changing the nature of the metal cations, the molar ratios of divalent/trivalent cations, and the types of interlayer anions. These compounds are composed of positively charged brucite-type octahedral sheets, interchanging with interlayers containing carbonate anions in the natural mineral or other exchangeable anions in the synthetic hydrotalcite (HT)-like materials, along with water molecules. The hydrogen bonding associated with the interlamellar water molecules serves as a driving force for the stacking of the clay layers (see Figure 4).

2.3. An overview of Preparation Methods of LDHs

In the last few decades, a number of studies associated with the synthesis of LDH have been reported and some are easy and simple to process for many industrial applications. Various kinds of low cost, environmentally and eco-friendly LDHs can be synthesized by using fundamental methods of choice. These commonly used methods include co-precipitation; ion exchange; reconstruction; sonochemical method; hydrothermal/solvochemical method; sol–gel method; induced hydrolysis method; and urea method [57,59–65].

2.3.1. Co-Precipitation Method

Co-precipitation method is also referred to as a large-scale and direct technique typically utilized for the synthesis of LDH platelets with different divalent and trivalent cations (M^{2+} and M^{3+}) coupled with many inorganic anions (Cl^-, NO_3^-, CO_3^{2-}) and organic molecules/outsized biomolecules [57]. In this co-precipitation method, a dropwise addition of alkali solution into the divalent and trivalent layer cations/mixed metal salts containing solution in a proper ratio resulted in formation of LDH as shown in Figure 5. During this method, a pH emerged as a very crucial factor which negatively influences both the structural and chemical properties of LDH component to a larger extent. In a dropwise addition, the pH of the reaction mixture is maintained constantly at the range of 8–10 and purged at N_2 atmosphere in an attempt to achieve high chemical homogeneity in LDH [61]. The resulting solution mixture is allowed to stay for a long period of time in order to obtain a reproducible and well-crystallized LDH structural material. The obtained precipitate is collected by filtration, washed thoroughly with deionized water and dried in an oven overnight. The underpinning principle of co-precipitation is based on a simple, economical and industrially feasible technique utilized for the synthesis of metal oxide materials in solution. This led to the brucite-like layers' formation, which uniformly dispersed metallic cations and inorganic anions.

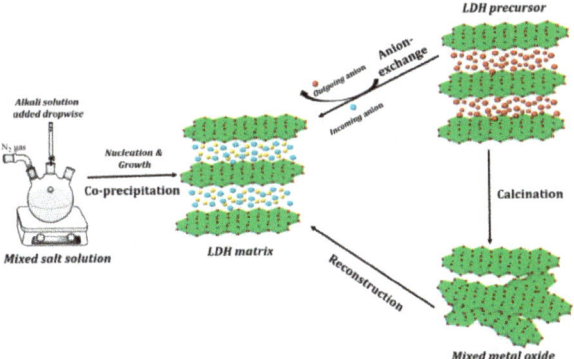

Figure 5. Schematic representation of co-precipitation, ion exchange and reconstruction methods LDH. Reproduced with permission from Reference [59]. Copyrights 2018, Elsevier Science Ltd.

2.3.2. Ion Exchange Method

Anion-exchange method is better for the incorporation of layered silicates into a solution containing anions species ready for exchange as compared to other methods. The ion exchange method (Figure 5) is based on the exchange of anions in interlayer space with other anionic species. In this method, the precursor LDH is suspended in an aqueous solution containing the anionic species to be exchanged.

The suspension is then stirred constantly for several hours at room temperature. The solid precipitate is then collected by filtration, washed several times with deionized water and dried in an oven overnight. Furthermore, the interlamellar region can also contain water molecules and is referred to as internal water dominant region where water molecules are organized by the inorganic layers via hydrogen bonding. The high anion exchange capacity of LDH matrix-like compounds produces their interlayer ion exchange by outgoing anions and incoming useful anions, easily accomplished and reflected in LDH precursor product formation [62] (Figure 5).

2.3.3. Reconstruction Method

As shown in Figure 5, this reconstruction method is a well-known regeneration or memory effect method of LDH. According to this method, the layered structure of brucite-like LDH with carbonate anion is used as a precursor during hydration and calcination due to its behavioral pattern. The reconstruction method is based on the memory effect or regeneration which is one of the unique properties of LDHs. In first step, the calcination of LDHs is performed at a particular temperature to obtain mixed metal oxides (Figure 5) and then subjected to rehydration in aqueous solution with the anion to be intercalated [63]. The solid precipitate is collected by filtration, washed several times with deionized water and dried in an oven overnight. The structural recovery, however, depends upon some experimental conditions such as, calcination temperature, duration and rate of heating. The reconstruction method is useful mainly in the preparation of large organic anions intercalated LDH [63,64].

2.3.4. Sonochemical Method

In sonochemical method, LDHs are prepared by co-precipitation method followed by sonochemical treatment. In first step, the co-precipitation method is performed to the latter as explained fully under Section 2.3.1 above. In the second step after successful completion of mixing, the resultant solution is subjected to ultrasound irradiation at a given time and temperature. The solid precipitate is filtered, washed thoroughly with deionized water and put in an oven overnight for further drying. This sonochemical method is best described as a synthetic and high intensity ultrasonic/three-fold acoustic cavitation phenomenon which assist in improving the crystallinity of LDH phases [61,65]. When the solution mixture is subjected to ultrasonic irradiation, rapid movement of the fluid leads to three-fold acoustic cavitation phenomenon (Figure 6) in which microbubbles undergo nucleation formation, growth and implosive collapse [65]. The formation of microbubbles produced a distinctive hot spot due to the compressional heating induced by collapsing of bubble and therefore yields the bubble with extremely high temperature, pressure and cooling rates [66].

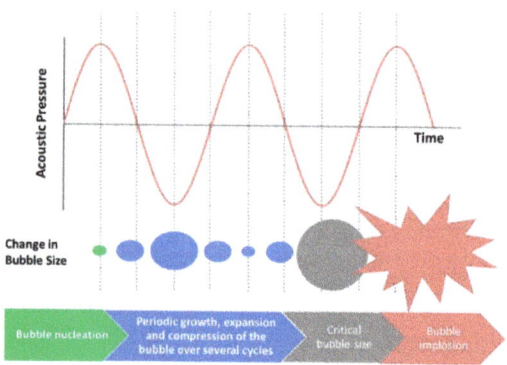

Figure 6. Schematic representation of three-fold acoustic cavitation phenomenon. Reproduced with permission from Reference [67]. Open Access 2020, MDPI Energies.

2.3.5. Hydrothermal/Solvothermal Method

The hydrothermal synthesis method illustrated in Figure 7a,b is similar to the co-precipitation method. In this method, two solutions containing M^{2+} and M^{3+} metal salts are added dropwise to another solution containing base under vigorous stirring at room temperature. Thus, the suspension is transferred into a Teflon-lined autoclave and heated at higher temperature (100–180 °C) for many hours (10–48 h) based on the metal ions [68]. The pH of the supernatant solution is in the threshold range of 8–10. The solid precipitate is collected by centrifugation washed thoroughly with deionized water and ethanol and dried in an oven overnight. The hydrothermal method is useful for synthesis of highly crystalline LDHs with uniform morphology compared to co-precipitation technique [69]. Solvothermal method is a synthesis method where a chemical reaction takes place in a closed solvent system at elevated temperatures above the boiling point and standard pressures. In a typical solvothermal synthesis, the amount of organic solvent such as glycerol or alcohol is used in a non-aqueous solution at somewhat high temperatures, while hydrothermal method refers to synthesis via chemical reactions in aqueous solution just above boiling point of water in a closed vessel. Many scientists realized the importance of preparing inorganic nanomaterials using hydrothermal and solvothermal reactions, upon which effective syntheses of novel high-technology and green materials would be established. The main reason for this remarkable and milestone achievement in preparing nanomaterials is the easy of processing which include low temperature process, low energy consumption, no harm to the environment and more importantly high degree of crystallinity of the material can easily be produced [68,69].

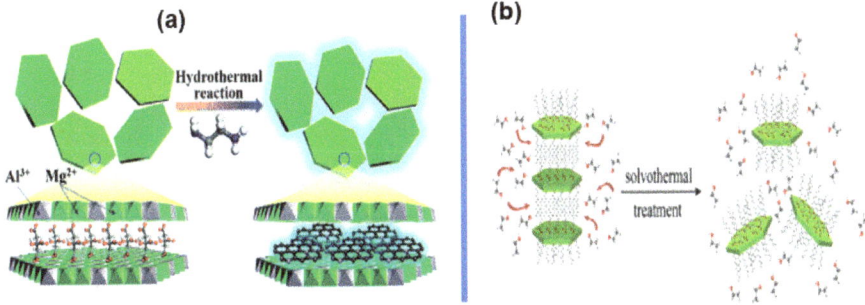

Figure 7. Schematic representation of (**a**) hydrothermal method and (**b**) solvothermal treatment of layered double hydroxide (LDH). Reproduced with permission from Reference [70,71]. Copyrights 2017 and 2019, RSC and Elsevier.

2.3.6. Adsorption and Layer-by-Layer Method

In this case, adsorption (Figure 8a) can be referred to as the adhesion of divalent and trivalent ions (M^{2+} and M^{3+}) from a liquid or dissolved solid to surface of the LDH adsorbent. This creates a film of the adsorbate over the surface in many processes such as chemical, physical, biological and natural systems and widely used in various industrial applications [72]. The adsorption process may occur through weak van der Waals forces (physisorption) or covalent bonding (chemisorption) and also may occur due to electrostatic attraction between the adsorbate and surface of the adsorbent. It is a surface phenomenon most widely adopted in wastewater treatment for removal of various organic contaminants from aqueous solution.

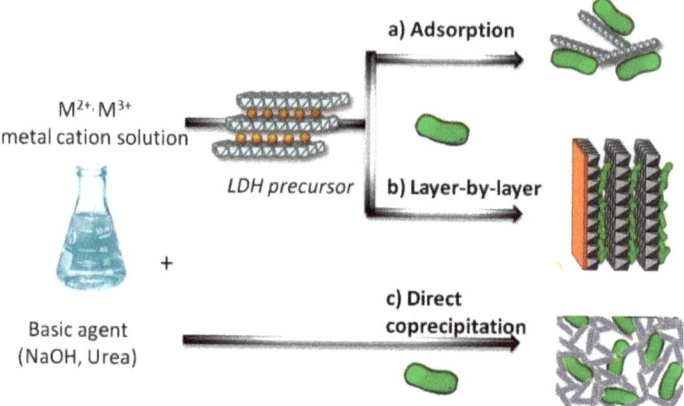

Figure 8. Schematic representation of (**a**) adsorption method, (**b**) layer-by-layer deposition and (**c**) direct co-precipitation method of layered double hydroxide (LDH). Reproduced with permission from Reference [73]. Copyrights 2018, John Wiley and Sons.

Layer-by-layer (LBL) assembly (Figure 8b) is a universal method for coating substrates with polymers, colloids, biomolecules, and even cells. This presents superior control and versatility when compared to other thin film deposition techniques in certain research and industrial applications. The LBL technique is known to support electrostatic interactions between positively charged layers and negatively-charged molecules and leads to nanostructured thin films [74]. This LBL deposition technique has three types of methods known as (i) the dipping layer-by-layer deposition technique (dipping-LBL); (ii) spray layer-by-layer deposition method (spray-LBL) and (iii) spin layer-by-layer deposition method (spin-LBL) method. Dipping-LBL is executed by chronologically adsorbing opposite charged materials onto a substrate via enthalpic and entropic driving forces [75]. In this method, the time depends on both the diffusion and adsorption of molecules, solutions or suspensions. Spray-LBL is a deposition technique where divalent and trivalent solutions are sprayed onto a vertical substrate, and the layer is formed after completion of drying in an oven overnight [58]. In spin-LBL method, the solutions or suspensions are deposited on a substrate attached to a spin coater, and the rotation speed generates a high centrifugal force. Thus, high rotational speed with high airflow rate at the surface leads to fast drying times of the liquid which in turn quickly and easily produce very uniform layers or thin films. In both spray-LBL and spin-LBL methods, the total time does not depend on the diffusion of molecular species. The co-precipitation method shown in Figure 8c has already been previously explained in Section 2.3.1.

2.3.7. Sol-Gel Method

The sol-gel method is a low-cost, simple preparation method and efficient wet-chemical method of high-purity metal oxide materials from LDH precursors through hydrolysis and condensation processes [76]. In this method (Figure 9), the mixed salt solution of $Al(NO_3)_3 \cdot 9H_2O$ and $Mg(NO_3)_2 \cdot 6H_2O$ and alkali solution of NH4OH were concurrently added to a beaker and heated under refluxed condition. The pH of the suspension is maintained at 8–10 by adding NH4OH base under magnetic stirring at ambient temperature until the gel formation is achieved. The resultant gel like product is filtered, washed properly with deionized water via re-dispersion/centrifugation and dried overnight. The formed gel was re-dispersed in water by ultra-sonication to produce LDH single layer nanosheets (SLNSs) dispersion. A portion of the LDH SLNS gel was further tried in an oven at 80 °C for 24 h to yield a well-crystallized LDH SLNS sol sample. The LDHs synthesized using sol-gel method is thermally very stable, but less crystalline than those synthesized via the co-precipitation method.

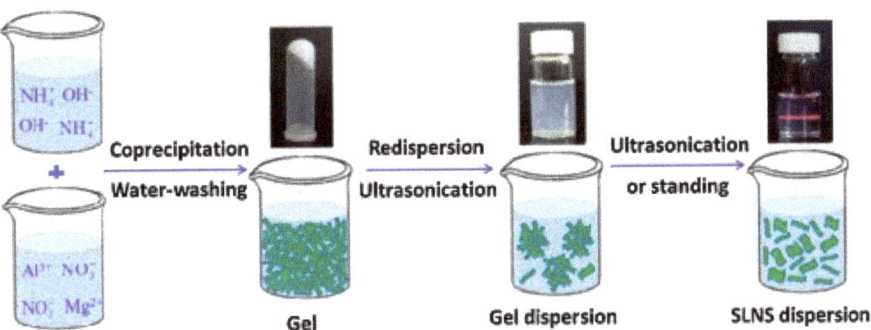

Figure 9. Schematic representation of sol-gel method of layered double hydroxide (LDH). Reproduced with permission from Reference [77]. Copyrights 2016, Elsevier.

2.3.8. Induced Hydrolysis Method

In this method, metal oxides are added dropwise to an acidic solution containing M^{3+} metal salts. The metal oxides are dissolved progressively in the acidic solution and precipitated into LDH. The pH is cushioned at 8–10 by the oxide suspension [59]. The obtained solid precipitate is collected, filtered, washed methodically with deionized water and dried at 80 °C for 24 h. This method of induced hydrolysis can also be used for synthesis of LDH with di-divalent, di-tetravalent and tri-trivalent systems.

2.3.9. Urea Method

In general, urea is added to an aqueous solution of preferred M^{2+} and M^{3+} metal salts and heated under reflux condition for several hours. The precipitate product is collected by filtration, washed thoroughly with deionized water and dried overnight. The rate of urea hydrolysis can possibly increase significantly with an increase in the reaction temperature to 100 °C [59]. The urea molecules undergo degradation to form ammonium carbonate, which initiates the precipitation into LDH with CO_3^{2-} as interlayer anion. This urea method provides high degree of crystallinity and a fine particle size distribution. Urea-based co-precipitation provided the better crystallinity and particle size due to thermal treatment and hydrolysis of urea which is proceeded in a very slow manner [57,59,60].

In comparison to many other nanoclays or layered materials, LDHs have compositional multiplicities in the cationic layers and in the hydrated interlayer of anions for charge balance which lead to some functional diversities. This implies that LDHs among layered materials have the great advantages and number of possible compositions, metal-anion combinations and morphologies useful for synthesis and processing methods. Apart from that, LDHs can be used in a variety of potential applications due to their anion exchangeability, compositional flexibility, good biocompatibility, low cost, facile synthesis, pH dependent solubility, thermal stability and high chemical versatility [78]. Due to their tunable chemistry and high charge density tailored properties, LDHs have attracted great attentions in various technologically significant fields and applications such as production of renewable energy [4,21,29], adsorbents [7,41,79], water purification [8–10,80,81], antimicrobial activities [10,24], sensors [29,82–84], flame resistance [48], drug delivery [85,86], cosmetics [87,88] and environmental catalysis [57,89,90]. In our previous work [91], detailed discussion about various applications were made and the current work focuses more on water purification.

2.4. Preparation Methods of Polymer-Clay Nanocomposites (PCNCs) and Surface Modification

The manufacturing of PCNCs depends mainly on a proper method selection which ensures acceptable level of dispersion of the nanofillers throughout the polymer matrix. Several processing methods were employed in preparing polymer-based clay nanocomposites such as in situ polymerization, the melt blending, and solution blending techniques [92,93] (see Figure 10). In each preparation method, an absolute goal is to achieve a desired uniform dispersion of nanoclays in the pristine polymer matrix. However, there are currently numerous interesting views about the applications and usage of these methods. According to the following studies [18,94,95], melt blending is regarded as a significantly, industrially viable and ecofriendly technique with high economic potential for preparation of polymer–clay nanocomposites. The in situ polymerization method is a commonly used synthesis technique and easy to modify by changing the polymerization conditions [96] and provides uniform dispersion. Both these methods require either a large amount of organic solvent or high viscosity or thermally unstable polymers at high temperatures. In comparison to melt blending, the solution-blending technique often produces pleasing dispersion of clay layers in the polymer matrix [31] due to its low viscosity and high agitation power. Each technique has its own relevant significance and limitations in relation to certain required industrial applications.

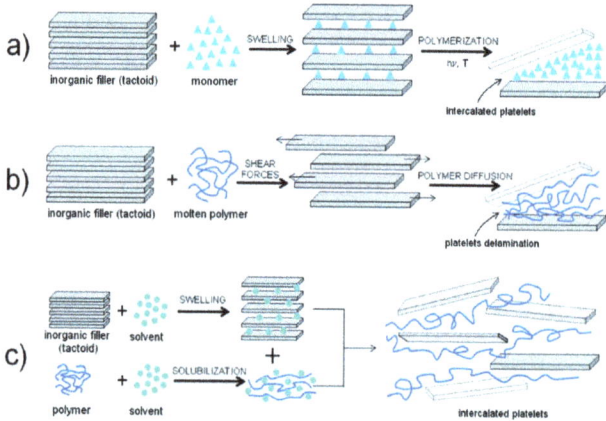

Figure 10. Illustration of (**a**) in situ polymerization, (**b**) melt intercalation and (**c**) solution intercalation. Reproduced with permission from Reference [97]. Open Access 2014, Royal Society of Chemistry.

2.4.1. In Situ Polymerization Technique

Due to the silicate dispersion deduced information, in situ polymerization is more effective in the preparation of composites and can sidestep the harsh thermodynamic requirements related to the polymer intercalation process [18,31,98] (Figure 10a). Furthermore, this polymerization technique (i) tolerates resourceful molecular strategies of the polymer matrix; (ii) it provides an effective approach to the synthesis of different polymer/nanoclay composites with prolonged property range and (iii) facilitates the development of the interface between the filler and the polymeric matrices by modification of the matrix composition and structure. Many studies focus on preparing novel polymer/nanoclay composites via the in situ polymerization method and demonstrate the benefits of this method in comparison with other types of synthesis methods [18,31,98,99]. For instance, Ozkose et al. [100] investigated the synthesis of poly(2-ethyl-2-oxazoline)/nanoclay composites for the first time using in-situ polymerization. In their finding, a ring-opening polymerization method was applied, which then initiated the delamination of clay layers in the polymer matrix and led to a composite formation.

2.4.2. Melt Blending Technique

Melt blending technique involves direct mixing of layered clay into the molten polymer matrix and can either be immobile or active. In an immobile melt blending (melt annealing), the process is performed under a vacuum at temperatures of approximately 50 °C above transition temperatures in the absence of mixing. In an active melt blending, the polymer melting is performed during a melt mixing in the presence of an inert gas [27,101]. As a result, the polymer clay nanocomposites are produced from the enthalpic driving force and influence of the polymer–organoclay interactions. The melt-mixing method (Figure 10b) provides better mixing of the polymer and nanoclay fillers and is well-suited with current industrially and ecofriendly viable processes such as extrusion and injection molding for thermoplastic and elastomeric material manufacturing. The absence of solvents reduces the environmental impact and minimizes potential interactions between the host and polymer solvents, which, in many cases, limits clay dispersion [18,102–104].

2.4.3. Solution Blending Technique

Solution-blending is a solvent based process in which the polymer and the prepolymer are soluble, which causes swelling of the clay layers, see Figure 10c. This technique involves thoroughly dispersing the layered silicate within appropriate solvents, which includes polymer/soluble prepolymer. These clay layers are dispersed into the solvent and further mixing with a dissolved polymer would be done to prepare the solution which allows polymer chains to be embedded into the exfoliated clay layers. Upon reaction completion stage, the solvent molecules would have evaporated, trapping the polymer chains intercalated into the gallery of clay interlayers [105,106] and the matrix segments combine with the dispersed clay layers.

The major driving force of intercalation process in solution mixing is the increased total disorder of the system referred to as desorption process of solvent molecules. This entire process normally consists of three stages known as (i) the dispersion of clay in a polymer solution, (ii) well-ordered solvent removal and (iii) lastly composite film casting [102,106,107]. The dispersion of clay in neat polymer necessitates active agitation such as stirring, reflux and shear mixing.

It is well documented that the morphology and dispersion of clay nanoplatelets in polymers is one of the key factors affecting their gas barrier properties [95,108]. One of the most vital challenges in the preparation of polymer/clay nanocomposites with improved barrier performances [95] is to achieve high level of exfoliation and orientation. In general, polymer/clay nanocomposites may result into three possible morphologies referred to as (i) phase-separated, intercalated and exfoliated structures (see Figure 11) [95]. For attainment of phase-separated nanocomposites, clay tactoids are formed throughout the pure polymer matrix, and no separation of clay nanoplatelets occurs. Polymer chains surround clay nanoplatelets but do not penetrate between the clay layers [109] and absence of platelets separation may result in large, micron-sized agglomerates. In intercalated nanocomposties, some of the polymer molecular chains have penetrated the interlayer galleries of the clay tactoids. Due to the penetration of polymer molecular chains, the spacing between individual clay platelets and the overall order of the clay layers is increased and maintained [110]. In exfoliated nanocomposite structures, the clay nanoplatelets are fully separated and dispersed uniformly within the continuous polymer matrix. Exfoliated nanocomposites produce the highest surface area interaction between clay nanoplatelets and neat polymer [111]. After a successful exfoliation, an enhancement in properties can be manifested in barrier properties, as well as improved mechanical properties, decreased solvent uptake, increased thermal stability and flame retardancy [112,113].

However, the main drawback to achieve homogeneous dispersion of most inorganic clays within organic polymers is closely related to the incompatibility between hydrophilic clay and hydrophobic polymer, which often causes agglomeration of clay mineral in the polymer matrix. Thus, surface modification of clay minerals for a good compatibility with the polymer is the most important step to achieve homogeneous dispersion of clay nanoplatelets in polymer matrix [29,31,76,109].

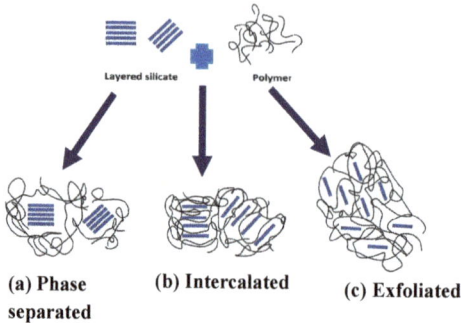

Figure 11. The main types of nanocomposites: (**a**) intercalated, (**b**) flocculated and (**c**) exfoliated. Reproduced with permission from Reference [113]. Open Access 2018, IntechOpen.

2.4.4. Surface Modification of Nanoclays and LDHs

Layered silicates including nanoclays and layered double hydroxides (LDHs) can be intercalated with hydrophilic polymers such as thermoplastic, thermosetting and elastomeric polymers. Most commonly used polymers are hydrophobic, while others such as poly(vinyl alcohol) (PVA), poly(ethylene glycol) (PEG), poly(acrylic acid) (PAA), poly(2-oxazoline) (POX), poly(methyl methacrylate) (PMMA), poly (ethylene-co-vinyl acetate) (EVA) are hydrophilic in nature. Despite their various applications, silicate layers also have one primary drawback due to the intrinsic incompatibility of hydrophilic silicate minerals and the hydrophobic polymer matrix. The incorporation of hydrophilic silicate minerals into a hydrophobic polymer causes agglomeration/aggregation, which lead to incompatibility between the components and weak extent of dispersion. Thus, it is indispensable to augment the degree of dispersion and the compatibility between the polymer matrix and the clay by surface modification [114]. The miscibility between layered silicates and the polymer matrices is enhanced as the clay becomes hydrophobic after surface modification using organic materials. For fabrication of layered silicates with engineering polymers such as thermoplastic or thermosetting, the surfaces of the layered silicate have to be modified by ion-exchange processes using cationic surfactants like quaternary alkylammonium salt, alkylphosphonium-based positively charged species or coupling agents [49]. The surface energy of layered silicates is reduced due to the modification, providing the efficiency and reinforcing characteristics in controlling the stability of the polar polymer matrix [105]. As a result, the interlayer spacing increases to high margins, producing better anchoring of the polymer chains for improvement of the overall properties of the system. The most preferential modification is the addition of coupling agent such as silane, which ensures good compatibility or chemical bonding with polymers, an exchange of the interlayer inorganic cations such as Na$^+$ with organic ammonium cations. In addition to ionic modifications, covalent and dual modifications (ionic and covalent are possible [115]. Other approaches, such as grafting polymer chains directly onto the surface of a nanoclay or using non-ionic surfactant have also been used [116]. There are two ways of ionic modification, called directly reacting anionic or cationic surfactants with the nanoclay or using ionic liquids. Imidazolium, pyridinium, trihexyltetradecylphosphonium tetrafluoroborate, and trihexyltetradecylphosphonium decanoate salts are commonly used for ionic liquid modification of nanoclays which show better properties [117–122]. The modified clay is commonly referred to as organoclay and the schematic illustration for the modification of clay particles is shown in Figure 12.

Chang et al. [119] prepared and characterized bio-oil phenolic foam (BPF) and surfactant modified bio-oil phenolic foam (MBPF) reinforced with Montmorillonite (MMT) as secondary phase. Their findings showed remarkably enhanced toughness as well as good flame resistance and improved the thermal stability of modified bio-oil phenolic foam (MBPF)-MMT nanocomposite foams compared to unmodified BPF-MMT nanocomposites. Covalently modified clay silicate is often synthesized via a step-reaction polymerization called condensation polymerization. During this process, the reaction is taking place between the hydroxyl groups from the surface of clays with mono- or tri-alkoxy silanes such as methoxy(dimethyl)octylsilane, tri-alkoxy silanes, trimethoxy(octyl)silane, (3-aminopropyl) triethoxysilane and others. The covalent modification renders the clay surface more hydrophobic [120]. Uwa et al. [121] studied the effect of nanoclay as reinforcing agent on the mechanical properties and thermal conductivity of polypropylene (PP) and maleic-anhydride-grafted-polypropylene (MAPP). The results of PP/MAPP/nanoclay composites exhibited a significant improvement in tensile strength and stiffness with low clay contents. Thermal conductivity analysis revealed that composites with high clay loadings have high resistance to heat. Twofold modifications can be done possibly by first covalently modified clay silicate followed by an ionic modification or vice versa. In comparison to single modifications (either ionic or covalent), dually modified clays show even more improved properties in terms of mechanics, thermal stability, dimensional stability, and viscoelastic characteristics.

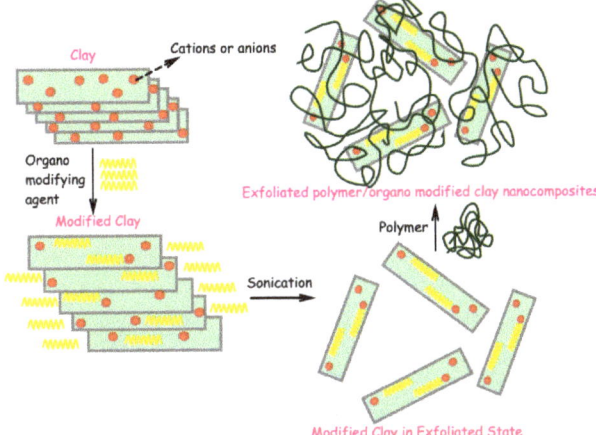

Figure 12. Schematic representation for the preparation of exfoliated polymer/organomodified clay nanocomposites. Reproduced with permission from Reference [122]. Copyrights 2017, Elsevier Ltd.

Various polymer-based layered silicates nanocomposite systems have been investigated, and their methods, structure and properties are compared and summarized in Table 3. The comparison of polymeric categories such as thermoplastic, elastomeric and thermoset matrices including thermoplastic polyurethane (TPU), polyisoprene (PIP), nacre-thermoset, poly(L-lactic acid) (PLLA), polypropylene (PP), polyamide 11 (PA11), nitrile butadiene rubber (NBR), styrene-butadiene rubber (SBR), Vinyl ester (VE), epoxy (EP), polylactic acid (PLA), Polybutylene terephthalate (PBT), polymethyl methacrylate (PMMA) reinforced with corresponding LDHs and other nanoclays are also included in Table 3 summary. Nevertheless, it is evident in the literature that the polymeric-thermoplastic matrices are utilized more preferentially over the thermosets because of their features such as light weight, can be re-melted/molded, and shaped. Recently, there is a growing demand to safeguard and deal with environmental contaminants and pollutants in preparation of biodegradable matrix/LDHs nanocomposites which are referred to as eco-friendly materials. Layered double hydroxides (LDHs) systems appeared to have better overall properties than most of other nanoclays due to their varied chemical compositions and methods of synthesis. LDHs possess higher layer charge densities and prefer multivalent anions within their interlayer space due to strong electrostatic interactions between the brucite-type sheets and the anions. Therefore, swelling is more difficult in LDHs than for other clay minerals. In short, LDHs containing monovalent anions like nitrate or chloride ions are viewed as good precursors for exchange reactions with charge balance, which lead to some functional diversities.

Table 3. Comparison of polymer-unmodified/modified layered double hydroxide and other nanoclays for water purification.

Polymer-Category	Layered Silicate	Processing Methods	Observed Morphology and Removal of Dyes or Other Heavy Metal Pollutants	References
Polyacrylamide (PAM)-thermoset	Sodium-montmorillonite (Na-MMT)	Free-radical cross-linking Polymerization (In situ polymerization)	A slightly intercalated MMT structure and incomplete exfoliation. PAM/Na-MMT nanocomposites efficiently removed the heavy metal ions such as Ni^{2+} and Co^{2+} wastewater with removal yield between 87.40% and 94.50%.	[123]
N-isopropylacrylamide (NIPAm) nanogel polymer-thermoset	Sodium-montmorillonite (Na-MMT)	Surfactant free dispersion radical crosslinking polymerization	Exfoliated structure. The Na-MMT nanogel composites showed drastic reduction in water surface tension and efficiently remove methylene blue (MB) dye. Co and Ni cations showed water within an hour. The prepared Na-MMT nanogels desorbed and reused four times to remove the heavy metal from water with the same efficiency.	[124]
Polyethylene (PE)-thermoplastic	Green clay	Solution mixing	An exfoliation nanocomposite morphology was achieved. The adsorption increased with increasing methylene blue concentration, the pH values and with increasing temperature due to the increased kinetic energies of the molecules. The removal of methylene blue (MB) from water solution was effectively achieved.	[125]
Cellulose-thermoplastic	Montmorillonite (MMT)	Aqueous solution method	The adsorption was not considerably affected by pH due to the presence of hydrophobic interaction between MB and hydrogels. The hydrogel samples containing intercalated clays showed high removal efficiency for MB aqueous solution with concentrations of 10 and 100 mg L^{-1}. The removal efficiency for MB increased with the clay contents of hydrogel networks and was reported as high as 97%.	[126]
PP-Thermoplastic Chitosan (CS)-thermoplastic	Montmorillonite Bentonite	Melt blending technique using twin-screw extruder Both melt compounding and crosslinking reaction between chitosan and glutaraldehyde	Structural morphology of intercalated PP/MMT is observed, while PP-g-MA/MMT appeared to have obtained an exfoliation morphology. Neat PP and synthesized PP-g-MA/MMT nanocomposites were used for removal of heavy metal adsorbent for adsorption of Pb(II) from aqueous solutions. The results revealed that adsorption efficiency of 96% for the removal of Pb(II) ion contaminant with neat PP and 0.5 wt% MMT were attained and conform the Langmuir isotherm. The PP-g-MA/MMT at 0.5 wt% nanocomposites showed can efficiently and effectively be used as super adsorbent for optimized removal percentage of contaminants like Pb(II) ions from wastewater. Intercalated structures and morphology were observed. The adsorption of an azo dye called Amido Black 10B (AB 10B) adsorbate onto the crosslinked chitosan (CCS)/Bentonite (BT) clay composites was reported to be optimal at high temperatures and low pH value of 2. CCS/BT composite is an effective biosorbent for the removal of AB10B from aqueous solutions.	[127,128]
Polyethylene glycol (PEG)-thermoplastic	Mg-Al-layered double hydroxides (LDHs)	Simple chemical precipitation method.	The morphology of the synthesized PEG-modified Fe_3O_4/Mg–Al-layered double hydroxides (LDHs) nanocomposites is heterogeneous and spherical with an average diameter of around 16–30 nm. It is a common knowledge that the morphologies of nanocomposites significantly influenced their adsorption capacity. Adsorbents exhibited a remarkable high adsorption capacity for the removal of methyl orange (MO) from water within a short time interval of 5 min and easy separation of adsorbents after successful adsorption process was achieved with the help of a magnet.	[129]
Polystyrene-thermoplastic	MgAl-LDH	Solution blending technique	The structural morphology of PS/LDH observed to be fibrous membranes. LDH-based sorbent showed a 67% adsorption efficiency of Cd^{2+} ion removal, while LDH-PS fibrous sorbents reached 10–15% adsorption efficiencies of Cd^{2+} ion removal based on the concentration of LDH in each of the sorbents. Since PS fibrous-based sorbents are hydrophobic, then the adsorption efficiency removal of Cd^{2+} ion can be attributed to the involvement of LDH-based sorbents which have better ion exchange capability.	[130]
Polyaniline (PANI)-conductive thermoset	Mg/Al Layered Double Hydroxide	Situ oxidative polymerization	A uniform fibrillar nanostructure is observed by SEM. The maximum adsorption efficiency of the PANI/LDHs is strongly affected by the initial solution pH for Cr(VI) wastewater treatment. Therefore, the adsorption efficiency removal of Cr(VI) decreases while the initial solution pH is above 7.0. This is probably due to the fact that the surface charge of PANI/LDHs was negative when pH > 7.0 and weakened electrostatic repulsion forces and significantly reduced adsorption efficiency/removal percentage of Cr(VI) from wastewater.	[131]

3. Properties of Polymer/Other Clays and LDHs Nanocomposites

The intention for the addition of clay minerals to the polymers is to improve the polymer properties and to produce the polymer/clay nanocomposites with desired applications. The key step is to prepare nanocomposites with highly preferred and value-added demand properties, which overcome downsides of polymers while maintaining their intrinsic advantages. Due to the low cost, availability, high aspect ratio as well as desirable nanostructure and interfacial interactions, clays can provide considerable improved properties at very low filler loadings, which help to obtain more useful properties. The nature and properties of constituents as well as preparation methods and conditions affect the final properties of polymer/clay nanocomposites. In this review, various improved properties of polymer/clay nanocomposites as well as the adsorption capacities and removal efficiency of dyes or heavy metal ions in water including morphology are discussed.

3.1. Morphology of Polymer/Other Clays and LDHs Nanocomposites

The morphology of the polymer/layered clay silicate nanocomposites significantly influenced their adsorption capacities and the removal efficiency of dyes or other heavy metal ions in water. The key aspect in nanocomposite structure is the clay–polymer interaction, which affects the dispersion level of clay in polymer matrix. Depending on the dispersion level of layered silicates, the structure can either be separated, intercalated or exfoliated structure [108,132–135]. Surface modification plays also important role in achieving good interaction between polymer and clay which affects the extent of dispersion and improves significantly the adsorption as well as removal of dyes or heavy metal ions from water. Thus, organically modified clay silicates such as montmorillonite (OMMT), kaolinite and LDH are mostly preferred nano reinforcement for proper selection of the functional groups and their abilities of ion-exchange. It is well-known that acid-modified clay resulted in higher rate of dye adsorption, an increased specific surface area and high porosity than in the case of base-modified clay [136].

Highly flame-retardant polymer/deoxyribonucleic acid (DNA)-modified clay nanocomposites were investigated by transmission electron microscopy (TEM) as shown in Figure 13a–f. The dark lines observed in Figure 13a,d are closely related to the layered silicate nanoclays and the light segments are associated to epoxy matrix. Naebe et al. [137] explained that intra-gallery reactions due to interfacial interactions made diffusion of more epoxy monomers within DNA-modified clay possible to enhance clay layers' separation and therefore induced formation of exfoliated structures as seen in Figure 13b. In addition, three individual intercalated ordered structures known as intercalated tactoids could be observed for epoxy nanocomposite at higher contents of clay (Figure 13e).

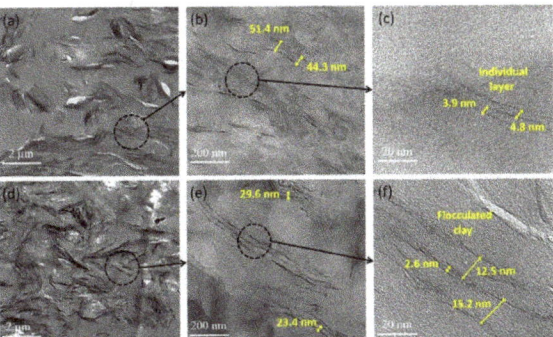

Figure 13. TEM micrographs of epoxy-2.5 wt% of DNA-clay (**a–c**) and epoxy-5 wt% of DNA-clay (**d–f**) nanocomposites. Reproduced with permission from Reference [137]. Open Access 2016, Springer Nature.

However, few clay layers possess thin and small tactoids which are uniformly and randomly dispersed in the epoxy resin. This indicates that the DNA-clay modification is an effective approach to improve both the exfoliation and dispersion of clay. In addition to achieved dispersion, most of microcracks under an effective load are initiated within the intra-layer of semi-stacked clay instead of epoxy-clay interfacial region. This phenomenon verifies that higher contents of clay (5 wt%) resulted in lower reinforcing impact in the overall mechanical properties.

Evidently, low clay content formed an exfoliated structural configuration with individual layers, well-dispersed with more homogeneous distribution (Figure 13c), while high content of clay resulted in flocculated structures. It can be seen in Figure 13f that phase-separated clay tactoid structures are formed and agglomeration at high content prevail over complete delamination of clay layers due to low penetration of epoxy monomers into stacked layers of modified clay [138].

Zubitur et al. [139] studied the poly(Lactic acid) (PLA)/modified drug 4-biphenyl acetic acid (Bph)-layered double hydroxide (LDH) nanocomposites. The nanocomposites were prepared by solvent casting with 5 wt% of drug-modified LDH, and the hydrolytic degradation was carried out in a Phosphate-buffered saline (PBS) solution at pH 7.2 and 37.8 °C. From their XRD results, PLA/LDH-Bph nanocomposites showed no peaks corresponding to LDH-Bph observed and this was attributed to an exfoliation or to the presence of entropic LDH layers. The degree of dispersion of acid/base-modified LDH was found to be good with small tactoids at low magnification, exfoliated layers were observed at higher magnification using TEM images. A number of studies reported the acid modification of clay silicate and layered double hydroxides (LDHs) reinforced with polymer matrices. Table 4 represents the summary of selective studies on acid/base-modification of polymer/clay and LDH nanocomposites for water purification.

Table 4. Summary of selective studies on acid/base-modification of polymer/clay and LDH nanocomposites for water purification.

Polymer Systems	Preparation Methods	Acid/Base-Modification	Dispersion and Structural Morphology Good for Removal of Dyes or Heavy Metal Ions.	References.
PLA/NiAl/LDHs	Melt mixing	SDBS	Good dispersion and predominantly exfoliated structures. Maximum adsorption capacity of dyes or heavy metal ions.	[140]
LDH/PEG$_{400}$	Physicochemical modification	PEG$_{400}$	Good dispersion and LDH layers were exfoliated in PEG$_{400}$ and showed higher dye adsorption capacity and effectively removed the azo dye, Acid Orange II (AO-II) in aqueous medium.	[141]
Polyethersulfone (PES)/AA-MMT	Phase inversion method	Acid Activated (AA)	The proper dispersion of nanoparticles in the membrane matrix was observed and PES/AA-MMT membranes showed better dye removal in the basic pH for MO and acidic pH for MB. The nanocomposite membranes exhibited considerably higher dye removal than neat PES. When the nano-clay content was increased, the dye removal percentage also increased in both dyes even in neutral pH due to the remarkable role of MMT particles in dye adsorption.	[142]
Ppy NF/Zn-Fe LDH	Interfacial polymerization of pyrrole	HCl as oxidant	The dispersion was observed to be weak, and structural morphology was separated grain or agglomerated particles-like stacked structure. Ppy NF/Zn-Fe LDH) composites enhanced adsorption capacity and high efficiency in the removal of safranin dye from raw water samples including tap water, groundwater, and sewage water.	[143]
Chitosan (CS)/laponite	Acid Activated (AA) aqueous solution	2-acrylamido-2-methyl-propanesulfonic acid (AMPS)	The degree of dispersion is better after being pre-adsorbed by AMPS and exfoliated microstructure morphology with high surface area, large pore volume and average pore size is observed. High surface area and large amount of micropores in adsorbent is suitable for the penetration of water and heavy metal ions into the interior and thus enhances the adsorption rate and removal efficiency. Thus, these nanocomposites showed an excellent adsorption capacity for removal of Cd(II), MB and CR from aqueous solution rapidly and efficiently.	[144]
Zeo/PVA/SA	Melt blending	The mixture of glutaraldehyde (GA) the cross-linking agent consists of 75 wt% (2% GA, 2% HCl and 71% acetone): 25 wt% DI water.	Zeo/PVA/SA NC beads have more pores and showed rougher and loose surfaces with porous structure. The dispersion of Zeo NPs had some agglomerations and exhibited an irregular inner morphology with stacks of tiny interspace structure with a very limited number of dents. The results revealed the removal efficiency of heavy metal ions such as Pb^{2+}, Cd^{2+}, Sr^{2+}, Cu^{2+}, Zn^{2+}, Ni^{2+}, Mn^{2+} and Li^{2+} using Zeo/PVA/SA NC modified beads reached the maximum at the pH value of 6.0, while the highest removal is achieved at pH = 5 for Fe^{3+} and Al^{3+} with 96.5 and 94.9%, respectively.	[145]
Chitosan/modified Bijoypur clay	Solution blending	HCl-purifier; Dodecylamine	The morphology of modified clay/chitosan showed smooth but discrete spherical particles with some dispersion. Chitosan/modified clay composite with high clay loading showed a better performance for cationic dye (MB) uptake, whereas heavy metals (Cr (VI) and Pb (II)) were better adsorbed on the composite, with high chitosan content.	[146]

3.2. Adsorption of Polymer/Nanoclay and LDH Systems

Various types of polymer/nanoclay composites are currently being explored for the primary usage in water purification and many other applications due to their unique properties, which are different from their counterparts. Polymer/clay nanocomposites used in water purifications technology and their corresponding applications as well as the removal of various heavy metal ions are shown in Figure 14. Adsorption is a removal of soluble material/adsorbed materials called adsorbate (heavy metal ions) present in water and clay nanomaterials (solid adsorbents) are used for the adsorption of contaminants. This process can either be physical (physisorption) or chemical (chemisorption) in nature. Nanoclay minerals have the high specific surface area and high sorption capacity giving high structural and chemical stability towards the adsorption of organic and inorganic contaminants.

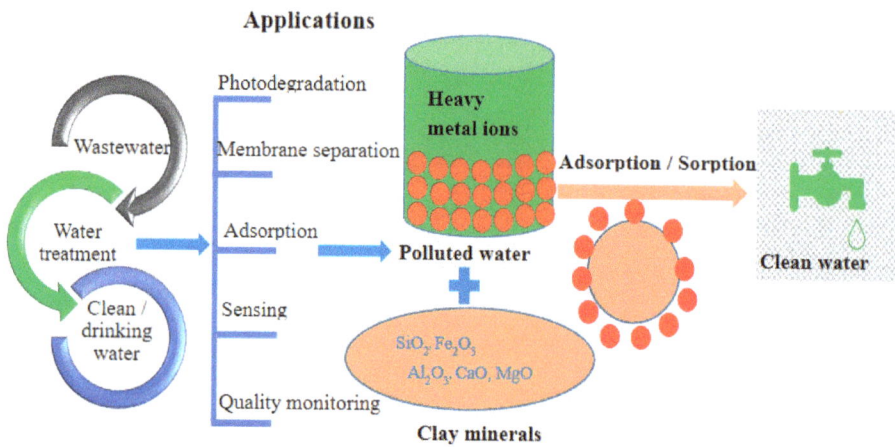

Figure 14. Role of nanocomposite in water purification and the removal of various heavy metal ions.

Due to its high internal surface area in the range of 500–1500 m^2/g, the most popular solid adsorbent material is activated carbon which is primarily used in large-industrial scale for water purification systems [41,49]. Thus, adsorption is a surface phenomenon commonly found in nature and plays a key role in water purification technology. Adsorption increases with the increase in the surface area of the adsorbent. This implies that more finely divided or rougher the surface of the adsorbent is, then the greater is the surface area and the adsorption. However, adsorption affinity of the surface area of the adsorbent is independent of the surface area and dependent on the favorable attractive interactions present at the pH value range below 7. When the pH is higher than 7, the electrostatic repulsion forces of the adsorbate-adsorbent are weakened, hence, the reduced adsorption efficiency or removal percentage of contaminants [41,49,131]. The adsorption of contaminants from water mostly depends on hydrophobic interactions between adsorbate and adsorbent, which make great contributions to the affinity of the organic anion to LDH. Therefore, the hydrophobic interactions interconnect many segments into a cluster in order for each natural organic anion to show a stronger affinity to LDH. The high hydrophilic nature of contaminants reduced significantly the adsorption capacity. The main reason for this is dominant force that decreases the surface tension between the matrix (adsorbent) and the solid adsorbed (adsorbate). The detailed discussion of the effect of pH as one of many factors affecting the adsorption, advantages and disadvantages of various methods of synthesis would also be outlined in the paragraphs or Table 5 below.

Table 5. Advantages and disadvantages of different methods of LDHs and clays in water purification.

Methods	Advantages	Disadvantages	References
Co-precipitation	Has high contaminant removal ability, applicable to communities and low reaction temperature and short reaction time.	Requires high maintenance and optimization of treatment is difficult.	[57,59,61]
Ion exchange	Able to effectively remove inorganic contaminants, has capacity to regenerate and inexpensive. No loss of sorbent on regeneration, effective.	It has high operating costs over a long period of time and cannot effectively remove pyrogens or bacteria. Causes economic limitations, not effective for dispersing and removing the dyes.	[62]
Sonochemical	Produces a better shear thickening transition at lower shear rate and significantly reduced the water content contamination. Improves reaction rate, involves high energies and pressures in a short time; no additives needed; reduced number of reaction steps.	Reactions need to be at certain temperature and there is not enough power to carry out the reaction. Extension of problems; inefficient energy; low yield.	[61,65–67]
Hydrothermal /Solvothermal	Has ability to synthesize large crystals of high quality and crystalline substances that are unstable near the melting point.	High cost of equipment.	[68,69]
Adsorption	The most effective adsorbent, great capacity, produce a high-quality treated effluent.	Ineffective against disperse and dyes, the regeneration is expensive and results in loss of the adsorbent, non-destructive process.	[58,72–75]
Sol-gel method	Can produce a thin coating to ensure excellent adhesion between the substrate and the top layer. It has the capacity of sintering at low temperatures, between 200–600 °C; simple, economical and efficient method to produce high quality coverage and high purity products.	The contraction that occurs during processing; long processing time; fine pores; use of organic solutions that can be toxic.	[76,77]
Urea/Induced hydrolysis	Cheap, easy to store.	Loss through leaching and volatilization, acidifying.	[57,59,80]
In situ polymerization	Easy processing method based on the dispersion of the filler in the polymer precursors.	Difficult control of intragallery polymerization and limited applications.	[18,31,98]
Melt mixing method	One step technique, economical and environmentally viable and easy to process and compatible with industrial polymer processes.	Sensitivity to reaction conditions, limited applications to low molecular weight polymers.	[102–104]
Solution mixing technique	Preparation of homogeneous dispersion of fillers.	The industries and sectors use a huge amount of solvents and this technique is very expensive.	[105,106]

In recent times, different synthesis and preparation methods of LDHs and other nanoclays have been applied as discussed in Sections 2.3 and 2.4, respectively. These different methods influence the adsorption performance of the nanocomposite and noticeably assist to reduce and remove the high concentration levels of contamination in water [147,148]. Individually, each method has some advantages and disadvantages as represented in Table 5. Amongst all the methods, adsorption is considered the most efficient, economical technique and easy processing to drive for contaminants removal from wastewater. Furthermore, the adsorption is reversible, and adsorbents can be regenerated. In general, there are three types of adsorption known as physisorption (the interaction between adsorbent-adsorbate), chemisorption (adsorbent-solvent) and electrostatic interactions (adsorbate-solvent) [149]. The underpinning principles/mechanisms involved in the adsorption material surface have been studied and reported in relation to factors affecting systems.

3.2.1. Factors Governing the Performance of Clays/LDH Based Adsorbents

In this review, the application of nanomaterials as adsorbents for removal of contaminants such as heavy metal ions and dyes from wastewater has been reviewed. It is important to understand how adsorbents interact with different adsorbates such as heavy metal ions and dyes in the laboratory small scale to determine their potential for application in water purification and their contribution to large scale [148,149]. However, the main challenges for adsorption process are waste products, non-selectivity, instability and low/poor heat transfer leading to long heating and cooling times. To address these challenges, the development and design of suitable and more effective nanoadsorbents with optimum adsorption efficiency for the removal of contaminants in water should be prioritized. In the field of wastewater treatment, different materials prepared through various methods bring about unique functionalities for adsorption efficiency for the removal of contaminants from industrial effluents, surface water, groundwater and tap/drinking water. As stated, many key factors affect the efficiency and performance of LDH/clay adsorbents including pH value, contact time, adsorbent dosage, initial ion concentration, temperature, coexisting ions, and sorption kinetics [148,149]. In this section, the influence of these factors on adsorption performance and capacity by LDH/clay adsorbents is explained as presented in Table 5.

Influence of pH Value

The pH of an adsorbate-adsorbent solution is most significant aspect for adsorptive removal of contaminants. This affects the type of the surface charge of the adsorbent during water purification by the adsorption technique. It is also the main factor taking care of the type of the surface of the adsorbent, degree of ionization and aqueous adsorbates [149]. The effect of solution pH was studied at different pH values from 3.0 to 11.0, and the results are shown in Figure 15a,b. The pH of solution noticeably changed by the addition of diluted HCl/NaOH. The maximum fluoride adsorptions of 98%, 94% and 91%, respectively for cerium bentonite clay-malic acid chitosan (CeBC-A@CS), lanthanum bentonite clay-malic acid chitosan (LaBC-A@CS) and aluminum bentonite clay-malic acid chitosan (AlBC-A@CS) adsorbents was attained at pH of 3.0. It was also reported that the minimum fluoride adsorption for these three different adsorbents was achieved as 43%, 37% and 35%, respectively at pH of 11. At neutral pH of 7.0, the maximum fluoride adsorption was 84%, 82% and 80% for, respectively. The maximum fluoride adsorption capacity can be attributed to the change in the surface charge of the adsorbent. The pH value at point of zero charge (pHzpc) for the adsorbents is 7.1 [150] as represented in Figure 15b. When the pH of solution was less than pHzpc, the fluoride ions moved towards the positively charged surface of the chitosan composites formed by the protonation of OH^- and carboxylic acid groups leading to the fluoride adsorption onto the surface [151]. At a pH value above the pHzpc, the fluoride adsorption was exceptionally low. This is probably because the composite surfaces were negatively charged due to deprotonation of the hydroxyl groups, ensuring mutual repulsion forces between the fluoride ions and the composite surfaces [152]. It can be observed that cerium bentonite clay-malic acid chitosan (CeBC-A@CS) adsorbent showed a higher fluoride adsorption capacity than lanthanum bentonite clay-malic acid chitosan (LaBC-A@CS) and aluminum bentonite clay-malic acid chitosan (AlBC-A@CS) adsorbents.

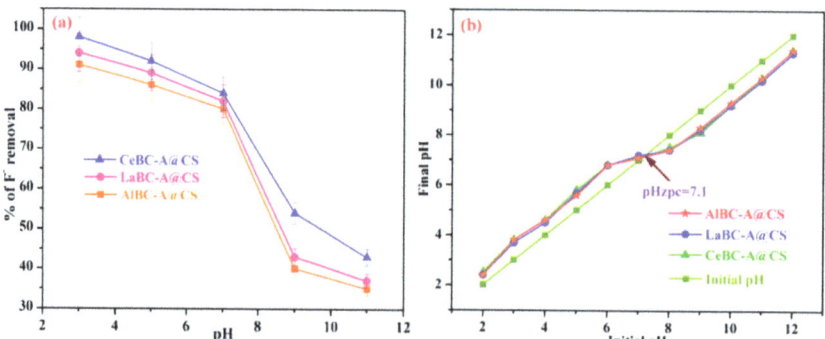

Figure 15. (a) Effect of pH and (b) pHzpc values of cerium bentonite clay-malic acid chitosan (CeBC-A@CS), lanthanum bentonite clay-malic acid chitosan (LaBC-A@CS) and aluminum bentonite clay-malic acid chitosan (AlBC-A@CS) adsorbents. Reproduced with permission from Reference [151]. Open Access 2020, RSC Advances.

Wei et al. [153] investigated the novel hydrotalcite-like material layered double hydroxide (FeMnMg-LDH) adsorbent synthesized by co-precipitation and its adsorption capacity for the removal of lead ions in water. In order to prevent the precipitation of Pb^{2+} at the high pH, the experiment pH was set below pH 6. The pH of 6 was maintained to prevent the precipitation of Pb^{2+}, and effects of pH on the adsorption are discussed. It was reported that the Pb^{2+} removal percentage by FeMnMg-LDH adsorbent mostly increased with the increasing pH value. The Pb^{2+} removal percentage higher than 97% at the pH range 3–6, was achieved. This is an indication of the outstanding efficiency and performance of FeMnMg-LDH adsorbent in Pb^{2+} adsorption except when pH is equal to 2. The suspension of the absorbent might take place at extremely low pH resulting into the collapse of the structure of FeMnMg-LDH and therefore reduce Pb^{2+} adsorption efficiency and capability. Comparing the bentonite clay and LDH-based adsorbent, it can be concluded that FeMnMg-LDH adsorbent has a higher adsorption capacity of more 97% than others except the cerium bentonite-malic acid chitosan, which appears to have more or less similar adsorption capacity percentage.

Influence of Contact Time

The contact time significantly affects the adsorption process and the economic efficiency of the process including the adsorption kinetics. Therefore, contact time is profoundly important and dependent factor for performance determination in adsorption process [154]. Figure 16a represents the fluoride adsorption capacity of the three adsorbents (cerium bentonite clay-malic acid chitosan (CeBC-A@CS), lanthanum bentonite clay-malic acid chitosan (LaBC-A@CS) and aluminum bentonite clay-malic acid chitosan (AlBC-A@CS)) at different contact times in the range of 10 to 90 min with neutral pH and initial concentration. The fluoride adsorption capacity of the adsorbents was gradually increased with an increase in contact time. Cerium bentonite clay-malic acid chitosan (CeBC-A@CS) adsorbent obtained the higher fluoride adsorption capacity of 84% than other adsorbents with 62%, 67%, 80%, 82%). Moreover, an equilibrium at 60 min and 45 min was achieved with different adsorbents which suggests the surfaces of the adsorbents CS and BC were completely covered with fluoride ions.

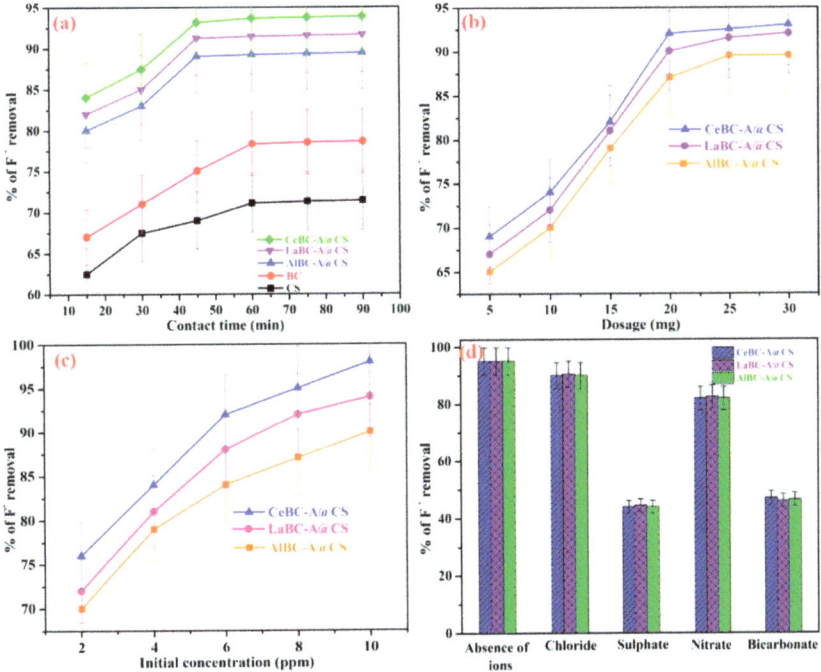

Figure 16. (**a**) Effect of contact time, (**b**) effect of dosage, (**c**) effect of initial fluoride concentration and (**d**) effect of co-ions on the fluoride adsorption of the adsorbents AlBC-A@CS, LaBC-A@CS and CeBC-A@CS at 303 K in neutral pH. Reproduced with permission from Reference [151]. Open Access 2020, RSC Advances.

Jaiswal and Chattopadhyaya [155] studied the effect of contact time on adsorption of Pb(II) on the Co/Bi-LDH synthesized by using co-precipitation method. Their impressive finding was that 90.0% of the adsorptive removal of contaminant, called heavy metal, was accomplished within 120 min of contact time. It was also observed that beyond 120 min, contact time has no effect in heavy metal removal percentage. At the beginning, very high adsorption rates were observed simply because of the larger number of vacancy sites available for the sorption and adsorption equilibria that were then steadily reached [139]. Effect of contact time for bentonite clay and LDH adsorbents using the same co-precipitation method were compared, and it can be concluded that LDH adsorbent seems to have an upper hand in terms of the adsorption capacity percentage for removal of contaminants. The reason for this is that LDH adsorbent has higher surface charge density and more ion exchange binding sites for good adsorption. This could also be attributed to the adsorption on the LDH layers via hydroxide precipitation or metal complexation.

Influence of Adsorbent Dosage

As matter of principle, the degree of adsorption of a solute increase with the increase in the content of an adsorbent. This can be attributed to the increase in adsorbent dosage, which indicates the increased active exchangeable adsorption surface vacancy sites. Nevertheless, the total solute adsorption per unit weight of an adsorbent can decline subsequent to the upsurge in adsorbent dosage because of meddling initiated by the interaction of active sites of an adsorbent [156]. In Figure 16b, the results showed the most favorable dose achieved as 25 mg for the adsorption of fluoride ions via three bentonite clay-based adsorbents. The resultant adsorption capacities of 87%, 90% and 92% were reported for aluminum bentonite clay-malic acid chitosan (AlBC-A@CS), lanthanum bentonite clay-malic acid chitosan (LaBC-A@CS) and cerium bentonite clay-malic acid chitosan (CeBC-A@CS) adsorbents, respectively. Beyond 25 mg, the results showed no significant increase in the fluoride removal limit due to the lower availability of active adsorption sites [157].

Li and co-authors [158] studied Mg–Al layered double hydroxides/MnO_2(Mg–Al LDHs/MnO_2) adsorbents for removal of Pb(II) from aqueous solutions synthesized by one-pot hydrothermal method. It was reported that the adsorbent dosage of Mg–Al LDHs/MnO_2 considerably influenced the adsorptive removal of contaminants like lead ions. It was also observed that the percentage adsorptive removal of Pb(II) contaminant increased fivefold from 18.48% to 99.56% with the adsorbent dosage increasing by a factor of 9 from 0.01 to 0.09 g. In addition, the higher adsorbent dose results in a reduced adsorption capacity of Mg–Al LDHs/MnO_2 at Pb(II) concentration of 50 mg/L. This observation can probably be associated with the low adsorbent dosage leading to the dispersion of Mg–Al LDHs/MnO_2 particles in aqueous solutions. The maximum adsorption efficiency and performance of LDH-based adsorbent and bentonite clay-based adsorbent was achieved at 99.56% and 92%, respectively. The LDH-based adsorbent achieved higher percentage removal efficiency of Pb(II) contaminant than bentonite clay-based adsorbent. This is attributed to the increase in the concentration of adsorption sites in aqueous solution, which enables the contaminants adsorption on a larger number of actives sites.

Influence of Initial Ion Concentration

The influence of the initial ion concentration of the contaminant on the adsorption is one of the most important factors to be studied. It can be seen from Figure 16c that the adsorption capacity of fluoride ions increased with increase in initial concentration. The initial concentrations improved from 2.0 mg per liter to 10 mg per liter where the adsorption capacity/proficiency of the aluminum bentonite clay-malic acid chitosan (AlBC-A@CS), lanthanum bentonite clay-malic acid chitosan (LaBC-A@CS) and cerium bentonite clay-malic acid chitosan (CeBC-A@CS) adsorbents moved from 70.1% to 98% [151]. Thus, the adsorption limit was observed to be straight forward undertaking related to the adsorption of fluoride ions. Mostafa et al. [159] investigated the effect of different Fe(II) concentrations on adsorption capacity of Co/Mo-LDH with carbonate $(CO_3)^{2-}$ as an interlayer anion prepared through co-precipitation method. Their results revealed that the Co/Mo-LDH seemingly removed a significant amount of Fe(II) contaminant from the aqueous solutions. The maximum adsorption efficiency improved to a 99.74%, and the saturation occurred when no more metal ions could be adsorbed on the surface of Co/Mo-LDH. A high efficiency for ferrous adsorption was obtained through a relatively short period of time up to 60 min at initial concentrations of 1.0, 2.0, 3.0 and 5.0 mg/L. The LDH-based adsorbent has 99.74% higher maximum adsorption capacity than bentonite clay-based adsorbent with 98%.

Influence of Co-Existing Ions

The adsorptive removal efficiency of contaminant is typically influenced by the presence of co-existing ions in solution leading to competitive adsorption on the adsorbent surface [160]. The influence of various negatively charged anions such as chloride (Cl^-), sulfate (SO_4^{2-}), nitrate (NO_3^-) and bicarbonate (HCO_3^-) ions on the adsorption of fluoride by the three adsorbents (aluminum bentonite clay-malic acid chitosan (AlBC-A@CS), lanthanum bentonite clay-malic acid chitosan (LaBC-A@CS) and cerium bentonite clay-malic acid chitosan (CeBC-A@CS)) was examined in Figure 16d. The findings suggest that the Cl^- and NO_3^- ions did not change the fluoride adsorption efficiency, while the SO_4^{2-} and HCO_3^- ions had an adverse effect. This can be associated with high coulombic repulsion forces in existence, which reduced the mobility of fluoride ions during interaction with the active sites of the adsorbent. The larger degree of interference of the SO_4^{2-} and HCO_3^- ions is due to the arrival of OH^- ions which induced the arrangement of sodium sulfate and sodium bicarbonate increasing the solution pH and thus made possible for these ions to compete with fluoride ions on the surface of the adsorbent [160].

Zhu et al. [161] studied the effect of calcined Mg/Al layered double hydroxides (Mg-Al LDH) as efficient adsorbents for polyhydroxy fullerenes (PHF) prepared by co-precipitation method. Naturally, PHF may co-exist with other inorganic anions, which may affect its adsorption on layered double oxide (LDO). The effect of selected various coexisting anions such as Cl^-, CO_3^{2-}, SO_4^{2-}, and HPO_4^{2-} at approximately pH of 10 was analyzed. Their results proved that Cl^-, CO_3^{2-}, and SO_4^{2-} slightly improved the adsorption of the PHF on LDO. The adsorption capacity of PHF on LDO lessened considerably in the entire concentration range of PHF with the presence of HPO_4^{2-} ions. The increase in PHF concentration implies that the negative effect of HPO_4^{2-} declined noticeably due to the improved competitive effect of PHF. The effect of co-existing anions on the adsorption of LDH adsorbent may influence the surface property of the adsorbents resulting into two effects known as inhibiting effect and synergistic effect. These happen by occupying some of the adsorption vacant sites (an inhibiting effect) and providing additional adsorption sites (a synergistic effect) [161]. This means that the properties of the adsorbates are influenced by promoting the aggregation or dispersion of adsorbates. Ultimately, the co-existing ions may have various effects such as promoting, inhibiting, or no effect at all on the adsorption of adsorbates of the adsorbents.

Influence of Temperature

The effect of temperature is typically one of the factors governing the adsorption efficiency and performance of adsorbents in water purification systems. An increase in temperature with an increase in the adsorption percentage enables the adsorption capacity of clay/LDH-based adsorbents at various temperatures [162]. This is due to the increase in the mobility of contaminants in aqueous solution, which leads to the improvement in the availability of adsorption vacant surface-active sites. Temperature parameter is well-known to have a strong effect on different chemical processes. It affects the adsorption rate by varying the molecular chain interactions and the solubility of the adsorbate. The effect of temperature on the adsorption of Pb(II) on clay/LDH was investigated by Ayawei et al. [162]. It was observed that on increasing the temperature, the percentage removal of metal ions increased (Figure 17). This presented sufficient evidence to conclusively refer to this adsorption process as an endothermic kind of the process.

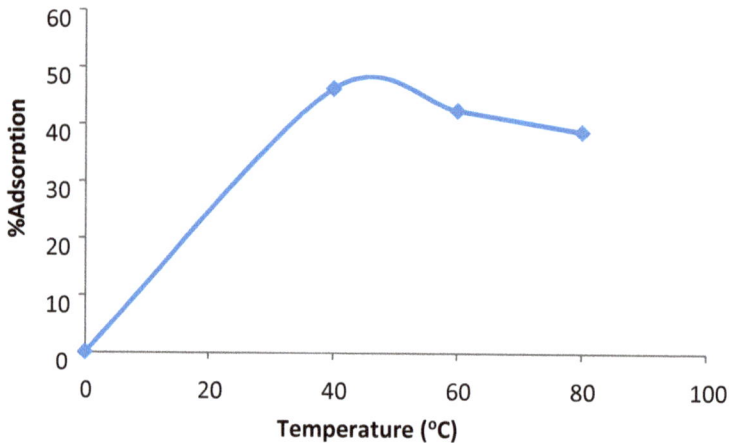

Figure 17. Adsorption capacity and percentage as a function temperature. Reproduced with permission from Reference [162]. Open Access 2015, International Journal of Chemistry.

Influence of Sorption Kinetics and Isotherms

The information on adsorption/desorption process and isotherms is one of the key factors required for proper analysis, structure and understanding of the adsorbent–adsorbate system [160]. In recent times, various models have been explored to explicitly explain the adsorption behavior including isotherm models. The latter provide valuable information about adsorption property of a system and the distribution of exchangeable sites on the surface of the adsorbent. In general sorption, diffusion, desorption, sorption isotherms will be discussed in this section [163].

Some of the factors that play a major role in the adsorption/sorption process are explained below. The nanocomposites usually exhibit a very high amount of barrier properties with even a small amount of layered silicate [29]. Some small molecules such as oxygen, carbon dioxide, water and nitrogen permeate through a polymer membrane due to a gas chemical potential gradient through the membrane. The chemical potential difference ($\Delta\mu$) acts as the driving force for the molecules to permeate from the high chemical potential side to the side of low chemical potential. The mode occurrence of permeant transport in polymers is described using the solution–diffusion model. According to this model, the permeation in polymers consists of three steps as illustrated in Figure 18a (i) sorption of the permeant from the high concentration side onto the membrane/film surface, (ii) diffusion of the permeant along the concentration gradient through the membrane and (iii) desorption through evaporation from the low concentration surface of the membrane. When the permeating molecule interacts with the polymer, then the deviations from a gradient with a straight line can be observed and is known as non-Fickian diffusion explained by the diffusion–relaxation model [164,165].

In this review, we discuss the tortuosity of the diffusive path for nanoplatelets for the gas barrier properties of nanocomposites in Figure 18b. Layered silicate nanoclays such as montmorillonite and kaolinite are the most innovative and promising nanofillers due to their ability to exfoliate to form single nanoplatelets when dispersed in a polymer matrix. The basic theory of the model is that the presence of impermeable clay platelets forces the permeant (gas) molecules to follow a longer diffusion path by traversing around the platelets. Therefore, this is also known as the tortuous path model as illustrated in Figure 18b. The nanoplatelets hinder the diffusion process of small gases through them and form a tortuous path which act as a barrier structure for gases. Tortuous paths explicitly explain the principle of the barrier behavior and its noticeable improvement in nanocomposites, which can be attributed to the impermeability of the layered clay silicate into the polymer matrix. Therefore, the intercalation molecules are placed in a wiggle shape around the nanoparticles in a random way [166,167] as seen in Figure 18b.

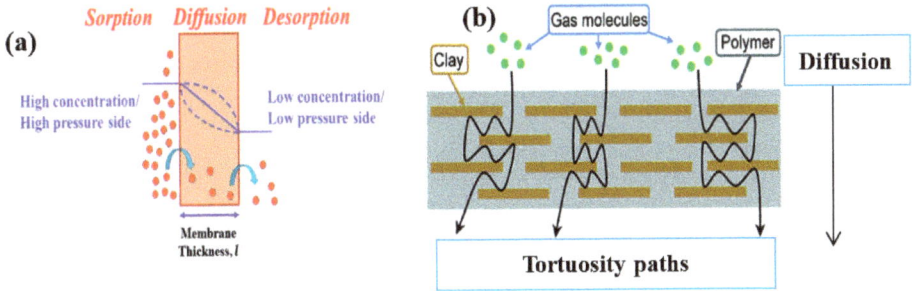

Figure 18. (a) Schematic illustration of the solution diffusion model. Reproduced with permission from Reference [168]. Open Access 2020, MDPI Polymers. (b) Schematic illustration of the tortuous path model. Reproduced with permission from Reference [95]. Open Access 2017, RSC Advances.

Follain et al. [169] investigated the water vapor transport properties by pervaporation and sorption measurements and evaluated nanoclay extent of dispersion in polymer matrix. They prepared nanocomposites based on poly(ethylene-co-vinyl acetate) (EVA)/organo-modified Cloisite clays at varying contents by melt blending. Their results showed a noticeable decrease in water permeation flux obtained when nanoplatelets are incorporated into the neat EVA matrix (Figure 19a). This barrier effect is typically attributed to the remarkable increase of the diffusion pathways due to organo-modified Cloisite clay-induced tortuosity effects. Furthermore, the water permeation flux seems to be proportional to the diffusion coefficient, which was found to be reduced due to a plasticization effect of water. The water-induced plasticization effect of sorbed water molecules was highlighted through sorption kinetics, and water barrier behavior was observed.

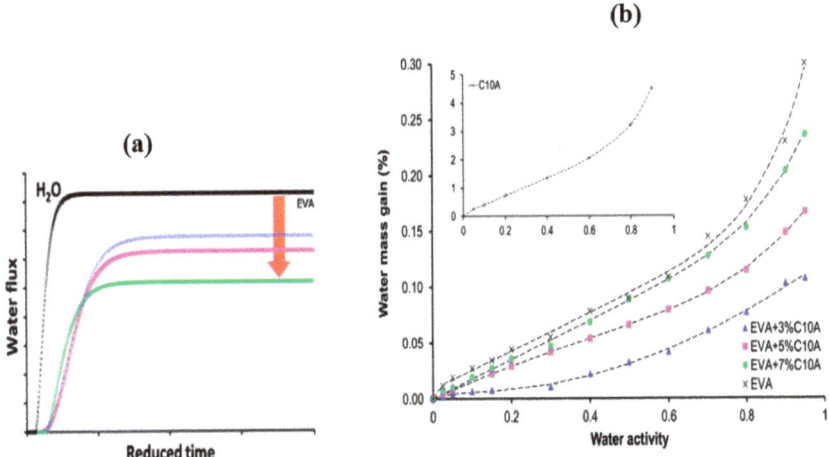

Figure 19. (a) Water permeation flux as a function reduced time. (b) Water vapor sorption isotherms of the neat EVA matrix and its nanocomposites with an inset of C10A nanoclay isotherm. The dotted lines correspond to the fitting of experimental data with Park's model. Reproduced with permission from Reference [169]. Open Access 2015, The Royal Society of Chemistry.

The water vapor sorption isotherm curves are also known as water mass gain in the equilibrium state versus water activity deduced from water sorption kinetics, reported in Figure 19b. The water sorption capacity of nanoclays is evidently higher than that of the EVA matrix. The incorporation of Cloisite (C10A) into the EVA matrix induced the shift of isotherm curves to lower values at a given activity (Figure 19b), reproducing a reduction in water mass gain. As a result, C10A sorbed higher water mass gain than the matrix, which contributed to increasing sorption capacity of nanocomposites, and this finding is in good agreement with the reduction of water permeability. The decrease in mass gain can be related to tortuosity effects induced by nanoclays in a matrix and which counterbalance the strong water sorption capacity of nanofillers [170]. An increase in nanoclay content, reproduced an increase in water mass gain for nanocomposite samples, which is measured for the whole water activity range due to a more hydrophilic character of nanoclays than the matrix one. This can be attributed to the increase of the polar site number of surfactant-modified nanoclays. The sorption process of the resulting nanocomposites is thus driven by the nanoclay sorption capacity, and the incorporation of nanoclay content into the nonpolar matrix. The nanocomposite with the highest nanoclay loading is categorized by the highest water mass gain. Water molecules are implicitly located in the matrix/nanoclay interfacial regions, and the conclusion is water absorption behavior of nanocomposites obeyed Fick's law. The increased tortuosity within the EVA matrix/nanoclays is therefore in opposite effect to the increased water solubility due to the hydrophilic character of nanoclays.

Rajan et al. [171] studied the role of dicarboxylic acid (malic acid (A)) in chitosan (CS)/modified-metal ion decorated bentonite clay (BC) and the defluoridation efficiency in fluoride contaminated groundwater. The samples were prepared through the solution mixing and their findings revealed the chemical changes of the adsorbent as shown in Figure 20. The bands observed in the range of 3413–3440 cm^{-1} in the spectra of chitosan confirm the presence of O–H bond stretching and N–H bond stretching frequency [172]. The bands at 3046 cm^{-1} and 2900 cm^{-1} related to aliphatic stretching vibrations of –CH, while the bands at 2927 and 2860 cm^{-1} can be attributed to the C–H stretching vibration of the –CH$_2$ groups in chitosan. The aliphatic stretching vibrations of –CH have very low intensity in chitosan [171].

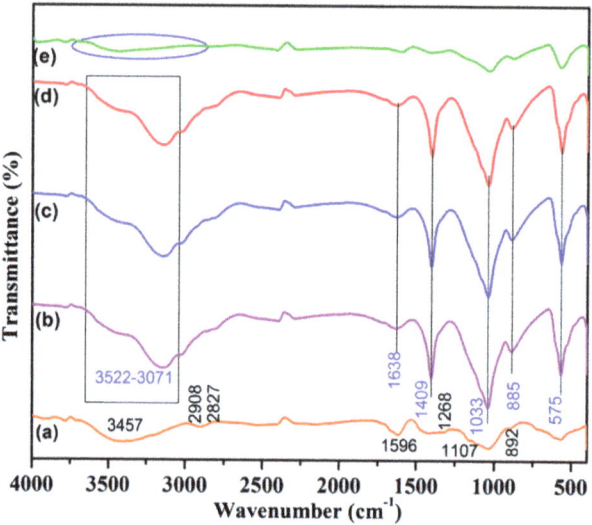

Figure 20. FTIR spectrum of (**a**) chitosan (CS), (**b**) aluminum bentonite clay-malic acid chitosan (AlBC-A@CS), (**c**) LaBC-A@CS, (**d**) cerium bentonite clay-malic acid chitosan (CeBC-A@CS) and (**e**) fluoride adsorbed CeBC-A@CS. Reproduced with permission from Reference [151]. Open Access 2020, RSC Advances.

The peak observed at 1596 cm^{-1} is related to N–H bond scission from the primary amine because of free amino groups in the crosslinked chitosan segments [173]. The peak which appeared at 1596 cm^{-1} indicates the aromatic ring stretching vibration. The C=O adsorption peak of secondary hydroxyl groups becomes more pronounced and shifts to 1107 cm^{-1} [174]. It can be seen in Figure 20e that the peak intensity of the hydroxyl group was remarkably reduced because of the fluoride adsorption. Furthermore, the nitrogen adsorption–desorption analysis was done by Rajan and co-authors [170] as evident in Figure 21. The surface area, pore width and pore volume of the synthesized CeBC-A@CS adsorbent was characterized by Brunauer–Emmett–Teller (BET) analysis and reported as 103 m$^2 \cdot$g^{-1}, 12.71 nm and 0.0321 cm$^3 \cdot$g^{-1}. The synthesized CeBC-A@CS adsorbent possessed high surface area, higher pore width and larger microspore volume, which confirmed that the synthesized adsorbent has maximum adsorption capacity.

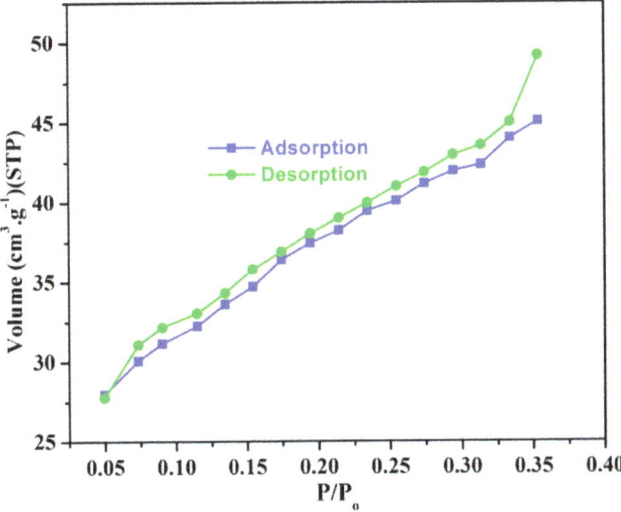

Figure 21. Nitrogen adsorption–desorption isotherm spectrum of cerium bentonite clay-malic acid chitosan (CeBCA@CS) adsorbent. Reproduced with permission from Reference [151]. Open Access 2020, RSC Advances.

4. Concluding Remarks and Future Prospects

In summary, polymer/layered clay nanocomposites are the most progressive and alternative procedures for water purification systems. Synthesis of nanomaterials such as nanoclays and/or LDHs play a pivotal role in improving the physicochemical properties of adsorbents for pollutant removal in water. Furthermore, layered double hydroxides (LDHs) materials have attracted considerable attention and are preferred candidates as sorbents in water treatment because of their remarkable ability to eliminate a variety of water contaminants instantaneously. Their unique properties as anion exchangers and their compositional versatility among other factors is an advantage over other types of clays in the field of water treatment.

Various modification techniques were discussed for the preparation of functionalized clay and LDH nanomaterials and showed influence on the dispersion of nanoclay fillers in the polymer matrices. The desired properties for the polymer/clay and LDH nanocomposites are primarily dependent on the type of modifying agents used for functionalization of layered silicates. Solution blending technique and in situ polymerization method seemed to provide good dispersion of clay layers in polymer matrix compared to melt blending technique. This is mainly because of the low viscosity and high agitation power associated with solution blending. However, melt blending is considered as an industrially viable as well as ecofriendly technique and shows high economic potential. The addition of stabilizers and/or compatibilizers during the processing stage is believed to lead to an improved number the properties of polymer/clay and LDH nanocomposites.

Development of an appropriate understanding of the structure, property and formulation relationship in both clay-based nanocomposites and LDH based hybrids needs to be further researched for better output in water purification. LDH-containing hybrids are the new emerging areas of research in water purification processes. Due to their nontoxicity, higher surface area than individual constituents and notable adsorption capacity, these LDH hybrids have attracted a considerable interest in water treatment applications. The polymer/modified-clay nanocomposites with exfoliated morphology disclosed strong reduction in water surface tension and efficiently remove methylene blue (MB) dye and other contaminants from water. The adsorption increased with increasing dye concentration, pH values and increasing temperature due to the increased kinetic energies of the molecules. The morphologies of polymer/LDH nanocomposites exhibited a remarkable high adsorption capacity for the efficient removal of dyes and heavy metal ions from water within a short time interval.

Author Contributions: M.M., J.S.S. and M.J.M. conceptualized, co-designed and steered the review as well as co-writing Sections 1, 2, 2.1–2.3 and 3. S.I.M. and K.L. co-wrote Sections 2.4 and 4, while M.M., J.S.S. and M.J.M. compiled the article together. J.S.S. and M.J.M. were responsible for funding acquisition of the manuscript. All authors have read and agreed to the published version of the manuscript.

Funding: This research was funded by the National Research Foundation (NRF) of South Africa, grant number (s) 114270 and 127278.

Acknowledgments: The National Research Foundation (NRF) of South Africa is highly acknowledged for financial support of this research work.

Conflicts of Interest: No conflicts of interest declared by the authors.

Abbreviations

LDHs	Layered double hydroxides	Cl^-	Chloride ion
2D	Two-dimensional	NO_3^-	Nitrate ion
WHO	World Health Organization	CO_3^{2-}	Carbonate ion
WASH	Water, sanitation and hygiene	N_2	Nitrogen
SDGs	Sustainable development goals	M^{2+}	Divalent cation
MF	Microfiltration	M^{3+}	Trivalent cation
UF	Ultrafiltration	LBL	Layer-by-layer
NF	Nanofiltration	NH_4OH	Ammonium hydroxide
RO	Reverse osmosis	SLNSs	Single layer nanosheets
SMNs	Surface modification of nanomaterials	PVA	Poly(vinyl alcohol)
PCNCs	Polymer clay nanocomposites	PEG	Poly(ethylene glycol)
PNCs	Polymer nanocomposites	HT	Hydrotalcite
MDPI	Multidisciplinary Digital Publishing Institute	PAA	Poly(acrylic acid)
NCs	Nanoclays	POX	Poly(2-oxazoline)
MMT	Montmorillonite	PP	Polypropylene
CEC	Cation exchange capacity	PIP	Polyisoprene
PMMA	Poly(methyl methacrylate)	PLLA	Poly(L-lactic acid)
$Mg(OH)_2$	Magnesium hydroxide	PA11	Polyamide
EVA	Poly(ethylene-co-vinyl acetate)	PAM	Polyacrylamide
$Al(NO_3)_3 \cdot 9H_2O$	Aluminum nitrate nonahydrate	PE	Polyethylene
$Mg(NO_3)_2 \cdot 6H_2O$	Magnesium nitrate hexahydrate	PLA	Poly(lactic acid)
BPF	Bio-oil phenolic foam	PS	Polystyrene
MBPF	Modified bio-oil phenolic foam	BT	Bentonite
MAPP	Maleic-anhydride-grafted polypropylene	MB	Methylene Blue
TPU	Thermoplastic polyurethane	CCS	Crosslinked chitosan
NBR	Nitrile butadiene rubber	MO	Methyl orange

PBT	Polybutylene terephthalate	PANI	Polyaniline
NIPAM	N-isopropylacrylamide	DNA	Deoxyribonucleic acid
SEM	Scanning electron microscopy	AA	Acid activated
OMMT	Organically modified montmorillonite	PES	Polyethersulfone
TEM	Transmission electron microscopy	AO-II	Acid Orange II
PBS	Phosphate-buffered saline	Ppy	Polypyrrole
SDBS	Sodium dodecylbenzene sulfonate	CS	Chitosan
AMPS	2-acrylsmido-2-methyl-propanesulfonic acid	SA	Sodium alginate
GA	Glutaraldehyde	NaOH	Sodium hydroxide
HCl	Hydrochloric acid	BC	Bentonite clay
CeBC-A@CS	Cerium bentonite clay-malic acid chitosan		
LaBC-A@CS	Lanthanum bentonite clay-malic acid chitosan		
AlBC-A@CS	Aluminum bentonite clay-malic acid chitosan		
pHzpc	pH zero charge	MnO_2	Manganese dioxide
FeMnMg-LDH	Iron Manganese Magnesium-layered double hydroxide		
DI	Deionized	Pb(II)	Lead ion
RSC	The Royal Society of Chemistry	HCO_3^-	Bicarbonate ion
Co/Mo-LDH	Cobalt/Molybdenum-layered double hydroxide	SO_4^{2-}	Sulphate ion
PHF	Polyhydroxy fullerenes	Cl^-	Chloride ion
SBR	Styrene-butadiene rubber	EP	Epoxy
VE	Vinyl ester	NO_3^-	Nitrate ion
LDO	Layered double oxide		
Mg-Al-LDH	Magnesium-Aluminum-layered double hydroxide	AB	Amido Black
XRD	X-ray Diffractometry	CH_2	Methylene
HPO_4^{2-}	Hydrogen phosphate ion	C=O	Carbon monoxide
N-H	Imidogen	C-H	Methylene group
C10A	Cloisite 10A		
Δμ	Chemical potential difference		
BET	Brunauer-Emmett-Teller		
NRF	National Research Foundation		

References

1. Briggs, A.M.; Cross, M.J.; Hoy, D.G.; Sanchez-Riera, L.; Blyth, F.M.; Woolf, A.D.; March, L. Musculoskeletal health conditions represent a global threat to healthy aging: A report for the 2015 World Health Organization world report on ageing and health. *Gerontologist* **2016**, *56*, S243–S255. [CrossRef] [PubMed]
2. Ortigara, A.R.C.; Kay, M.; Uhlenbrook, S. A review of the SDG 6 synthesis report 2018 from an education, training, and research perspective. *Water* **2018**, *10*, 1353. [CrossRef]
3. Ait-Kadi, M. Water for development and development for water: Realizing the sustainable development goals (SDGs) vision. *Aquat. Procedia* **2016**, *6*, 106–110. [CrossRef]
4. Le, N.L.; Nunes, S.P. Materials and membrane technologies for water and energy sustainability. *Sustain. Mater. Techno.* **2016**, *7*, 1–28. [CrossRef]
5. Anjum, M.; Miandad, R.; Waqas, M.; Gehany, F.; Barakat, M.A. Remediation of wastewater using various nano-materials. *Arab. J. Chem.* **2019**, *12*, 4897–4919. [CrossRef]
6. Krstić, V.; Urošević, T.; Pešovski, B. A review on adsorbents for treatment of water and wastewaters containing copper ions. *Chem. Eng. Sci.* **2018**, *192*, 273–287. [CrossRef]
7. Warsinger, D.M.; Chakraborty, S.; Tow, E.W.; Plumlee, M.H.; Bellona, C.; Loutatidou, S.; Karimi, L.; Mikelonis, A.M.; Achilli, A.; Ghassemi, A.; et al. A review of polymeric membranes and processes for potable water reuse. *Prog. Polym.* **2018**, *81*, 209–237. [CrossRef]
8. Pandey, N.; Shukla, S.K.; Singh, N.B. Water purification by polymer nanocomposites: An overview. *Nanocomposites* **2017**, *3*, 47–66. [CrossRef]
9. Tlili, I.; Alkanhal, T.A. Nanotechnology for water purification: Electrospun nanofibrous membrane in water and wastewater treatment. *J. Water Reuse Desal.* **2019**, *9*, 232–248. [CrossRef]

10. Yaqoob, A.A.; Parveen, T.; Umar, K.; Mohamad Ibrahim, M.N. Role of nanomaterials in the treatment of wastewater: A review. *Water* **2020**, *12*, 495. [CrossRef]
11. Guerra, F.D.; Attia, M.F.; Whitehead, D.C.; Alexis, F. Nanotechnology for environmental remediation: Materials and applications. *Molecules* **2018**, *23*, 1760. [CrossRef] [PubMed]
12. Selatile, M.K.; Ray, S.S.; Ojijo, V.; Sadiku, R. Recent developments in polymeric electrospun nanofibrous membranes for seawater desalination. *RSC Adv.* **2018**, *8*, 37915–37938. [CrossRef]
13. Liao, Y.; Loh, C.H.; Tian, M.; Wang, R.; Fane, A.G. Progress in electrospun polymeric nanofibrous membranes for water treatment: Fabrication, modification and applications. *Prog. Polym.* **2018**, *77*, 69–94. [CrossRef]
14. Umar, K.; Parveen, T.; Khan, M.A.; Ibrahim, M.N.M.; Ahmad, A.; Rafatullah, M. Degradation of organic pollutants using metal-doped TiO_2 photocatalysts under visible light: A comparative study. *Desal. Water Treat.* **2019**, *161*, 275–282. [CrossRef]
15. Tang, K.; Gong, C.; Wang, D. Reduction potential, shadow prices, and pollution costs of agricultural pollutants in China. *Sci. Total Environ.* **2016**, *541*, 42–50. [CrossRef] [PubMed]
16. Richter, K.E.; Ayers, J.M. An approach to predicting sediment microbial fuel cell performance in shallow and deep water. *Res. J. Appl. Sci.* **2018**, *8*, 2628. [CrossRef]
17. Sizmur, T.; Fresno, T.; Akgül, G.; Frost, H.; Moreno-Jiménez, E. Biochar modification to enhance sorption of inorganics from water. *Bioresour. Technol.* **2017**, *246*, 34–47. [CrossRef]
18. Wang, J.; Wang, Z.; Vieira, C.L.; Wolfson, J.M.; Pingtian, G.; Huang, S. Review on the treatment of organic pollutants in water by ultrasonic technology. *Ultrason. Sonochem.* **2019**, *55*, 273–278. [CrossRef]
19. Liu, C.; Hong, T.; Li, H.; Wang, L. From club convergence of per capita industrial pollutant emissions to industrial transfer effects: An empirical study across 285 cities in China. *Energy Policy* **2018**, *121*, 300–313. [CrossRef]
20. Bayoumi, T.A.; Saleh, H.M. Characterization of biological waste stabilized by cement during immersion in aqueous media to develop disposal strategies for phytomediated radioactive waste. *Prog. Nucl. Energy* **2018**, *107*, 83–89. [CrossRef]
21. Ma, H.; Guo, Y.; Qin, Y.; Li, Y.Y. Nutrient recovery technologies integrated with energy recovery by waste biomass anaerobic digestion. *Bioresour. Technol.* **2018**, *269*, 520–531. [CrossRef] [PubMed]
22. Hlongwane, G.N.; Sekoai, P.T.; Meyyappan, M.; Moothi, K. Simultaneous removal of pollutants from water using nanoparticles: A shift from single pollutant control to multiple pollutant control. *Sci. Total Environ.* **2019**, *656*, 808–833. [CrossRef]
23. Rajasulochana, P.; Preethy, V. Comparison on efficiency of various techniques in treatment of waste and sewage water–A comprehensive review. *Res. Effic. Technol.* **2016**, *2*, 175–184. [CrossRef]
24. Jeevanandam, J.; Barhoum, A.; Chan, Y.S.; Dufresne, A.; Danquah, M.K. Review on nanoparticles and nanostructured materials: History, sources, toxicity and regulations. *Beilstein J. Nanotechnol.* **2018**, *9*, 1050–1074. [CrossRef]
25. Palmero, P. Structural ceramic nanocomposites: A review of properties and powders' synthesis methods. *Nanomaterials* **2015**, *5*, 656–696. [CrossRef]
26. Chen, J.; Liu, B.; Gao, X.; Xu, D. A review of the interfacial characteristics of polymer nanocomposites containing carbon nanotubes. *RSC Adv.* **2018**, *8*, 28048–28085. [CrossRef]
27. Kumar, S.; Nehra, M.; Dilbaghi, N.; Tankeshwar, K.; Kim, K.H. Recent advances and remaining challenges for polymeric nanocomposites in healthcare applications. *Prog. Polym. Sci.* **2018**, *80*, 1–38. [CrossRef]
28. Lofrano, G.; Carotenuto, M.; Libralato, G.; Domingos, R.F.; Markus, A.; Dini, L.; Gautam, R.K.; Baldantoni, D.; Rossi, M.; Sharma, S.K.; et al. Polymer functionalized nanocomposites for metals removal from water and wastewater: An overview. *Water Res.* **2016**, *92*, 22–37. [CrossRef] [PubMed]
29. Müller, K.; Bugnicourt, E.; Latorre, M.; Jorda, M.; Echegoyen Sanz, Y.; Lagaron, J.M.; Miesbauer, O.; Bianchin, A.; Hankin, S.; Bölz, U.; et al. Review on the processing and properties of polymer nanocomposites and nanocoatings and their applications in the packaging, automotive and solar energy fields. *Nanomaterials* **2017**, *7*, 74. [CrossRef]
30. Ashraf, M.A.; Peng, W.; Zare, Y.; Rhee, K.Y. Effects of size and aggregation/agglomeration of nanoparticles on the interfacial/interphase properties and tensile strength of polymer nanocomposites. *Nanoscale Res. Lett.* **2018**, *13*, 214. [CrossRef]
31. Guo, F.; Aryana, S.; Han, Y.; Jiao, Y. A review of the synthesis and applications of polymer–nanoclay composites. *Appl. Sci.* **2018**, *8*, 1696. [CrossRef]

32. Crucho, C.I.; Barros, M.T. Polymeric nanoparticles: A study on the preparation variables and characterization methods. *Mater. Sci. Eng. C* **2017**, *80*, 771–784. [CrossRef] [PubMed]
33. Tessema, A.; Zhao, D.; Moll, J.; Xu, S.; Yang, R.; Li, C.; Kumar, S.K.; Kidane, A. Effect of filler loading, geometry, dispersion and temperature on thermal conductivity of polymer nanocomposites. *Polym. Test.* **2017**, *57*, 101–106. [CrossRef]
34. Ghadimi, M.; Zangenehtabar, S.; Homaeigohar, S. An Overview of the water remediation potential of nanomaterials and their ecotoxicological impacts. *Water* **2020**, *12*, 1150. [CrossRef]
35. Nasir, A.; Masood, F.; Yasin, T.; Hameed, A. Progress in polymeric nanocomposite membranes for wastewater treatment: Preparation, properties and applications. *J. Ind. Eng. Chem.* **2019**, *79*, 29–40. [CrossRef]
36. Schaming, D.; Remita, H. Nanotechnology: From the ancient time to nowadays. *Found. Chem.* **2015**, *17*, 187–205. [CrossRef]
37. Usmani, M.A.; Khan, I.; Ahmad, N.; Bhat, A.H.; Sharma, D.K.; Rather, J.A.; Hassan, S.I. Modification of Nanoclay Systems: An Approach to Explore Various Applications. In *Nanoclay Reinforced Polymer Composites*; Springer: Singapore, 2016; pp. 57–83.
38. Uddin, M.K. A review on the adsorption of heavy metals by clay minerals, with special focus on the past decade. *Chem. Eng. J.* **2017**, *308*, 438–462. [CrossRef]
39. Nadziakiewicza, M.; Kehoe, S.; Micek, P. Physico-chemical properties of clay minerals and their use as a health promoting feed additive. *Animals* **2019**, *9*, 714. [CrossRef]
40. Jawaid, M.; Qaiss, A.; Bouhfid, R. *Nanoclay Reinforced Polymer Composites: Nanocomposites and Bionanocomposites*; Springer: Singapore, 2016; p. 391.
41. Awasthi, A.; Jadhao, P.; Kumari, K. Clay nano-adsorbent: Structures, applications and mechanism for water treatment. *SN Appl. Sci.* **2019**, *1*, 1076. [CrossRef]
42. Jlassi, K.; Krupa, I.; Chehimi, M.M. Overview: Clay preparation, properties, modification. In *Clay-Polymer Nanocomposites*; Elsevier: Amsterdam, The Netherlands, 2017; pp. 1–28.
43. Ghadiri, M.; Chrzanowski, W.; Rohanizadeh, R. Biomedical applications of cationic clay minerals. *RSC Adv.* **2015**, *5*, 29467–29481. [CrossRef]
44. Lázaro, B.B. Halloysite and kaolinite: Two clay minerals with geological and technological importance. *J. Acad. Exact Phys. Chem. Nat. Sci. Zaragoza* **2015**, *70*, 7–38.
45. Yu, F.; Deng, H.; Bai, H.; Zhang, Q.; Wang, K.; Chen, F.; Fu, Q. Confine clay in an alternating multilayered structure through injection molding: A simple and efficient route to improve barrier performance of polymeric materials. *ACS Appl. Mater. Inter.* **2015**, *7*, 10178–10189. [CrossRef] [PubMed]
46. Ferrari, P.C.; Araujo, F.F.; Pianaro, S.A. Halloysite nanotubes-polymeric nanocomposites: Characteristics, modifications and controlled drug delivery approaches. *Cerâmica* **2017**, *63*, 423–431. [CrossRef]
47. Yusoh, K.; Kumaran, S.V.; Ismail, F.S. Surface modification of nanoclay for the synthesis of polycaprolactone (PCL)–clay nanocomposite. In *MATEC Web of Conferences*; EDP Sciences: Pahang, Malaysia, 2018; Volume 150, p. 02005.
48. Irshidat, M.R.; Al-Saleh, M.H. Thermal performance and fire resistance of nanoclay modified cementitious materials. *Constr. Build. Mater.* **2018**, *159*, 213–219. [CrossRef]
49. Murugesan, S.; Scheibel, T. Copolymer/Clay Nanocomposites for Biomedical Applications. *Adv. Funct. Mater.* **2020**, *30*, 1908101. [CrossRef]
50. Wang, W.; Wang, A. Nanoscale Clay Minerals for Functional Ecomaterials: Fabrication, Applications, and Future Trends. In *Handbook of Ecomaterials*; Springer: Cham, Germany, 2019; pp. 1–82.
51. Satish, S.; Tharmavaram, M.; Rawtani, D. Halloysite nanotubes as a nature's boon for biomedical applications. *Nanobiomedicine* **2019**, *6*, 1–16. [CrossRef]
52. Gaaz, T.S.; Sulong, A.B.; Kadhum, A.A.H.; Al-Amiery, A.A.; Nassir, M.H.; Jaaz, A.H. The impact of halloysite on the thermo-mechanical properties of polymer composites. *Molecules* **2017**, *22*, 838. [CrossRef]
53. Lazzara, G.; Cavallaro, G.; Panchal, A.; Fakhrullin, R.; Stavitskaya, A.; Vinokurov, V.; Lvov, Y. An assembly of organic-inorganic composites using halloysite clay nanotubes. *Curr. Opin. Colloid Interface Sci.* **2018**, *35*, 42–50. [CrossRef]
54. Ursino, C.; Castro-Muñoz, R.; Drioli, E.; Gzara, L.; Albeirutty, M.H.; Figoli, A. Progress of nanocomposite membranes for water treatment. *Membranes* **2018**, *8*, 18. [CrossRef]
55. Sajid, M.; Basheer, C. Layered double hydroxides: Emerging sorbent materials for analytical extractions. *Trac-Trend Anal. Chem.* **2016**, *75*, 174–182. [CrossRef]

56. Mir, Z.M.; Bastos, A.; Höche, D.; Zheludkevich, M.L. Recent Advances on the Application of Layered Double Hydroxides in Concrete—A Review. *Materials* **2020**, *13*, 1426. [CrossRef] [PubMed]
57. Mohapatra, L.; Parida, K. A review on the recent progress, challenges and perspective of layered double hydroxides as promising photocatalysts. *J. Mater. Chem.* **2016**, *4*, 10744–10766. [CrossRef]
58. Jaśkaniec, S.; Hobbs, C.; Seral-Ascaso, A.; Coelho, J.; Browne, M.P.; Tyndall, D.; Sasaki, T.; Nicolosi, V. Low-temperature synthesis and investigation into the formation mechanism of high quality Ni-Fe layered double hydroxides hexagonal platelets. *Sci. Rep.* **2018**, *8*, 1–8. [CrossRef] [PubMed]
59. Mishra, G.; Dash, B.; Pandey, S. Layered double hydroxides: A brief review from fundamentals to application as evolving biomaterials. *Appl. Clay Sci.* **2018**, *153*, 172–186. [CrossRef]
60. Barahuie, F.; Hussein, M.Z.; Gani, S.A.; Fakurazi, S.; Zainal, Z. Synthesis of protocatechuic acid–zinc/aluminium–layered double hydroxide nanocomposite as ananticancer nanodelivery system. *J. Solid State Chem.* **2015**, *221*, 21–31. [CrossRef]
61. Yu, J.; Wang, Q.; O'Hare, D.; Sun, L. Preparation of two dimensional layered double hydroxide nanosheets and their applications. *Chem. Soc. Rev.* **2017**, *46*, 5950. [CrossRef]
62. He, X.; Qiu, X.; Hu, C.; Liu, Y. Treatment of heavy metal ions in wastewater using layered double hydroxides: A review. *J. Dispers. Sci. Technol.* **2018**, *39*, 792–801. [CrossRef]
63. Sokol, D.; Klemkaite-Ramanauske, K.; Khinsky, A.; Baltakys, K.; Beganskiene, A.; Baltusnikas, A.; Pinkas, J.; Kareiva, A. Reconstruction effects on surface properties of Co/Mg/Al layered double hydroxide. *Mater. Sci.* **2017**, *23*, 144–149. [CrossRef]
64. Abo El-Reesh, G.Y.; Farghali, A.A.; Taha, M.; Mahmoud, R.K. Novel synthesis of Ni/Fe layered double hydroxides using urea and glycerol and their enhanced adsorption behavior for Cr(VI) removal. *Sci. Rep.* **2020**, *10*, 587. [CrossRef]
65. Pahalagedara, M.N.; Samaraweera, M.; Dharmarathna, S.; Kuo, C.H.; Pahalagedara, L.R.; Gascón, J.A.; Suib, S.L. Removal of azo dyes: Intercalation into sonochemically synthesized NiAl layered double hydroxide. *J. Phys. Chem. C* **2014**, *118*, 17801–17809. [CrossRef]
66. Skorb, E.V.; Möhwald, H.; Andreeva, D.V. Effect of cavitation bubble collapse on the modification of solids: Crystallization aspects. *Langmuir* **2016**, *32*, 11072–11085. [CrossRef] [PubMed]
67. Altay, R.; Sadaghiani, A.K.; Sevgen, M.I.; Sisman, A.; Koşsar, A. Numerical and experimental studies on the effect of surface roughness and ultrasonic frequency on bubble dynamics in acoustic cavitation. *Energies* **2020**, *13*, 1126. [CrossRef]
68. Daud, M.; Kamal, M.S.; Shehzad, F.; Al-Harthi, M.A. Graphene/layered double hydroxides nanocomposites: A review of recent progress in synthesis and applications. *Carbon* **2016**, *104*, 241–252. [CrossRef]
69. Zhao, X.; Cao, J.-P.; Zhao, J.; Hu, G.-H.; Dang, Z.-M. A hybrid Mg–Al layered double hydroxide/graphene nanostructure obtained via hydrothermal synthesis. *Chem. Phys. Lett.* **2014**, *605–606*, 77–80. [CrossRef]
70. Liu, W.; Xu, S.; Liang, R.; Wei, M.; Evans, D.G.; Duan, X. In situ synthesis of nitrogen-doped carbon dots in the interlayer region of a layered double hydroxide with tunable quantum yield. *J. Mater. Chem. C* **2017**, *5*, 3536–3541. [CrossRef]
71. Cermelj, K.; Ruengkajorn, K.; Buffet, J.-C.; O'Hare, D. Layered double hydroxide nanosheets via solvothermal delamination. *J. Ener. Chem.* **2019**, *35*, 88–94. [CrossRef]
72. Wang, L.; Shi, C.; Wang, L.; Pan, L.; Zhang, X.; Zou, J.-J. Rational design, synthesis, adsorption principles and applications of metal oxide adsorbents: A review. *Nanoscale* **2020**, *12*, 4790–4815. [CrossRef]
73. Forano, C.; Bruna, F.; Mousty, C.; Prevot, V. Interactions between biological cells and layered double hydroxides: Towards functional materials. *Chem. Rec.* **2018**, *18*, 1–18. [CrossRef]
74. Larocca, N.M.; Filho, R.B.; Pessan, L.A. Influence of layer-by-layer deposition techniques and incorporation of layered double hydroxides (LDH) on the morphology and gas barrier properties of polyelectrolytes multilayer thin films. *Surf. Coat. Technol.* **2018**, *349*, 1–12. [CrossRef]
75. Shao, M.; Zhang, R.; Li, Z.; Wei, M.; Evans, D.G.; Duan, X. Layered double hydroxides toward electrochemical energy storage and conversion: Design, synthesis and applications. *Chem. Commun.* **2015**, *51*, 15880–15893. [CrossRef]
76. Danks, A.E.; Hall, S.R.; Schnepp, Z. The evolution of 'sol–gel' chemistry as a technique for materials synthesis. *Mater. Horiz.* **2016**, *3*, 91–112. [CrossRef]
77. Zhang, Y.; Li, H.; Du, N.; Zhang, R.; Hou, W. Large-scale aqueous synthesis of layered double hydroxide single-layer nanosheets. *Colloids and Surfaces A: Physicochem. Eng. Asp.* **2016**, *501*, 49–54. [CrossRef]

78. Sarma, G.K.; Rashid, M.H. Synthesis of Mg/Al layered double hydroxides for adsorptive removal of fluoride from water: A mechanistic and kinetic study. *J. Chem. Eng.* **2018**, *63*, 2957–2965. [CrossRef]
79. Sajid, M.; Nazal, M.K.; Baig, N.; Osman, A.M. Removal of heavy metals and organic pollutants from water using dendritic polymers based adsorbents: A critical review. *Sep. Purif. Technol.* **2018**, *191*, 400–423. [CrossRef]
80. Maziarz, P.; Matusik, J.; Leiviskä, T. Mg/Al LDH Enhances Sulfate removal and Clarification of AMD Wastewater in Precipitation Processes. *Materials* **2019**, *12*, 2334. [CrossRef]
81. Daud, M.; Hai, A.; Banat, F.; Wazir, M.B.; Habib, M.; Bharath, G.; Al-Harthi, M.A. A review on the recent advances, challenges and future aspect of layered double hydroxides (LDH)–Containing hybrids as promising adsorbents for dyes removal. *J. Mol. Liq.* **2019**, *288*, 110989. [CrossRef]
82. Li, M.; Tian, R.; Yan, D.; Liang, R.; Wei, M.; Evans, D.G.; Duan, X. A luminescent ultrathin film with reversible sensing toward pressure. *ChemComm* **2016**, *52*, 4663–4666. [CrossRef]
83. Asif, M.; Aziz, A.; Azeem, M.; Wang, Z.; Ashraf, G.; Xiao, F.; Chen, X.; Liu, H. A review on electrochemical biosensing platform based on layered double hydroxides for small molecule biomarkers determination. *Adv. Colloid Interface Sci.* **2018**, *262*, 21–38. [CrossRef]
84. Baig, N.; Sajid, M. Applications of layered double hydroxides based electrochemical sensors for determination of environmental pollutants: A review. *Trends Environ. Anal.* **2017**, *16*, 1–15. [CrossRef]
85. Guan, S.; Liang, R.; Li, C.; Yan, D.; Wei, M.; Evans, D.G.; Duan, X. A layered drug nanovehicle toward targeted cancer imaging and therapy. *J. Mater. Chem. B* **2016**, *4*, 1331–1336. [CrossRef]
86. Yan, L.; Gonca, S.; Zhu, G.; Zhang, W.; Chen, X. Layered double hydroxide nanostructures and nanocomposites for biomedical applications. *J. Mater. Chem. B* **2019**, *7*, 5583–5601. [CrossRef] [PubMed]
87. Bastianini, M.; Faffa, C.; Sisani, M.; Petracci, A. Caffeic Acid-layered Double Hydroxide Hybrid: A New Raw Material for Cosmetic Applications. *Cosmetics* **2018**, *5*, 51. [CrossRef]
88. Amberg, N.; Fogarassy, C. Green consumer behavior in the cosmetics market. *Resources* **2019**, *8*, 137. [CrossRef]
89. Caminade, A.M.; Ouali, A.; Laurent, R.; Turrin, C.O.; Majoral, J.P. Coordination chemistry with phosphorus dendrimers. Applications as catalysts, for materials, and in biology. *Coord. Chem. Rev.* **2016**, *308*, 478–497. [CrossRef]
90. Jing, M.; Hou, H.; Banks, C.E.; Yang, Y.; Zhang, Y.; Ji, X. Alternating voltage introduced NiCo double hydroxide layered nanoflakes for an asymmetric supercapacitor. *ACS Appl. Mater. Inter.* **2015**, *7*, 22741–22744. [CrossRef]
91. Mochane, M.J.; Magagula, S.I.; Sefadi, J.S.; Sadiku, E.R.; Mokhena, T.C. Morphology, Thermal Stability, and Flammability Properties of Polymer-Layered Double Hydroxide (LDH) Nanocomposites: A Review. *Crystals* **2020**, *10*, 612. [CrossRef]
92. Salavagione, H.J.; Diez-Pascual, A.M.; Lázaro, E.; Vera, S.; Gomez-Fatou, M.A. Chemical sensors based on polymer composites with carbon nanotubes and graphene: The role of the polymer. *J. Mater. Chem. A* **2014**, *2*, 14289–14328. [CrossRef]
93. Bhattacharya, M. Polymer nanocomposites—A comparison between carbon nanotubes, graphene, and clay as nanofillers. *Materials* **2016**, *9*, 262. [CrossRef]
94. Madhumitha, G.; Fowsiya, J.; Mohana Roopan, S.; Thakur, V.K. Recent advances in starch–clay nanocomposites. *Int. J. Polym. Anal. Charact.* **2018**, *23*, 331–345. [CrossRef]
95. Cui, Y.; Kumar, S.; Kona, B.R.; van Houcke, D. Gas barrier properties of polymer/clay nanocomposites. *RSC Adv.* **2015**, *5*, 63669–63690. [CrossRef]
96. Hammad, S.; Noby, H.; Elkady, M.F.; El-Shazly, A.H. In-situ of polyaniline/polypyrrole copolymer using different techniques. *Mater. Sci. Eng.* **2017**, *290*, 012001. [CrossRef]
97. Unalan, I.U.; Cerri, G.; Marcuzzo, E.; Cozzolino, C.A.; Farris, S. Nanocomposite films and coatings using inorganic nanobuilding blocks (NBB): Current applications and future opportunities in the food packaging sector. *RSC Adv.* **2014**, *4*, 29393–29428. [CrossRef]
98. Abedi, S.; Abdouss, M. A review of clay-supported Ziegler–Natta catalysts for production of polyolefin/clay nanocomposites through in situ polymerization. *Appl. Catal. A Gen.* **2014**, *475*, 386–409. [CrossRef]
99. Reddy, K.R.; El-Zein, A.; Airey, D.W.; Alonso-Marroquin, F.; Schubel, P.; Manalo, A. Self-healing polymers: Synthesis methods and applications. *Nano Struct. Nano Objects* **2020**, *23*, 100500. [CrossRef]

100. Ozkose, U.U.; Altinkok, C.; Yilmaz, O.; Alpturk, O.; Tasdelen, M.A. In-situ preparation of poly (2-ethyl-2-oxazoline)/clay nanocomposites via living cationic ring-opening polymerization. *Eur. Polym. J.* **2017**, *88*, 586–593. [CrossRef]
101. Cruz, S.M.; Viana, J.C. Structure–Properties Relationships in Thermoplastic Polyurethane Elastomer Nanocomposites: Interactions between Polymer Phases and Nanofillers. *Macromol. Mater. Eng.* **2015**, *300*, 1153–1162. [CrossRef]
102. Jafarbeglou, M.; Abdouss, M.; Shoushtari, A.M.; Jafarbeglou, M. Clay nanocomposites as engineered drug delivery systems. *RSC Adv.* **2016**, *6*, 50002–50016. [CrossRef]
103. Zabihi, O.; Ahmadi, M.; Nikafshar, S.; Preyeswary, K.C.; Naebe, M. A technical review on epoxy-clay nanocomposites: Structure, properties, and their applications in fiber reinforced composites. *Compos. B. Eng.* **2018**, *135*, 1–24. [CrossRef]
104. Ercan, N.; Durmus, A.; Kaşgöz, A. Comparing of melt blending and solution mixing methods on the physical properties of thermoplastic polyurethane/organoclay nanocomposite films. *J. Compos. Mater.* **2017**, *30*, 950–970. [CrossRef]
105. Mohd Zaini, N.A.; Ismail, H.; Rusli, A. Short review on sepiolite-filled polymer nanocomposites. *Polym. Plast. Technol. Eng.* **2017**, *56*, 1665–1679. [CrossRef]
106. Saeed, K.; Khan, I. Characterization of clay filled poly (butylene terephthalate) nanocomposites prepared by solution blending. *Polímeros* **2015**, *25*, 591–595. [CrossRef]
107. Luecha, W.; Magaraphan, R. A novel and facile nanoclay aerogel masterbatch toward exfoliated polymer-clay nanocomposites through a melt-mixing process. *Adv. Mater. Sci. Eng.* **2018**, *2018*, 1–14. [CrossRef]
108. Kong, J.; Li, Z.; Cao, Z.; Han, C.; Dong, L. The excellent gas barrier properties and unique mechanical properties of poly (propylene carbonate)/organo-montmorillonite nanocomposites. *Polym. Bull.* **2017**, *74*, 5065–5082. [CrossRef]
109. Fu, S.; Sun, Z.; Huang, P.; Li, Y.; Hu, N. Some basic aspects of polymer nanocomposites: A critical review. *Nano Mater. Sci.* **2019**, *1*, 2–30. [CrossRef]
110. Pesetskii, S.S.; Bogdanovich, S.P.; & Aderikha, V.N. Polymer/clay nanocomposites produced by dispersing layered silicates in thermoplastic melts. In *Polymer Nanocomposites for Advanced Engineering and Military Applications*; IGI Global: Pennsylvania, PA, USA, 2019; pp. 66–94.
111. Bai, C.; Ke, Y.; Hu, X.; Xing, L.; Zhao, Y.; Lu, S.; Lin, Y. Preparation and properties of amphiphilic hydrophobically associative polymer/montmorillonite nanocomposites. *R. Soc. Open Sci.* **2020**, *7*, 200199. [CrossRef]
112. Szadkowski, B.; Marzec, A.; Rybiński, P.; Żukowski, W.; Zaborski, M. Characterization of Ethylene–propylene Composites Filled with Perlite and Vermiculite Minerals: Mechanical, Barrier, and Flammability Properties. *Materials* **2020**, *13*, 585. [CrossRef]
113. Al-Shahrani, A.; Taie, I.; Fihri, A.; Alabedi, G. Polymer-Clay Nanocomposites for Corrosion Protection. In *Current Topics in the Utilization of Clay in Industrial and Medical Applications*; IntechOpen: London, UK, 2018; pp. 61–79.
114. Alvi, M.U.; Zulfiqar, S.; Sarwar, M.I.; Kidwai, A.A. Preparation and properties of nanocomposites derived from aromatic polyamide and surface functionalized nanoclay. *Chem. Eng. Commun.* **2016**, *203*, 242–250. [CrossRef]
115. Chen, H.H.; Thirumavalavan, M.; Lin, F.Y.; Lee, J.F. A facile approach for achieving an effective dual sorption ability of Si/SH/S grafted sodium montmorillonite. *RSC Adv.* **2015**, *5*, 57792–57803. [CrossRef]
116. Haider, S.; Kausar, A.; Muhammad, B. Overview on polystyrene/nanoclay composite: Physical properties and application. *Polym. Plast. Technol. Eng.* **2017**, *56*, 917–931. [CrossRef]
117. Grishina, E.P.; Ramenskaya, L.M.; Kudryakova, N.O.; Vagin, K.V.; Kraev, A.S.; Agafonov, A.V. Composite nanomaterials based on 1-butyl-3-methylimidazolium dicianamide and clays. *J. Mater. Res. Technol.* **2019**, *8*, 4387–4398. [CrossRef]
118. Bischoff, E.; Simon, D.A.; Liberman, S.A.; Mauler, R.S. Adsorption of ionic liquid onto halloysite nanotubes: Thermal and mechanical properties of heterophasic PE-PP copolymer nanocomposites. In *AIP Conference Proceedings*; AIP Publishing LLC: Jeju Island, Korea, 2016; Volume 1713, pp. 1–5.
119. Xu, P.; Yu, Y.; Chang, M.; Chang, J. Preparation and Characterization of Bio-oil Phenolic Foam Reinforced with Montmorillonite. *Polymers* **2019**, *11*, 1471. [CrossRef] [PubMed]

120. Chanra, J.; Budianto, E.; Soegijono, B. Surface modification of montmorillonite by the use of organic cations via conventional ion exchange method. In *IOP Conference Series: Materials Science and Engineering*; IOP Science: Semarang, Indonesia, 2019; Volume 509, p. 012057.
121. Uwa, C.A.; Jamiru, T.; Sadiku, E.R.; Huan, Z.; Mpofu, K. Polypropylene/nanoclay Composite: A solution to refrigerated vehicles. *Procedia Manuf.* **2019**, *35*, 174–180. [CrossRef]
122. Valapa, R.B.; Loganathan, S.; Pugazhenthi, G.; Thomas, S.; Varghese, T.O. An overview of polymer–clay nanocomposites. In *Clay-Polymer Nanocomposites*; Elsevier: Amsterdam, The Netherlands, 2017; pp. 29–81.
123. Moreno-Sader, K.; García-Padilla, A.; Realpe, A.; Acevedo-Morantes, M.; Soares, J.B.P. Removal of Heavy Metal Water Pollutants (Co2+ and Ni2+) Using Polyacrylamide/Sodium Montmorillonite (PAM/Na-MMT) Nanocomposites. *ACS Omega* **2019**, *4*, 10834–10844. [CrossRef]
124. Atta, A.M.; Al-Lohedan, H.A.; ALOthman, Z.A.; Abdel-Khalek, A.A.; Tawfeek, A.M. Characterization of reactive amphiphilic montmorillonite nanogels and its application for removal of toxic cationic dye and heavy metals water pollutants. *J. Ind. Eng. Chem.* **2015**, *31*, 374–384. [CrossRef]
125. Şen, F.; Demirbaş, Ö.; Çalımlı, M.H.; Aygün, A.; Alma, M.H.; Nas, M.S. The dye removal from aqueous solution using polymer composite films. *Appl. Water Sci.* **2018**, *8*, 206. [CrossRef]
126. Peng, N.; Hu, D.; Zeng, J.; Li, Y.; Liang, L.; Chang, C. Superabsorbent cellulose–clay nanocomposite hydrogels for highly efficient removal of dye in water. *ACS Sustainable Chem. Eng.* **2016**, *4*, 7217–7224. [CrossRef]
127. Moja, T.N.; Bunekar, N.; Mojaki, S.; Mishra, S.B.; Tsai, T.-Y.; Hwang, S.S.; Mishra, A.K. Polypropylene–Polypropylene-Grafted-Maleic Anhydride–Montmorillonite Clay Nanocomposites for Pb(II) Removal. *J. Inorg. Organomet. Polym. Mater.* **2018**, *28*, 2799–2811. [CrossRef]
128. Liu, Q.; Yang, B.; Zhang, L.; Huang, R. Adsorption of an anionic azo dye by cross-linked chitosan/bentonite composite. *Int. J. Biol. Macromol.* **2015**, *72*, 1129–1135. [CrossRef]
129. Natarajan, S.; Anitha, V.; Gajula, G.P.; Thiagarajan, V. Synthesis and characterization of magnetic superadsorbent Fe$_3$O$_4$-PEG-Mg-Al-LDH nanocomposites for ultrahigh removal of organic dyes. *ACS Omega* **2020**, *5*, 3181–3193. [CrossRef]
130. Alnaqbi, M.A.; Samson, J.A.; Greish, Y.E. Electrospun polystyrene/LDH fibrous membranes for the removal of Cd^{2+} ions. *J. Nanomater.* **2020**, *12*, 5045637. [CrossRef]
131. Quispe-Dominguez, R.; Naseem, S.; Leuteritz, A.; Kuehnert, I. Synthesis and characterization of MgAl-DBS LDH/PLA composite by sonication-assisted masterbatch (SAM) melt mixing method. *RSC Adv.* **2019**, *9*, 658. [CrossRef]
132. Devi, K.U.; Ponnamma, D.; Causin, V.; Maria, H.J.; Thomas, S. Enhanced morphology and mechanical characteristics of clay/styrene butadiene rubber nanocomposites. *Appl. Clay Sci.* **2015**, *114*, 568–576. [CrossRef]
133. Sari, M.G.; Ramezanzadeh, B.; Shahbazi, M.; Pakdel, A.S. Influence of nanoclay particles modification by polyester-amide hyperbranched polymer on the corrosion protective performance of the epoxy nanocomposite. *Corros. Sci.* **2015**, *92*, 162–172. [CrossRef]
134. Tsai, T.Y.; Bunekar, N.; Liang, S.W. Effect of Multiorganomodified LiAl-or MgAl-Layered Double Hydroxide on the PMMA Nanocomposites. *Adv. Polym. Technol.* **2018**, *37*, 31–37. [CrossRef]
135. Nagendra, B.; Mohan, K.; Gowd, E.B. Polypropylene/layered double hydroxide (LDH) nanocomposites: Influence of LDH particle size on the crystallization behavior of polypropylene. *ACS Appl. Mater. Inter.* **2015**, *7*, 12399–12410. [CrossRef]
136. Adeyemo, A.A.; Adeoye, I.O.; Bello, O.S. Adsorption of dyes using different types of clay: A review. *Appl. Water Sci.* **2017**, *7*, 543–568. [CrossRef]
137. Zabihi, O.; Ahmadi, M.; Khayyam, H.; Naebe, M. Fish DNA-modified clays: Towards highly flame retardant polymer nanocomposite with improved interfacial and mechanical performance. *Sci. Rep.* **2016**, *6*, 38194. [CrossRef]
138. Bashar, M.; Mertiny, P.; Sundararaj, U. Effect of nanocomposite structures on fracture behavior of epoxy-clay nanocomposites prepared by different dispersion methods. *J. Nanomater.* **2014**, *70*, 1–12. [CrossRef]
139. Oyarzaba, A.; Mugica, A.; Müller, A.J.; Zubitur, M. Hydrolytic degradation of nanocomposites based on poly(L-lactic acid) and layered double hydroxides modified with a model drug. *J. Appl. Polym. Sci.* **2016**, *133*, 43648.

140. Leng, J.; Kang, N.; Wang, D.Y.; Falkenhagen, J.; Thünemann, A.F.; Schönhals, A. Structure–property relationships of nanocomposites based on Polylactide and layered double hydroxides–comparison of MgAl and NiAl LDH as Nanofiller. *Macromol. Chem. Phys.* **2017**, *218*, 1700232. [CrossRef]
141. Mandal, S.; Kalaivanan, S.; Mandal, A.B. Polyethylene glycol-modified layered double hydroxides: Synthesis, characterization, and study on adsorption characteristics for removal of acid orange II from aqueous solution. *ACS Omega* **2019**, *4*, 3745–3754. [CrossRef] [PubMed]
142. Mahmoudian, M.; Balkanloo, P.G.; Nozad, E. A facile method for dye and heavy metal elimination by pH sensitive acid activated montmorillonite/polyethersulfone nanocomposite membrane. *Chin. J. Polym. Sci.* **2018**, *36*, 49–57. [CrossRef]
143. Mohamed, F.; Abukhadra, M.R.; Shaban, M. Removal of safranin dye from water using polypyrrole nanofiber/Zn-Fe layered double hydroxide nanocomposite (Ppy NF/Zn-Fe LDH) of enhanced adsorption and photocatalytic properties. *Sci. Total Environ.* **2018**, *640–641*, 352–363. [CrossRef] [PubMed]
144. Xu, G.; Zhu, Y.; Wang, X.; Wang, S.; Cheng, T.; Ping, R.; Cao, J.; Kaihe, L. Novel chitosan and Laponite based nanocomposite for fast removal of Cd(II), methylene blue and Congo red from aqueous solution. *e-Polymers* **2019**, *19*, 244–256. [CrossRef]
145. Isawi, H. Using Zeolite/Polyvinyl alcohol/sodium alginate nanocomposite beads for removal of some heavy metals from wastewater. *Arab. J. Chem.* **2020**, *13*, 5691–5716. [CrossRef]
146. Biswas, S.; Rashid, T.U.; Debnath, T.; Haque, P.; Rahman, M.M. Application of chitosan-clay biocomposite beads for removal of heavy metal and dye from industrial effluent. *J. Compos. Sci.* **2020**, *4*, 16. [CrossRef]
147. Sharma, S.; Bhattacharya, A. Drinking water contamination and treatment techniques. *Appl. Water Sci.* **2017**, *7*, 1043–1067. [CrossRef]
148. Sarma, G.K.; Sen Gupta, S.; Bhattacharyya, K.G. Nanomaterials as versatile adsorbents for heavy metal ions in water: A review. *Environ. Sci. Pollut. Res.* **2019**, *26*, 6245–6278. [CrossRef]
149. Crini, G.; Lichtfouse, E.; Wilson, L.; Morin-Crini, N. Adsorption-oriented processes using conventional and non-conventional adsorbents for wastewater treatment. In *Green Adsorbents for Pollutant Removal*; Environmental Chemistry for a Sustainable World; Springer Nature: Besançon, France, 2018; pp. 23–71. ISBN 978-3-319-92111-2.
150. Nagaraj, A.; Munusamy, M.A.; Ahmed, M.; Kumar, S.S.; Rajan, M. Hydrothermal synthesis of a mineral-substituted hydroxyapatite nanocomposite material for fluoride removal from drinking water. *New J. Chem.* **2018**, *42*, 12711–12721. [CrossRef]
151. Nagaraj, A.; Pillay, K.; Kumar, S.K.; Rajan, M. Dicarboxylic acid cross-linked metal ion decorated bentonite clay and chitosan for fluoride removal studies. *RSC Adv.* **2020**, *10*, 16791. [CrossRef]
152. Schneckenburger, T.; Riefstahl, J.; Fischer, K. Adsorption of aliphatic polyhydroxy carboxylic acids on gibbsite: pH dependency and importance of adsorbate structure. *Environ. Sci. Eur.* **2018**, *30*, 1. [CrossRef] [PubMed]
153. Zhou, H.; Jiang, Z.; Wei, S. A new hydrotalcite-like absorbent FeMnMg-LDH and its adsorption capacity for Pb2+ ions in water. *Appl. Clay Sci.* **2018**, *153*, 29–37. [CrossRef]
154. Srivastava, V.; Sharma, Y.; Sillanpää, M. Green synthesis of magnesium oxide nanoflower and its application for the removal of divalent metallic species from synthetic wastewater. *Ceram. Int.* **2015**, *41*, 6702–6709. [CrossRef]
155. Jaiswal, A.; Chattopadhyaya, M.C. Synthesis and characterization of novel Co/Bi-layered double hydroxides and their adsorption performance for lead in aqueous solution. *Arab. J. Chem.* **2017**, *10*, S2457–S2463. [CrossRef]
156. Xie, J.; Lin, Y.; Li, C.; Wu, D.; Kong, H. Removal and recovery of phosphate from water by activated aluminum oxide and lanthanum oxide. *Powder Technol.* **2015**, *269*, 351–357. [CrossRef]
157. Rout, T.K.; Verma, R.; Dennis, R.V.; Banerjee, S. Study the removal of fluoride from aqueous medium by using nanocomposites. *J. Encapsulation Adsorpt. Sci.* **2015**, *5*, 38–52. [CrossRef]
158. Bo, L.; Li, Q.; Wang, Y.; Gao, L.; Hu, X.; Yang, J. One-pot hydrothermal synthesis of thrust spherical Mg–Al layered double hydroxides/MnO$_2$ and adsorption for Pb(II) from aqueous solutions. *J. Environ. Chem. Eng.* **2015**, *3*, 1468–1475. [CrossRef]
159. Mostafa, M.S.; Bakr, A.A.; Eshaq, G.; Kamel, M.M. Novel Co/Mo layered double hydroxide: Synthesis and uptake of Fe(II) from aqueous solutions (Part 1). *Desalin. Water Treat.* **2014**, *56*, 239–247. [CrossRef]
160. Zhang, L.; Zeng, Y.; Cheng, Z. Removal of heavy metal ions using chitosan and modified chitosan: A review. *J. Mol. Liq.* **2016**, *214*, 175–191. [CrossRef]

161. Zhu, Y.; Zhu, R.; Chen, Q.; Laipan, M.; Zhu, J.; Xi, Y.; He, H. Calcined Mg/Al layered double hydroxides as efficient adsorbents for polyhydroxy fullerenes. *Appl. Clay Sci.* **2018**, *151*, 66–72. [CrossRef]
162. Ayawei, N.; Ekubo, A.T.; Wankasi, D.; Dikio, E.D. Synthesis and Application of Layered Double Hydroxide for the removal of Copper in Wastewater. *Int. J. Chem.* **2015**, *7*, 122–132.
163. Zhou, Q.; Zhu, R.; Parker, S.C.; Zhu, J.; He, H.; Molinari, M. Modelling the effects of surfactant loading level on the sorption of organic contaminants on organoclays. *RSC Adv.* **2015**, *5*, 47022. [CrossRef]
164. El Afif, A. Flow and non-Fickian mass transport in immiscible blends of two rheologically different polymers. *Rheol. Acta* **2015**, *54*, 929–940. [CrossRef]
165. Hairch, Y.; El Afif, A. Mesoscopic modeling of mass transport in viscoelastic phase-separated polymeric membranes embedding complex deformable interfaces. *J. Membr. Sci.* **2020**, *596*, 117589. [CrossRef]
166. Akin, O.; Tihminlioglu, F. Effects of organo-modified clay addition and temperature on the water vapor barrier properties of polyhydroxy butyrate homo and copolymer nanocomposite films for packaging applications. *J. Polym. Environ.* **2018**, *26*, 1121–1132. [CrossRef]
167. Barik, S.; Badamali, S.K.; Behera, L.; Jena, P.K. Mg–Al LDH reinforced PMMA nanocomposites: A potential material for packaging industry. *Compos. Interfaces* **2018**, *25*, 369–380. [CrossRef]
168. Singha, S.; Hedenqvist, M.S. A Review on Barrier Properties of Poly (Lactic Acid)/Clay Nanocomposites. *Polymers* **2020**, *12*, 1095. [CrossRef]
169. Wilson, R.; Follain, N.; Tenn, N.; Kumar, A.; Thomas, S.; Marais, S. Tunable water barrier properties of EVA by clay insertion? *Phys. Chem. Chem. Phys.* **2015**, *17*, 19527–19537. [CrossRef]
170. Trifol, J.; Plackett, D.; Szabo, P.; Daugaard, A.E.; Giacinti Baschetti, M. Effect of Crystallinity on Water Vapor Sorption, Diffusion, and Permeation of PLA-Based Nanocomposites. *ACS Omega* **2020**, *5*, 15362–15369. [CrossRef]
171. Smit, W.J.; Tang, F.; Nagata, Y.; Sánchez, M.A.; Hasegawa, T.; Backus, E.H.; Bonn, M.; Bakker, H.J. Observation and identification of a new OH stretch vibrational band at the surface of ice. *J. Phys. Chem. Lett.* **2017**, *8*, 3656–3660. [CrossRef]
172. Lichawska, M.E.; Bodek, K.H.; Jezierska, J.; Kufelnicki, A. Coordinative interaction of microcrystalline chitosan with oxovanadium (IV) ions in aqueous solution. *Chem. Cent. J.* **2014**, *50*, 1–9. [CrossRef] [PubMed]
173. Ivanova, O.P.; Krinichnaya, E.P.; Morozov, P.V.; Zav'yalov, S.A.; Zhuravleva, T.S. The Effect of Filler Content on the IR Spectra of Poly (p-xylylene)–Sulfide Nanocomposites. *Nanotechnol. Russ.* **2019**, *14*, 7–15. [CrossRef]
174. Homaeigohar, S. The nanosized dye adsorbents for water treatment. *Nanomaterials* **2020**, *10*, 295. [CrossRef] [PubMed]

Publisher's Note: MDPI stays neutral with regard to jurisdictional claims in published maps and institutional affiliations.

© 2020 by the authors. Licensee MDPI, Basel, Switzerland. This article is an open access article distributed under the terms and conditions of the Creative Commons Attribution (CC BY) license (http://creativecommons.org/licenses/by/4.0/).

Review

Morphology, Thermal Stability, and Flammability Properties of Polymer-Layered Double Hydroxide (LDH) Nanocomposites: A Review

Mokgaotsa Jonas Mochane [1,*], Sifiso Innocent Magagula [1], Jeremia Shale Sefadi [2,*], Emmanuel Rotimi Sadiku [3] and Teboho Clement Mokhena [3]

[1] Department of Life Sciences, Central University of Technology, Free State, Private Bag X20539, Bloemfontein 9300, South Africa; smagagula@cut.ac.za
[2] Department of Physical and Earth Sciences (PES), Sol Plaatje University, Kimberley 8301, South Africa
[3] Department of Chemical, Institute of NanoEngineering Research (INER), Metallurgical and Materials Engineering, Tshwane University of Technology, RSA, Pretoria 0001, South Africa; SadikuR@tut.ac.za (E.R.S.); mokhenateboho@gmail.com (T.C.M.)
* Correspondence: mochane.jonas@gmail.com or mmochane@cut.ac.za (M.J.M.); jeremia.sefadi@spu.ac.za (J.S.S.)

Received: 15 May 2020; Accepted: 1 July 2020; Published: 14 July 2020

Abstract: The utilization of layered nanofillers in polymer matrix, as reinforcement, has attracted great interest in the 21st century. This can be attributed to the high aspect ratios of the nanofillers and the attendant substantial improvement in different properties (i.e., increased flammability resistance, improved modulus and impact strength, as well as improved barrier properties) of the resultant nanocomposite when compared to the neat polymer matrix. Amongst the well-known layered nanofillers, layered inorganic materials, in the form of LDHs, have been given the most attention. LDH nanofillers have been employed in different polymers due to their flexibility in chemical composition as well as an adjustable charge density, which permits numerous interactions with the host polymer matrices. One of the most important features of LDHs is their ability to act as flame-retardant materials because of their endothermic decomposition. This review paper gives detailed information on the: preparation methods, morphology, flammability, and barrier properties as well as thermal stability of LDH/polymer nanocomposites.

Keywords: nanocomposites; nanofillers; thermal stability; flammability; polymer matrix

---

1. Introduction

Polymer matrices are normally reinforced with inorganic fillers in order to improve their properties and widen their applications [1]. The well-known fillers include silicate, carbon based, calcium carbonate, fibres, etc. It is apparent that the incorporation of filler requires high content in order to have any significant influence of the properties of the polymer matrices. A higher composition of fillers, in most cases, results in increased weight of the resultant composites, which limits the applications of such systems. In order to solve the problem of weight, nanoparticles have recently emerged as the filler of choice to enhance the properties of the resultant polymer matrix. This is due to the ability of nanoparticles to influence the properties of a polymer matrix with considerably low contents, thereby allowing the nanocomposites to maintain low density of the polymer matrix.

The incorporation of layered inorganic fillers into polymer matrices to form polymer/layered inorganic nanocomposites has attracted a lot attention due to their distinctive properties [2]. Layered double hydroxides/polymer nanocomposites belong to an important class of polymer/layered inorganic nanocomposites because they have shown significant improvement in the composites' thermal stability, flame retardancy, and improvement in overall physical properties [3,4]. Due to high pressure that can limit or prohibit the use of halogen flame retardant materials because of environmental concerns, LDHs have emerged as a suitable candidate for halogen-free flame-retardant material [5–8]. It is well documented in the literature that the methods used for the synthesis of LDH include: urea hydrolysis, hydrothermal synthesis, co-precipitation, and ion exchange [9,10].

From a chemistry point of view, the structure of LDH can be presented by the following formula: $[M^{II}_{1-x} M_x^{III} (OH)_2]_{intra} [A_{x/m}^{m-} \cdot nH_2O]_{inter}$.

In the formula, *inter* and *intra* are the intralayer crystalline domain and interlayer spaces, respectively. The layers of the LDHs are positively charged edge-shared octahedral coordinated metal hydroxide crystal structures, sandwiched by charge compensating interlayer anions with optional solvation in water. Furthermore, M^{II} (M^{2+}) is the divalent cation, whereas M^{III} (M^{3+}) is the trivalent cation and A is described as an anion with the valency m. Nevertheless, it is apparent that LDHs consist of high charge density in the interlayer and they seem to have an impenetrable action between the hydroxides when compared with the well-known layered silicates, which makes exfoliation very difficult [2]. Furthermore, the fact that polymers are hydrophobic results in further hindrance of the polymer chains into the LDHs. It is very clear that there is the need to incorporate anionic materials in order to improve the intercalation of polymers into the LDHs layers. The easiest and convenient route for fabrication of polymer/LDHs with improved properties is to modify the clay with surfactant or other materials with the aim of preparing a stable LDHs/polymer nanocomposite system. This review paper discusses the different modifications of LHDs and the preparation of polymer nanocomposites with enhanced properties, i.e., better dispersion, flammability resistance, and thermal stability.

2. History of Layered Double Hydroxides (LDHs)

The existence of LDHs dates back to 1842, where minerals consisting of LDHs were discovered in Sweden. The laboratory synthesis of LDHs began in 1942 and was based on the reaction of dilute metal solutions with bases. Due to their structural similarities to the hydrotalcites, LDHs were referred to as hydrotalcite-like compounds (HTLCs). Hydrotalcites are compounds that exist as hydroxycarbonates of magnesium and aluminium or magnesium and iron (pyroaurite). These hydroxycarbonates are found in nature, in the form of foliated and twisted plates [11]. In the early 1970s, hydrotalcites began to be used as catalysts and precursors of various catalysts. This triggered a lot of interest towards the research of LDHs [11,12]. The first studies on the single crystal X-ray diffraction of minerals revealed that LDHs possessed a layered structure. Each layer consists of two cations and the interlayer space was filled with water and carbonate ions. However, at first, this was debated by several researchers. The reason for this was that even though the main components of the LDH structure had been identified, some researchers still felt that the intrinsic details associated with the structural components of LDHs were not yet understood [13]. LDHs can be formed naturally through natural processes and synthetically in the laboratory. In nature, LDHs are formed naturally by natural processes, such as the weathering of basalt rocks [14,15] and the precipitation [16] of saline water. As mentioned earlier, the structure of LDHs resembles that of naturally occurring hydrotalcites with the formula, $[Mg_6Al_2(OH)_{16}]CO_3 \cdot 4H_2O$ and a general formula, $[M(II)_{1-x} M(III)_x (OH)_2](Y^{n-})_{x/n} \cdot YH_2O$, where M(II) and M(III) are divalent and trivalent metals, respectively; $0.2 < x < 0.33$ and Y^{n-} are exchangeable interlayer anions [17–19]. Synthetically formed LDHs have a highly hydrophilic nature with an amorphous or semicrystalline hexagonal structure. The structure of the LDH layers is based on the brucite compound $[Mg(OH)_2]$. The layers that are adjacent to each other are usually tightly bound together [11]. The structure of LDHs is shown in Figure 1.

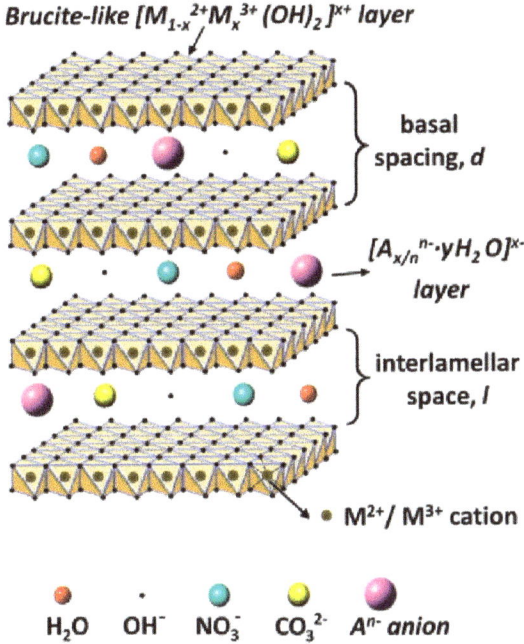

Figure 1. General chemical structure of layered double hydroxides (LDHs) [10].

During the synthesis of LDHs, many different combinations of divalent and trivalent metal cations are used; these include: magnesium, aluminium, zinc, nickel, chromium, iron, copper, indium, gallium, and calcium [17,20–22].

LDHs are usually preferred over clays or other layered materials. This is because the synthesis of LDHs has the potential of forming LDHs with a wide range of compositions and metal-ion combinations. Their preference over clay is also due to the fact that LDHs have high charge density. The charge density of the LDHs is determined by the ratio of the divalent and trivalent metal cations. If the divalent/trivalent ratio is low, the charge density increases [11]. LDHs have unique physical and chemical properties that are closely related to those of clays. The positively charged layered structure of LDHs induces properties, such as anion mobility, surface basicity, and anion exchangeability. The water and the anions found between the layers of LDHs are labile. Therefore, exchange reactions can be used to replace these interlayer anions with various inorganic or organic anions [23]. When LDHs are calcinated, mixed metal oxides with properties, such as large surface area and surface basicity are obtained. At elevated temperatures, the metal oxides formed also form a homogenous mixture with small crystallite sizes [11]. The LDHs and metal oxides formed during calcination also have a high catalytic activity.

LDHs possess a structural reconstruction or memory effect property that is only unique to them. Structural reconstruction is a property that is induced by the calcination of LDHs and the treatment of metal oxides with a specific anionic solution [24]. These materials can easily adsorb anions and cations [25,26]. The magnetic properties of LDHs are usually regulated by the chemical nature of the interlayer spaces in them. The chemical environment in these spaces can be modified by the intercalation with organic anions of different chain lengths. This results in hybrid materials with tunable magnetic properties [27]. The intercalation of LDHs with long-chain surfactants, e.g., dodecyl sulphates, forms hybrid materials that swell in organic solvents. This swelling characteristic is usually exploited in the fabrication of monolayers used in nanohybrid and nanocomposite synthesis [28].

The chemical environment in between the layers of LDHs is usually altered by the exchange of anions. The exchange of anions follows the following order of preference:

$$NO_3^- < Br^- < Cl^- < F^- < OH^- < SO_4^{2-} < CO_3^{2-}$$

Here, the NO_3^- anion can be easily replaced by the CO_3^{2-} anion. Therefore, in the preparation of a precursor for interaction, the NO_3^- anion is preferred over the CO_3^{2-} anion. This is because the interaction has to happen in such a manner that the introduction of the guest molecule does not change the structure of the host. During this interaction, the existing ion is replaced by the guest molecule. The anions that are weakly bonded to the hydroxide layers are usually the most vulnerable for replacement by other ions [29–31].

Many different methods are used for the synthesis of LDHs. The type of method used depends on the required characteristics and applications of the resultant material. The most commonly used methods/techniques are methods such asco-precipitation, hydrothermal synthesis, urea hydrolysis, sol-gel, ion-exchange, and rehydration. There also are other methods often employed, such as self-oxide method, template synthesis method, and surface synthesis method. The co-precipitation method is one of the commonly used methods. In this method, the LDH structure is formed by the mixing of aqueous solutions of M(II), M(III), and interlayer anions. This method gives the liberty to prepare LDHs that consist of a wide range of anions and cations. The co-precipitation method is uniquely used to prepare organic-anion LDHs [13]. The co-precipitation method can be further subdivided into three other methods, viz co-precipitation by filtration, co-precipitation at lower supersaturation, and co-precipitation at higher supersaturation methods.

The hydrothermal synthetic method is usually employed to regulate particle size and particle size distribution. The hydrothermal synthesis method follows two synthesis routes. The first route is where the materials are prepared at temperatures above 373 K in a pressured autoclave. Here, the LDHs are synthesized from MgO and Al_2O_3 precursors or from mixtures formed through the decomposition of the precursor nitrate compounds [11,32,33]. In the other synthesis route, LDHs are prepared at low temperatures and are also subjected to a process of aging. During the aging process, the LDH precipitate is refluxed at a specific temperature for 18 h.

In the urea hydrolysis method, urea is used as a precipitation agent in the synthesis of LDHs, at specific temperatures. The degree of crystallinity of LDHs depends on the synthesis temperature and decomposition rate. At low temperatures, large particles are formed due to the slow nucleation and slow decomposition rates of the urea [34,35]. The urea hydrolysis method is uniquely used in the synthesis of LDHs with a high charge density [34].

In the sol-gel method, LDHs are produced by first forming a sol via the hydrolysis and partial condensation of a metallic precursor, followed by the formation of a gel. Here, the hydrolysis and condensation rates of the metallic precursors determine the properties of the resultant LDHs [36]. The condensation and hydrolysis rates of the metallic precursors are also susceptible to modification by various reaction parameters, such as pH, type and concentration of the precursor, synthesis temperature, and solvent used. The sol-gel method forms LDHs with a larger surface area than those formed by the co-precipitation method [37–39]. However, properties such as basicity as well as the divalent and trivalent metal ion molar ratios of LDHs synthesized with the sol-gel method are still not understood [38–40].

The ion-exchange method involves the exchange of interlayer anions with other guest anions introduced into the LDH structure in order to obtain the LDH-guest compound. Several factors, such as affinity towards the guest anions, the medium of exchange, pH, and the chemical nature of the brucite layers affect the ion-exchange in LDHs [41].

In the rehydration method, the mixed metal oxides formed after the calcination of LDHs at high temperatures between 500–800 °C are rehydrated and formed into an LDH structure in the presence of the desired anions [42–47].

3. Selective Polymer Matrices for Fabrication LDHs

Polymer matrices are frequently reinforced by different nanofillers to reduce the limitations of polymers and widen their applications. LDHs nanofillers have been incorporated into polymer matrices to improve the mechanical properties, flammability resistance, and the overall physical properties of polymers. Polymer matrices in nanofiller polymer nanocomposites are used to provide shape and durability in the nanocomposites.

Different polymer matrices have been used for fabrication of polymer-LDHs nanocomposites. The polymer matrices include polymethyl methacrylate (PMMA), epoxy, polylactic acid (PLA), polyvinyl chloride (PVC), polypropylene (PP), poly(ethylene terephthalate)(PET), polystyrene (PS), polyaniline, low-density polyethylene (LDPE), acrylonitrile-butadiene-styrene (ABS), ethylene vinyl acetate (EVA), linear low density polyethylene (LLDPE), cellulose, and poly(caprolactone) [48–60]. Based on the polymer matrices mentioned above, it is clear that thermoplastic and thermosets were used as host matrices for preparation of LDHs-nanocomposites. However, it is apparent that thermoplastic matrices were preferred over the thermosets because the former are lightweight, can be re-melted, and shaped. Due to an environmental protection, there is a slight shift towards fabrication of biopolymer matrix/LDHs nanocomposites. The resultant nanocomposite is termed "green nanocomposite", and these nanocomposites are favourites to replace petroleum-based plastics. Table 1 summarizes selective polymer matrices for preparation of LDHs nanocomposites.

Table 1. LDHs nanocomposites with selective polymer matrices.

Nanocomposites	Type of Polymer (Thermoset or Thermoplastic)	Typical Example of Nanocomposite	References
Polyaniline (PANI)/LDHs	Thermoset	PANI/Mg–Al-LDH	[48]
Polyaniline (PAn)/LDHs	Thermoset	PAn/(3:1; Zn/Al-LDHs)	[49]
Polypropylene (PP)/LDHs	Thermoplastic, polyolefin	PP/MgAl-layered double hydroxides	[50]
PLA/LDHs	Thermoplastic, polyester	PLA/(Mg-Al-LDH-C_{12}) and PLA/Mg- Al -LDH-CO_3	[51]
Epoxy (EP)/LDHs	Thermoset	EP/Zn-Al-CO_3 -HA LDH and EP/Mg-Al-CO_3-HA LDH, HA = Hydroxyapatite	[52]
Ethylene propylene diene (EPDM)/LDH	Thermoplastic, elastomer	EPDM/Cu–Al –LDHs	[53]
LLDPE/LDH	Thermoplastic	LLDPE/Zn Al-LDH	[54]
LDPE/LDHs	Thermoplastic	LDPE with (i) Mn_2Al-LDH-stearate and (ii) Co_2Al-LDH-stearate	[55]
poly(3-hydroxybutyrate-co-3-hydroxyvalerate) (PHBV)	Thermoplastic, polyester	PHBV/Mg- Al layered double hydroxide	[56]
Thermoplastic Polyurethane	Thermoplastic	TPU/CoAl-LDH and TPU/APP@ Co Al –LDH APP = ammonium polyphosphate	[57]
Poly(ε-caprolactone) (PCL)/LDH	Thermoplastic, polyester	PCL/Zn Al -LDH	[58]
Highly amorphous vinyl alcOHol (HAVOH)/LDH	Thermoplastic	HAVOH/Zn Al -LDH-CNTs CNTs = carbon nanotubes	[59]
Polybutylene succinate (PBS)/LDH	Thermoplastic, polyester	PBS/Mg Al-LDH	[60]

4. Preparation and Morphology of Polymer-LDHs Nanocomposites

In most studies, it was explained that dispersion of the filler/nanofiller within a polymer matrix plays a very important role in the properties of the fabricated polymer nanocomposite. In an LDHs/polymer system, the LDHs are normally modified with organic anions in order to enhance their interactions with polymer matrices. The main reason for improving the interaction between the two components, i.e., polymer and LDHs, is because the polymer is hydrophobic and LDHs is hydrophilic [61]. Furthermore, the method of preparation was also found to have a huge impact on the dispersion of the LDHs within the polymer matrix. The literature has shown that in most cases, LDHs are fabricated with different metals, depending on the desired applications. Leng and co-workers investigated the structure-property relationship of composites formed from polylactic acid (PLA) and layered double hydroxides and the comparison of MgAl and NiAl LDH as nanofillers [62]. The metal (s) LDHs were organically modified with sodium dodecyl benzene sulfonate (SDBS). LDHs and PLA were prepared by a melt mixing process at a temperature of 190 °C (463 K). The morphological comparison was based on a 3 wt.% content of LDH. Large particles with an estimated size of 100 nm with less agglomerates were obtained for MgAl LDH/PLA (Figure 2A). According to the authors, this arrangement will allow partial exfoliation of the metal Al/LDH layers. However, in the system fabricated from NiAl/LDH-PLA, more aggregates were reported. The structures obtained are more favourable to intercalation than exfoliation. With the findings obtained above, it can be summarized that the system of NiAl/LDH-PLA favoured intercalation, whereas the MgAl/LDH-PLA nanocomposites is a typical exfoliated system.

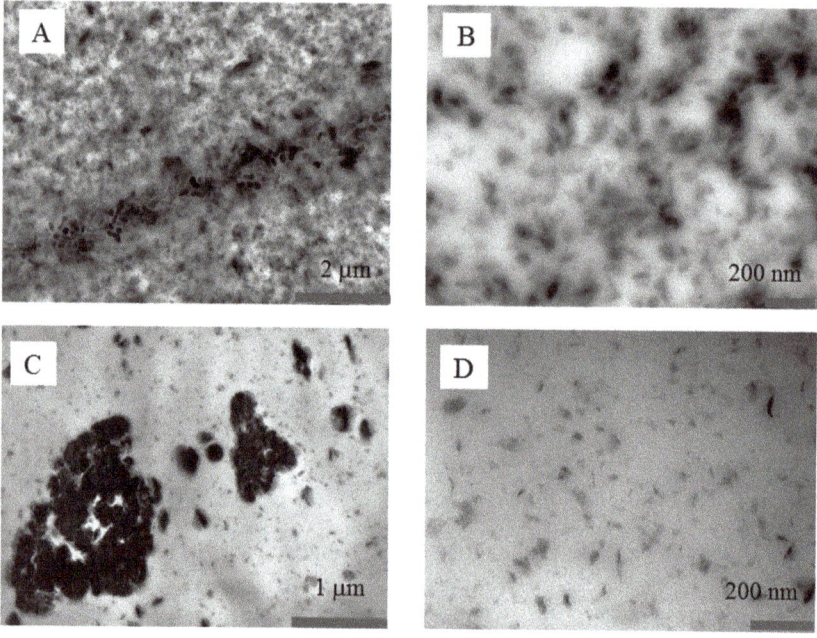

Figure 2. TEM picture for nanocomposites with an LDH concentration of 3 wt.%. (**A**) MgAl/LDH–PLA size bare 2 µm; (**B**) MgAl/LDH–PLA size bare 200 nm; (**C**) NiAl/LDH–PLA size bare 1 µm; (**D**) NiAl/LDH–PLA size bare 200 nm [62].

Quispe-Dominguez et al. [63] investigated two types of mixing methods in order to compare which method provides better dispersion of MgAl-DBS LDH in a PLA matrix. LDH was modified with sodium dodecylbenzene sulfonate (DBS). The methods employed in this study include: (i) sonication-assisted masterbatch melt mixing and (ii) direct melt mixing. Sonication-assisted masterbatch melt mixing was undertaken by dissolving PLA and MgAl-DBS in methylene chloride as a solvent. After 12 h and at a temperature of 80 °C, the solvent was evaporated, with the masterbatch formed, processed at 170 °C in a twin-screw compounder for 10 min. However, for direct mixing, there was no need for solvent mixing; PLA and MgAl-DBS were melt-mixed directly at 170 °C for 10 min. It was observed that the sonication-assisted masterbatch melt mixing method resulted in better dispersion, intercalation, and exfoliation of LDH when compared to direct mixing for all investigated compositions. The particle size of the metal (s)-LDHs nanofiller has been proven to have an influence on the dispersion of the nanofiller within a polymer matrix [64]. It was reported that the fabrication of LDH with gel resulted in large particle sizes (of between 3–4 µm), whereas the preparation of LDH nanoparticles through sonication produced smaller nanoparticles (of between 50–200 nm) (Figure 3). The nanocomposites with compositions of between 1–10 wt.% of Mg–Al LDH were prepared by the modified solution mixing method. The authors reported better dispersion for smaller sonicated LDH nanoparticles within isotactic polypropylene (iPP) matrix, as confirmed by wide angle x-ray diffraction (WAXD) and atomic force microscopy. Table 2 illustrates the summary of some selective studies on the preparation and morphology of polymer-LDHs nanocomposites.

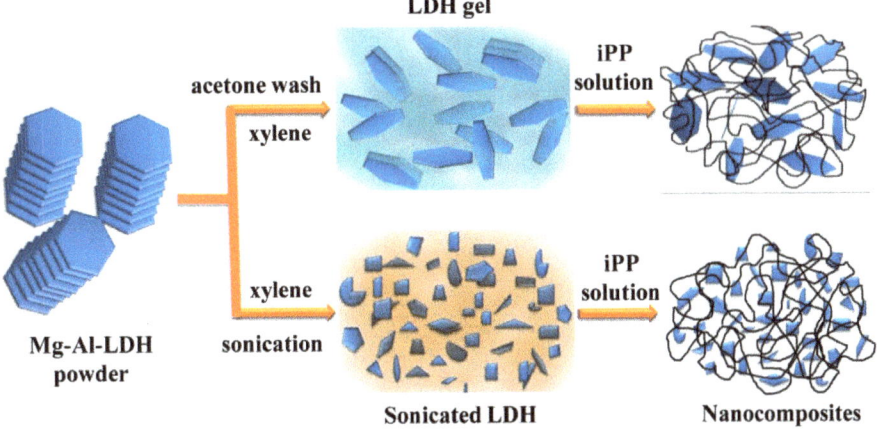

Figure 3. Graphic representation of the preparation of isotactic polypropylene/Mg–Al LDH layered double hydroxide nanocomposites [64].

Table 2. Selective studies on preparation and morphology of polymer-LDHs nanocomposites.

Polymer/LDHs System	Synthesis of the Metal-LDH Nanofiller	Preparation Method of the Nanocomposites	Summary of the Resultant Morphology	References
Polyurethane (Pu)/CoAl-LDH	Urea hydrolysis	In-situ intercalation polymerization	The exfoliation of the nanofiller within a matrix was reported.	[65]
Functionalized Poly (vinyl chloride) (PVC)/Mg-Al LDH	Co-precipitation method	Solution intercalation method	Four different nanocomposites were prepared depending on the chemical functionalizing of PVC: (i) PVC/Mg-Al LDH (ii) PVC+thiosulfate)/Mg-Al LDH (iii) (PVC+sulfate)/Mg-Al LDH (iv) (PVC+thiourea)/Mg-Al LDH. Amongst all formed nanocomposites, more exfoliated structures were observed for (PVC+thiourea)/Mg-Al.	[66]
Polyacrylonitrile (PAN)/Zn-Al LDH	Co-precipitation	In-situ polymerisation technique	Different LDH content (viz 2, 4, 6, 8%) were synthesized with PAN. Transmission electron microscopy (TEM) showed disordered dispersion of nanofiller in the PAN matrix. However, at higher content of the nanofiller (viz 8%), more agglomerates were obtained.	[67]
Poly(methyl methacrylate)/Mg–Al LDH (PMMA/LDH)	Co-precipitation	In-situ polymerisation	The PMMA nanocomposites consisting of 2, 4, 6, 8% composition of LDH was prepared by in situ polymerisation. The authors reported a random dispersion of the nanofiller with a polymer matrix. There was an observation of exfoliation of the nanofiller with partial intercalation at higher content of the nanofiller.	[68]
Poly (ethylene terephthalate) (PET)/CaAl-LDH and MgAl-LDH	The urea-assisted hydrothermal method was used for preparation of MgAl-LDH, while the co-precipitation method was used for fabrication of CaAl-LDH. The co-precipitation method was employed for MgAl-LDH with stearic acid (MgAl-LDH-SA)	Solution blending method	Scanning electron microscopy (SEM) and TEM showed homogenous distribution of both CaAl-LDH and MgAl-LDH within a PET matrix.	[69]
Linear Low Density Polyethylene/ZnAl-PDP LDH	High-energy ball milling	Melt blending and blowing	The nanofiller in this study were modified by Potassium dodecyl phosphate (PDP). SEM pictures showed uniform dispersion of the nanofiller into the polymer matrix. The results were supported by XRD, which indicated no diffraction peak for LDH.	[70]
Poly(methyl methacrylate)/Co–Al LDH (PMMA/LDH)	Instinctive self-assembly approach	Solvent blending technique	The PMMA nanocomposites were prepared by the solvent blending technique. The authors reported a wide dispersion of the nanofiller within the PMMA matrix. There was an observation of exfoliation of the nanofiller with partial intercalation at higher content of the nanofiller.	[71]
Linear Low Density Polyethylene (LLDPE)/LDH	Anion exchange method	Solution intercalation method	The nanofiller in this study was modified by dodecyl sulfate (DS). TEM results showed uniform dispersion of the nanofiller into the polymer matrix. Moreover, XRD and TEM results showed the formation of a mixture of intercalated-exfoliated structures in the LLDPE/LDH composites.	[72]

5. Thermal Stability of Polymer/LDHs Nanocomposites

Different studies have investigated the thermal stability of LDH/polymer nanocomposites [63,65,73–84]. Various thermal stability results (decrease or increase in the thermal stability of the nanocomposites) were recorded, depending on the LDH/polymer nanocomposites system. Lee et al. [84] reported an increase in the thermal stability of a composite that had the addition of the nanofiller into an ethylene vinyl acetate (EVA) matrix. In this study, an LDH nanofiller was modified with anionic surfactant, such as sodium dodecyl sulfate (DS), sodium dodecylbenzene sulfonate (DBS), and stearate (SA). When comparing the thermal stability (at 50% weight loss) of the three modifications, 6 phr of DS-LDH/EVA nanocomposites showed 19 °C increment, while both the 6 phr of DBS-LDH/EVA and SA-LDH/EVA nanocomposites recorded 12 °C increase from the pristine EVA (Figure 4). However, Quispe-Dominguez et al. [63] reported a decrease in thermal stability with the incorporation of magnesium-aluminium layered double hydroxides modified with dodecylbenzene sulfonate (DBS). This was attributed to the catalytic effect of the modified nanofiller in the polymer matrix and as a result, reducing the thermal stability of the overall nanocomposites. Due to the two different mixing methods (viz: sonication-assisted masterbatch melt mixing and direct mixing) employed in this study for preparation of the nanocomposites, differences in thermal stabilities were observed, depending on the mixing method employed. It was reported that as much as the thermal stability decreases with the addition and increases in nanofiller content, the nanocomposites prepared by masterbatch melt mixing exhibited enhancement in thermal stability when compared with direct mixing, especially when considering same nanofiller content. For example, at 1.25% of the nanofiller, masterbatch melt mixing showed higher thermal stability than direct mixing at the same content. This behaviour was ascribed to a finer dispersion and better intercalation of the nanofiller during masterbatch melt mixing. Better dispersion and intercalation emphasize the fact that the PLA chains were more intercalated in the LDHs layers, in case of the PLA nanocomposites prepared by the SMA melt mixing method, resulting in an enhancement of thermal stability than the poorly dispersed and agglomerated nanofiller in case of the direct mixing method. Zoromba et al. [83] modified both copper-aluminium LDH and nickel-aluminium LDH with sodium stearate modifier and melt-mixed them with PP in an extruder. The authors reported better improvements in the thermal stability of the modified nickel-aluminium LDH when compared with neat PP, the unmodified nickel-aluminium LDH/PP, and copper-aluminium LDH/PP nanocomposites. The improvement was attributed to better interfacial interaction between the nanofiller and the polymer matrix.

In this study, PMMA nanocomposites were prepared by solution mixing by adding flame resistant materials, such as intumescent flame retardant (IFR) (i) 1,2-Bis(5,5-dimethyl-1,3,2 -dioxyphospacyclOHexane phosphoryl amide) ethane (BPEA), (ii) graphene (reduced graphene oxide), and (iii) magnesium aluminium-layered double hydroxide modified with sodium dodecyl sulfate. It was reported that in the absence of IFR (BPEA), there was an increase in the thermal stability of PMMA nanocomposites with reduced graphene oxide (rGO), LDH, and LDH+graphene as nanofillers when compared with pure PMMA. The synergistic effect of LDH and LDH+graphene showed more delay in the thermal decomposition of PMMA in comparison to PMMA/LDH and PMMA/graphene alone. This is an indication that the synergistic effect of nanofillers can form a better protective heat barrier and therefore delay the decomposition of the polymer, improving the thermal stability. Interestingly, it became apparent from the same study that the synergy of LDH and graphene further enhanced the thermal stability of PMMA/IFR composites. On the contrary, the synergistic effect of two LDH nanofillers i.e., MgAl-layered double hydroxides (MgAl-LDH) and NiCo-layered double hydroxides (NiCo-LDH) exhibited a decrease in the thermal decomposition of the nanocomposites, in comparison to neat epoxy (EP) at the initial decomposition temperature ($T_{5\%}$) [85]. This was ascribed to the catalytic effect of the metals, which speed-up the polymer degradation. However, as much as the onset of degradation temperature decreased with the addition of the nanolayers, a combination of 2.5% of MgAl@NiCo exhibited a slightly higher thermal stability when compared with 2.5% of MgAl in the epoxy matrix. Therefore, a general remark can be made such that the synergy of LDH and nanofillers tend to produce a more compact char content than sole LDH in the polymer matrix.

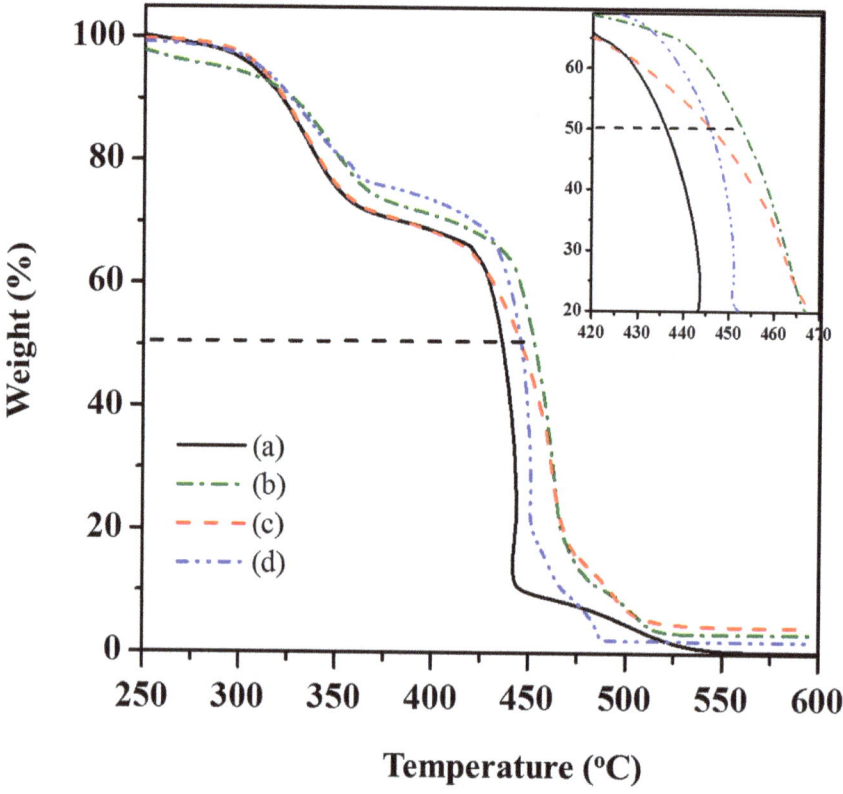

Figure 4. Thermogravimetric analysis curves of (a) neat EVA, (b) 6-phr DS-LDH/EVA, (c) DBS-LDH/EVA, and (d) SA-LDH/EVA. The zoomed area corresponds to the TGA curves at temperatures from 420 to 470 °C [84].

6. Flammability Properties of Polymer/LDHs Nanocomposites

Layered double hydroxide nanofillers have proven to be good flame-retardant materials for protection of polymers against heat [86]. A summarized version of the flame-retardant mechanism of LDH occurs through the endothermic process with the generation of water and metal oxide char. A number of authors have investigated the flammability properties of LDH/polymer and/or polymer/LDHs+another nanofiller in order to enhance the flammability resistance of the nanocomposites [85–89]. In most of these studies, the flammability properties of the nanocomposites were investigated by cone calorimetry and limiting oxygen index methods. Cone calorimetry parameters include: heat release rate (HRR), total heat release (THR), time to ignition (TTI), mass loss rate (MLR), and smoke production rate (SPR). Amongst the cone calorimetry parameters, the HRR has emerged as an important parameter since it measures the intensity of fire. A decrease in HRR peak symbolizes an improvement in flammability resistance of the material under investigation. Furthermore, an increase in limiting oxygen index (LOI) values is an indication of improvement in flammability of the nanocomposites. It was reported that the synergistic effect of LDHs with another nanofiller and/or an intumescent flame-retardant material significantly improved the flammability resistance of the polymer matrix when compared with LDH alone. Li et al. [87] investigated the flammability properties of the three LDHs (viz: MgFe-LDHs, MgAl-LDHs and MgAlFe-LDHs) incorporated into an EVA matrix. In this study, the LDHs nanofillers were produced from bittern by the co-precipitation method. Generally, there was a decrease in the heat release rate (HRR) of the EVA matrix with the addition of the LDHs nanofillers. The HRR peak of pristine EVA was recorded as 1645.8 kW/m^2. It became apparent that with the addition of the three LDHs, the HRR peaks were found to have reduced, i.e., EVA 1 (MgAl-LDHs) recorded a value of 222.65 kW/m^2, while EVA 2 (MgFe-LDHs) and EVA 3 (MgAlFe-LDHs) showed values of 311.87 and 286.96 kW/m^2, respectively. It can be concluded that the reduction in HRR is attributed to the formation of char residues in the presence of nanofillers, which acted as a protective barrier against heat. Since ≥90% of the layered double hydroxides must be modified in order to improve the dispersion of the nanoplatelets within the polymer matrix, it became apparent that the surfactants played a key role in the flammability properties of the resultant nanocomposite. Qiu et al. reported on the effect of surfactant (viz: sodium dodecyl sulfate (DDS) and stearic) on the flammability of PP and Mg$_3$Al LDHs nanocomposites [88]. Mg$_3$AlLDHs were prepared with surfactants i.e., sodium dodecyl sulfate (DBS) and stearic by employing the aqueous miscible organic solvent modification technique, whereas the nanocomposites were prepared by the solvent mixing technique. The addition of both sodium dodecyl sulfate and stearic modified-LDH resulted in a decrease in the peak heat release rate (PHRR), with the stearic-based LDH nanofiller showing moderate improvement in the flame resistance of PP than the sodium dodecyl sulfate (DDS) modified-LDH. At both 20 wt.% of stearic modified-LDH and sodium dodecyl sulfate modified-LDH, the pHRR decreased by 61% and 58%, respectively, in comparison to the neat PP. However, as much as there is an improvement in the flammability resistance of the LDHs/nanocomposites, there is a need for the formation of strong char layers during the process of burning. This can be achieved by combining two or more nanofillers with LDHs. Another study [85] reported on the flammability properties of the synergistic effect between NiCo-LDH and MgAl-LDH incorporated in an epoxy matrix. ZIF-67, a type of metal organic framework (MOF), was used as a precursor to tie up the Co^{2+} on the layered double hydroxide in order to convert more of it into NiCo-LDH platelets. Various weight percentages (wt.%) of MgAl@NiCO (viz 2, 2.5 and 3) as well as MgAl-LDH (2.5 and 3 wt.%.) were mixed with the epoxy matrix. MgAl@ZIF-67 was used as the benchmark to compare its flammability with neat EP, EP/2.5% MgAl, and EP/2.5% MgAl@NiCo. The addition of 2.5% MgAl reduced the PHRR when compared to neat EP, with the peak decreasing further with the addition of EP/2.5% MgAl@NiCo (Figure 5). The reduction of PHRR in the presence of 2.5% MgAl@NiCo was attributed to the formation of more compact char than the MgAl, which can delay the entrance of heat and oxygen into the system and as a result, enhance the flammability resistance.

It was further proven that the addition of three metal LDHs improved the flammability resistance of the polymer matrix more than two metal LDHs [89]. The authors investigated the flammability properties of neat iPP, 6% Co-Al LDH, 6% Zn-AlLDH, 6% Co-Zn-AlLDH, and 10% Co-Zn-Al LDH polypropylene nanocomposites. The three metal LDH nanocomposites showed better reduction in flammability than the two metal LDH nanocomposites, at the same content. The behaviour was attributed to the effective char formation for the 6% Co-Zn-Al LDH when compared with 6% Co-Al LDH and 6% Zn-Al LDH. This thermally stable char can prevent volatile gases/products from escaping out of the system and acting as a protective heat barrier, thereby preventing heat from entering the system, as a result improving the flammability resistance of the overall system. It is well known that a protective char will become fragile when more gaseous volatile products escape the system, which will provide easier passage of oxygen and heat into the material. Hence, it is important for flame retardant materials to form a stable and compact char in order to prevent the escape of volatiles gases and entering of heat into the system. Table 3 summarizes selective studies on the flammability of the polymer/LDH system.

Table 3. Selective studies on the flammability properties of a polymer/LDHs system.

Polymer/LDHs System	Synthesis of the Metal-LDH Nanofiller	Preparation Method of the Nanocomposites	Summary of the Flammability Results	References
Polypropylene (PP)/(AMO-LDHs) and (O-CNT), AMO = Aqueous miscible organic o = oxidized	Hydrothermal method	Solution mixing	It was reported that the hybrid mixture of AMO-LDH-OCNT improved the flammability resistance of PP better than AMO-LDHs alone. A synergy fabricated from 10 wt.% AMO-LDH+1 wt.% oCNT showed 40% reduction in peak heat release rate (pHRR) when compared with 20 wt.% AMO-LDH (31% reduction in pHRR).	[90]
Polypropylene (PP)/LDHs and PP/M-LDHs, M-LDHs = Mg-Al-H$_2$PO$_4$ LDHs	An ion exchange method was used for fabrication of dihydrogen phosphate intercalation (Mg-Al- H$_2$PO$_4^-$)	Melt mixing by using an extruder	The addition of LDH (Mg$_2$-Al- CO$_3^{-2}$ LDH) reduced the pHRR from 1032 kW/m^2 (pristine PP) to 837 kW/m^2 with the pHRR decreasing (534 kW/m^2) further with the addition M- LDHs (Mg-Al-H$_2$PO$_4^-$ LDHs).	[81]
Polypropylene (PP)/APP-LDHs and ZB, APP = Ammonium polyphosphate, ZB = Zinc borate	Solvent treatment (Aqueous miscible organic) was used for treatment of LDHs. Mg$_3$Al-APP LDH and Mg$_3$Al-CO$_3$ LDH were synthesized by the hydrothermal method	Solvent mixing	The addition of 10 and 20 wt.% of APP-LDH resulted in an improvement in flammability resistance in comparison to carbonate LDH (Mg$_3$Al-CO$_3$ LDH). The synergistic effect of ZB and APP-LDH improved the flammability resistance further in comparison to APP-LDH alone.	[91]
Thermoplastic polyurethane/LDHs-graphene oxide (TPU/LDHs-GOs)	NO$_3^-$-LDHs-GO was fabricated by the co-precipitation method	Melt mixing method	The investigated samples include: TPU (100%), TPU/NO$_3^-$-LDHs-GO (80/20 mass %), TPU/SDS-LDHs (80/20 mass %), TPU/SDS-LDHs-1%GO (80/20 mass %), TPU/SDS-LDHs-3%GO (80/20 mass %) and TPU/SDS-LDHs-5%GO (80/20 mass %). The addition of the nanofillers into the TPU matrix reduced the peak heat release rate (PHRR), which is an indication of flammability resistance. The LDHs nanofiller modified with sodium dodecyl sulfate (SD) showed better flammability resistance properties when compared with NO$_3^-$-LDHs. This was attributed to a better exfoliation of the SDS-modified nanofiller, which provided a better chance of a char formation. The synergy between GO and LDHs improved the flammability resistance more than NO$_3^-$-LDHs-Go and SDS-LDHs samples, with TPU/SDS-LDHs-5%GO (80/20 mass %) showing more reduction in PHRR. This is due to the ability of Go to be an effective flame-retardant material.	[92]
Acrylonitrile-Butadiene-Styrene/(CaMgAl -Layered Double Hydroxides) (CaMgAl-LDHs)	CaMgAl-LDHs was fabricated by Co-precipitation method. Borated CaMgAl-LDHs was prepared by dissolving CaMgAl-LDHs into boric acid solution. In order to form O-CaMgAl-LDHs, B-CaMgAl-LDHs was dissolved in sodium oleate solution.	Melt blending by two-roll mix	The addition of 10, 20, 30 and 40% of O-CaMgAl-LDHs increased with the addition and increasing in o-CaMgAl-LDHs content. The synergistic of O-CaMgAl-LDHs, ammonium polyphosphate (APP) and graphite (EG) showed higher values than O-CaMgAl-LDHs.	[93]

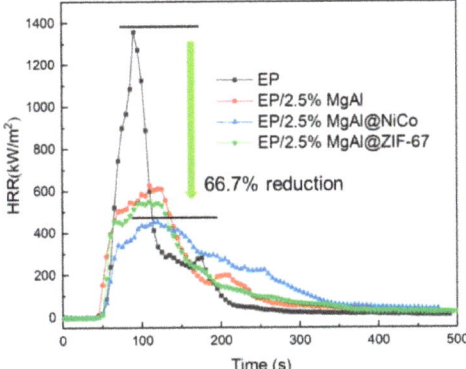

Figure 5. Heat release rate (HRR) of neat, EP, EP/2.5% MgAl, EP/2.5% MgAl@NiCo, EP/2.5% MgAl@ZIF-67 [85].

7. Barrier Properties of Layered Double Hydroxides (LDH)

Layered double hydroxides (LDHs) are usually combined with polymer matrices to form films with good gas barrier properties, for applications in food packaging and encapsulation of electronic devices [94,95]. LDHs consist of crystalline layered structures [96–99] that induce gas barrier properties by increasing the diffusion length of gases, therefore resisting permeating gases in the resultant polymer film. The development of organic-inorganic composite materials with gas barrier properties, e.g., LDH/polymer films, still faces a few challenges. One of them is that the diffusion of gas molecules is only suppressed in the direction of the film, i.e., the gas molecules flowing parallel to the inorganic structural layers are not resisted [100–102]. In such a case, the gas barrier properties of the film are usually improved by increasing the content of the crystalline layered structures. However, the loading with these inorganic crystalline structures reduces the flexibility and toughness of the barrier materials. Another challenge is that the incorporation of highly oriented inorganic structural layers into polymer matrices also causes polymer aggregation. This aggregation creates voids through which gas molecules permeate easily. This results in a film with compromised gas barrier properties [103]. In a quest to curb the challenges faced by the development of LDH-based gas barrier materials, considerable efforts have been directed towards developing the structure of LDHs. Dou et al. [104] discovered that the incorporation of plate-like LDH (P-LDH) into a polymer matrix, e.g., chitosan (CTS), improved the oxygen barrier properties of the resultant films. However, the challenge was that the oxygen barrier properties were not that good for very thin films. Hence, in another study [95], the authors converted the LDH with a plate-like structure to one with a hierarchical structure (H-LDH). The oxygen transmission rate of the resultant H-LDH/CTS films was reduced by almost 37% when compared to that of the original P-LDH/CTS films, indicating that the structural conversion of the LDHs improved their gas barrier properties. The synthesis of H-LDH was performed via the continuous calcination-rehydration treatment of P-LDH. The resultant H-LDH was then used as a scaffolding material for the fabrication of chitosan multi-layered films via an alternate spin-coating process. This process led to the formation of (H-LDH/CTS)$_n$ films with excellent oxygen barrier properties. The (H-LDH/CTS)$_n$ films exhibited an oxygen transmission rate (OTR) that was below the detection limit of commercial instruments (<0.005 cm^3/m^2 day atm). This was attributed to the capabilities of H-LDH to resist the migration of oxygen molecules from multiple directions by creating a longer diffusion pathway. Large amounts of oxygen molecules were also absorbed by the large surface area of the H-LDH. The large surface area of the H-LDH filled-in all the gaps between H-LDH and the polymer matrix, thus closing-up the space for oxygen permeation [95]. During the development of LDH/polymer gas barrier films, improving the durability of the films is quite important, especially for applications such as food packaging and encapsulation of electronic devices. In order to achieve this, LDH/polymer films with self-healing properties are designed and fabricated. This results not only in a film with excellent gas barrier properties, but also one with the ability to repair itself after damage by external stimuli. In a study by Dou et al. [94], (LDH/PSS)$_n$-PVA films with self-healing properties triggered by humidity were fabricated via the layer-by-layer assembling of layered double hydroxide nanoplatelets and poly(sodium styrene-4-sulfonate) (PSS), followed by the subsequent incorporation of poly(vinyl alcohol) (PVA). The even distribution of the highly oriented LDH nanoplatelets in the film was responsible for the resistance of permeating gases by creating a long diffusion pathway. The PVA was responsible for the humidity-stimulated self-healing properties of the films. When the films were exposed to humidity after the development of the stimuli crack, the water molecules triggered the formation of hydrogen bonds among the hydroxyl groups of PVA, thus causing the stimuli crack to close. Hydrogen is usually used as an alternate fuel to fossil fuels because its combustion only produces water, which implies less air pollution. However, amongst many other methods of storing hydrogen, LDHs have been considered as the best substances for the storage of hydrogen gas. This is achieved through the conversion of LDHs into microporous materials through intercalation with other anionic substances. Huang and Cheng [105] intercalated Li-Al layered double hydroxides with various organic anions via a co-precipitation method. The maximum hydrogen absorption per micropore surface area of the LDHs prepared in this study was higher than that of metal organic frameworks (MOFs) reported in the literature, hence confirming the microporous nature of the prepared LDHs. Table 4 summarizes selective studies on the barrier properties of polymer-LDHs nanocomposites.

Table 4. Selective studies for the preparation and barrier properties of polymer-LDHs nanocomposites.

Polymer/LDHs System	Synthesis of the Metal-LDH Nanofillers	Preparation Method of the Nanocomposites	Summary of the Resultant Barrier Properties	References
Cellulose nanofibrils (CNFs)/MgAl-CO$_3$-LDHs	LDHs were synthesized via the hydrothermal method	CNFs/MgAl-CO$_3$-LDHs with different LDH ratios were prepared by a filtering/evaporation process that induced barrier and strengthening properties in the composite films.	The resultant composite films exhibited improved gas-barrier properties. The water vapour penetration of the films decreased significantly at low MgAl-CO$_3$-LDHs concentrations of: 5 wt.% and 10 wt.%. At 5 wt.% MgAl-CO$_3$-LDHs, the water vapour penetration was reduced by 50 % as compared to the pure CNFs film. When the concentration was increased to 10 wt.%, the water vapour penetration decreased further, reaching a low of 1927 g/m^2·24 h, which was the lowest content of penetrated water vapour compared to all the MgAl-CO$_3$-LDHs loadings.	[106]
Nitrile butadiene rubber (NBR)/polyvinyl pyrrolidone modified ultrathin LDH nanoplatelets (U-mLDHs).	The U-mLDHs nanoplatelets were prepared by a slightly improved co-precipitation method.	The NBR/U-mLDH composites were prepared by a layer-by-layer spin-coating assembly technique.	The oxygen transmission rate (OTR) of the films with a higher aspect ratio, (U-mLDH/NBR)$_{30}$, was reduced by 92.2% compared with the pure NBR film. The improved gas barrier properties were due to the decreased diffusion pathway of the oxygen molecules. The free space between the U-mLDH and NBR was due to the large aspect ratio of the U-mLDH and the improved interfacial adhesion at the LDH-polymer interface.	[107]
Linear low density polyethylene (LLDPE)/LDH composite films.	LDH intercalated with an aliphatic long-chain anion was prepared by a single pot high-energy balling method.	The films were prepared by melt blending and blow processing.	The water vapour barrier properties of the LLDPE composite films with 1% LDH were enhanced by 60.36%. This was attributed to the LDH inducing a longer diffusion pathway for the water molecules.	[70]
Poly(vinyl alcohol)(PVA)/hybrid layered double hydroxides (LDHs)-reduced graphene oxide (rGO) (LDH-rGO).	MgAl-LDH-rGO hybrids were prepared by the co-precipitation method.	The PVA/LDH-rGO hybrid films were prepared by the solution casting method.	The oxygen transmission rate (OTR) of PVA/LDH-rGO films was decreased by 86% at 1% LDH-rGO loading. The improved barrier properties were attributed to the uniformly dispersed LDH-rGO hybrids in PVA.	[108]

8. Mechanical Properties of Polymer-LDHs Systems

Mechanical properties are used to determine the strength and ductility of polymer nanocomposites. The mechanical properties of LDH-based polymer nanocomposites were investigated by different studies [109–114]. Numerous factors were found to affect the mechanical properties of LDHs nanocomposites, including the content of LDHs, the type of LDHs, polymer matrix, dispersion of LDH in polymer matrix, and the preparation method. Botan et al. investigated the mechanical properties of polyamide 6 incorporated with two types of LDHs (viz Zn/Cr-L and Zn/Cr-P, with Zn = Zinc, Cr = Chromium, L = lauric acid and P = palmitic acid) [109]. The LDHs nanocomposites were fabricated by in situ polymerization with various composition of the filler (viz 1, 2, and 3 wt.%). Mechanical properties of nanocomposites were investigated with tensile tester. The modulus of elasticity (E) decreased at lower content of Zn/Cr-L i.e., 1 wt.%, with the E values increasing at higher compositions of Zn/Cr-L (viz 2, and 3 wt.%). The decrease in E values at lower content was attributed to the plasticizing effect due to the absorption of the water at lower content. Similarly, the addition of Zn/Cr-P increased the E values of the polyamide 6 nanocomposites in all investigated filler composition. The optimum composition for both fillers was obtained at 2 wt.%, with the Zn/Cr-P nanofiller showing higher E values when compared with Zn/Cr-L counterpart. This was attributed to the bilayer structures, which allowed an efficient stress-transfer. The effect of stearate intercalated LDH on the properties of PU was investigated in the literature [110]. The polyurethane (PU)/Stearate-intercalated LDH was fabricated by solution intercalation. There was an enhancement in tensile strength (TS) with the incorporation of stearate-LDH (viz 1, 3, 5, and 8 wt.%) into PU matrix, when compared with the neat PU. The tensile strength was observed to increase with decreasing in stearate-LDH nanoparticles content. A higher tensile strength (TS) at 1 wt.% of stearate-LDH was ascribed to a better exfoliation at this content. There was a reported increase in elongation at break with addition of stearate-LDH nanoparticles into the PU. The behaviour was ascribed to the plasticization of the alkyl chain intercalated into the LDH in the PU/stearate-LDH system. Feng and co-workers investigated the properties of the LDHs reinforced peroxide-cured acrylonitrile butadiene rubber [111]. The LDHs were organically modified with sodium dodecylbenzene (SDBS) and sodium styrene sulfonate (SSS). The LDH modified with sodium styrene sulfonate (SSS) composites showed better mechanical properties than neat acrylonitrile butadiene rubber (NBR) and LDH modified with sodium dodecylbenzene composites. This behaviour was attributed to a better chemical bond between the organically modified LDHs sodium styrene sulfonate (SSS) and NBR. Based on the above study, it became apparent that the type of organic modifier may influence the overall properties of the LDHs-polymer nanocomposites. Suresh and co-workers [112] investigated the Co-Al layered double hydroxide reinforced polystyrene nanocomposites. In this study, Co-Al layered double hydroxide was organically modified with sodium dodecyl sulfate (SDS) and the composites were prepared by melt compounding. It was observed that at 1 wt.% of Co-Al LDH nanocomposite, there is an enhancement in both tensile and tensile modulus when compared with neat PS and 3, 5 as well as 7 wt.% Co- Al LDH nanocomposites, respectively (Figure 6). At lower content, i.e., 1 wt.%, there is a better exfoliation of the nanoparticles in the PS matrix as well as enhanced interfacial interaction between the two phases, which resulted in better mechanical properties. At higher content of the nanofiller, there is a probability of agglomerated Co-Al LDH, which formed defects, and as a result lowering the mechanical properties. Table 5 summarizes selective studies on the mechanical properties of LDH-polymer nanocomposites.

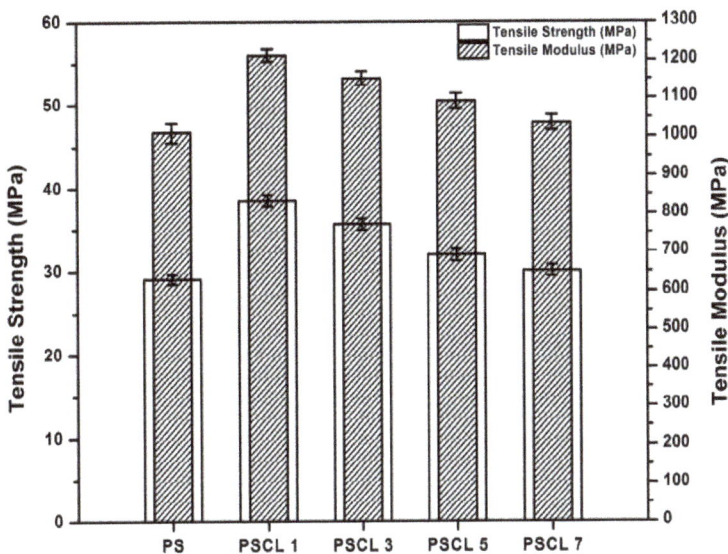

Figure 6. Tensile strength and tensile modulus of polystyrene and its Co-Al layered double hydroxide nanocomposites [112].

Table 5. Selective studies on mechanical properties of LDH-polymer nanocomposites.

Polymer-LDHs Nanocomposites	Preparation of Polymer-LDHs Nanocomposites	Summary of the Mechanical Properties	References
Epoxy (EP)/Mg-Al LDH intercalated with ammonium alcOHol polyvinyl phosphate (AAPP)	Solution intercalation	The addition of 10 and 20 mass% of intercalated LDHs into EP showed higher tensile strength than neat EP.	[113]
Poly(ε-caprolactone) (PCL)/Silver-LDH (Ag-LDH). LDH = Mg-Al LDH Type of silver-LDHs used are: Ag-LDHs@PDA PDA = polydopamine Ag-LDHs@TA-Fe (III) TA = tannic acid Fe (III) = Iron (III) Ag-LDHs (PVP) PVP = pyrrolidone	Solution casting method	It was reported that when the composition of Ag-LDHs was 0.5 wt.%, the tensile strength of the Ag-LDHs@TA-Fe(III)/PCL system decreased by 11%, while the LDHs@PDA/PCL nanocomposite reduced by 4% when compared with neat PCL. The 0.5 wt.% of Ag-LDHs(PVP) showed 26% reduction in tensile strength.	[114]
Aromatic Polyimide (PI)/Zn/Cr-LDH Zn = Zinc, Cr = Chromium	In situ polymerization	The study investigated 1, 2, and 4% of LDH incorporated into PI. The 2% of LDH showed higher tensile strength value when compared with neat PI, 1% LDH/PI and 4% LDH/PI nanocomposites. This was attributed to a better dispersion LDH in a polymer matrix at low content. However, at higher content, i.e., 4 wt.%, there was formation of an aggregate, which resulted in defect in the nanocomposite.	[6]
Cellulose nanofibrils (CNF)/Mg Al-CO$_3$-LDHs	Filtering/evaporation method	According to the study, 5 wt.% of the LDH showed higher tensile strength in comparison to neat CNF, 10, 15 and 25% of LDHs. Higher tensile strength at low content was ascribed to a better dispersion of the nanofiller in a matrix.	[106]
Poly(ethylene-co-vinyl alcOHol) (EVAL)/LDHs, LDH was organically modified with: Stearate (SA), to form SA-LDH	Melt compounding	Mechanical properties of EVAL/LDH composites were compared with neat EVAL. The authors reported that the charpy notched impact strength of the composites was twice that of the neat EVAL polymer. This was attributed to the extensive internal micro-cavitation of the highly dispersed and randomly dispersed LDH platelets during impact loading. The large surface area created by the micro-cavitation enhanced the requisite energy dissipation mechanism.	[115]

9. Selective Applications of Different Polymer-LDHs Systems

Layered double hydroxides nanocomposites have been prepared for various applications including energy, food packaging, water purification, gas sensing, biomedical, flame retardant, and agricultural applications [68,116–119]. Qin and co-workers investigated the LDHs nanocomposites based on PANI/ZnTI-LDHs for sensing ammonia gas (NH_3) [116]. The LDHs (ZnTi-LDHs) nanofiller was prepared by hydrothermal method and the nanocomposites were fabricated by in situ chemical oxidative polymerization. The ability of the LDH nanocomposites in sensing NH_3 was compared with neat PANI and ZnTi-LDHs. The LDHs nanocomposites were found to exhibit a significant NH_3 sensing ability with good lengthy stability when compared with neat polymer and LDHs, respectively. The results were ascribed to a more-loose architecture structure of the nanocomposite, which improves the adsorption site as well as facilitation of the gas adsorption. The system consisting of sulfonated polyaniline (SPAN) reinforced with graphene oxide (GO)-LDHs was investigated for extraction phthalates in drinking water and distilled herbal beverages [117]. The study compared the extraction efficiency of GO-LDH@SPAN nanocomposite with LDH and GO-LDH for extraction of phthalates from aqueous solution. The extraction efficiency of the LDH@SPAN nanocomposite for phthalate was higher than neat LDH and GO-LDH, with LDH showing lower extraction than GO-LDH. The reason LDH showed lower extraction for phthalates was due to a possible weak hydrogen bond between LDH and phthalates. The higher extraction efficiency of phthalates by GO-LDH was attributed to a π-π bond between GO and aromatic ring of phthalates. The GO-LDH@SPAN nanocomposite showed higher extraction of the analyte because SPAN can promote more π-π interaction with the phthalate. Furthermore, as SPAN has O=S=O in the matrix, it presents more active sites for extraction of the analytes. PMMA/Mg- Al LDH nanocomposites was fabricated by in situ polymerization for a possible packaging application [68]. Two key results were significant in determining a possible packaging application, i.e., thermal stability and gas permeability. The fabricated nanocomposites were reinforced with 2, 4, and 8% of Mg-Al LDH. There was a decrease in the oxygen flow rate of the nanocomposites in comparison to neat PMMA. This behavior was attributed to a dispersion of LDH in a polymer matrix, which acts as protective barrier for oxygen permeability; as for neat PMMA, the presence of voids resulted in oxygen penetration within the matrix. The addition of the LDH into the PMMA enhanced the thermal stability of PMMA matrix when compared with neat PMMA. The reduction in oxygen permeability and the enhancement in thermal stability of the PMMA/Mg-Al LDH system suggest that the nanocomposite may be suitable for packaging application.

In recent times, the demand for highly flexible, durable, and lightweight piezoelectric nanogenerators has led to the fabrication of piezoelectric and dielectric electrospun nanofabrics of poly (vinylidene fluoride) (PVDF)/Ca-Al LDH composites. During the fabrication of the PVDF/Ca-Al LDH composite nanofabrics, the Ca Al- LDH nanosheets were first synthesized via a modified coprecipitation method before they were incorporated as filler into the PVDF matrix. The composite nanofabrics of PVDF/Ca-Al LDH were finally obtained via the electrospinning of the composite solutions. The synergy between the PVDF-LDH interaction and the in situ stretching, which was attributed to the electrospinning, enhanced the nucleation of the electroactive β phase up to 82.79%. This was an indication that these composite nanofabrics are suitable for piezoelectric-based nanogenerators. The hand slapping and frequency-dependent mechanical vibration mode methods showed that the piezoelectric performance of the PVDF/Ca-Al LDH composite nanofabrics can reach a maximum open circuit output voltage of 4.1 and 5.72 V. The composite nanofabrics also had a high dielectric constant and a low dielectric loss, which were attributed to high interfacial polarization at low frequencies with increasing LDH loading. This showed that these materials have the potential to be used in electronic devices [118].

Wang et al. [119] prepared Zn Al -LDH/polycaprolactone (PCL) nanocomposites for use in drug delivery systems. The Zn Al-LDH was synthesized via the co-precipitation method while the ZnAl -LDH/PCL nanocomposites were prepared by the solution intercalation method. The ZnAl-LDH/PCL composite nanofabrics exhibited a higher weight loss and drug release amount when compared to neat

PCL. The drug release kinetics followed the first-order kinetic model for the Zn Al-LDH/PCL composites, indicating that the drug release was content-dependent. However, for the neat PCL, the drug release kinetics followed the Ritger-Peppas kinetic model, indicating that the release followed the Fikian mechanism. Table 6 summarizes selective studies on the applications of LDH-polymer nanocomposites.

Table 6. Summary of selective studies on the application of LDH-polymer nanocomposites.

Polymer-LDH System	Preparation Method for the Nanocomposites	Intended Application	References
Poly(lactide-co-glycolic acid) (PLGA)/Mg Al-LDH	Solution mixing and casting into thin films	Drug delivery applications	[120]
Polypropylene-grafted maleic anhydride (PP-g-MA)/Dye structure-intercalated layered double hydroxide (d-LDH)	co-precipitation	Flame retardant applications	[121]
Waterborne polyurethane (WPU)/LDH LDH = $(M_xAl/CO_3^{2-}$, M = Mg and/or Zn, and x = 2, 3 and 4)	Solution mixing and casting into thin films	Coatings applications	[122]
Chitosan (CS)/NiFe-LDH	Low-saturation co-precipitation method.	Catalytic applications	[123]
Polylactic acid (PLA)/Intumescent flame retardant (IFR)/Phosphotungstic acid intercalated Mg Al-LDH (PWA-LDH)	Melt blending and hot pressing into films	Flame retardant applications	[124]
CoNiMn-LDH/Polypyrrole (PPy)/Reduced graphene oxide (RGo)	One-step route in which the co-precipitation reaction of metal ions (Co^{2+}, Ni^{2+} and Mn^{2+}) was used to prepare LDH and the polymerization of pyrrole (Py) was used to prepare PPy. Modified Hummer's method was used to prepare graphene oxide	Electrocatalytic applications	[125]
Isotactic polypropylene (iPP)/ZnAl-LDH	Solvent mixing method	Flame retardant applications	[89]
Polystyrene (PS)/MgAl-LDH	Solution mixing	Removal of Cd^{2+} ions from aqueous media	[126]
Poly(vinyl chloride) (PVC)/MgAl LDH	Solution intercalation method	Biomedical applications	[127]

10. Conclusions and Future Recommendations

Layered double hydroxides (LDHs) have been the nanofiller of choice in terms of improving the flame retardancy and barrier properties of polymer matrices. However, the improvement in properties of the LDHs/polymer systems depended on the dispersion of the nanofiller within the polymer matrices. It is apparent that LDHs nanofillers in most cases has been organically modified in order to improve the exfoliation of the nanofillers in polymer matrices. The dispersion of LDHs/polymer system varied depending on the method of preparation, type of modifier, and the type of LDHs/polymer system. Generally, there was an improvement in the flammability resistance of polymer matrices with the addition of LDHs with different compositions. The flammability resistance improved more in the presence of LDHs and other flame-retardant fillers. The synergy of LDHs and nanofillers improved the flame retardancy more because it is able to form a strong and compact char layer, which can inhibit the entry of heat into the system, thereby improving the overall flammability resistance of the system. The gas barrier properties of LDH/polymer films were dependent on the orientation and distribution of the LDH nanoplatelets. Highly oriented and evenly distributed LDH nanoplatelets created a resistance for the migration of gas molecules from multiple directions by increasing the diffusion pathway of the permeating gases. The intercalation of LDHs with organic anions altered their gas barrier properties by converting the LDHs into microporous materials. However, although LDH improved the properties of various polymer matrices, there is still a huge gap in terms of the thermal conductivity of the LDHs-polymer nanocomposites. In the past, LDHs nanofillers were fabricated with conductivity polymers such as polyaniline and polypyrrole; however, there are a few studies investigating the thermal conductivity of LDHs in combination with well-known conductive fillers such as expanded graphite, carbon nanotubes, carbon black, and carbon fiber to widen the applications of LDHs nanocomposites. The fabricated hybrid LDH/conductive filler/polymer nanocomposite may be used in applications such as flame retardant, supercapacitors, and batteries. Furthermore, there is also less investigation on the effect of LDHs in natural fiber reinforced biopolymer composite to form a "green" LDH/natural fiber/biopolymer hybrid composite. The resultant hybrid material (viz LDH/natural fiber/biopolymer hybrid composite) may exhibit enhanced mechanical properties, thermal stability, and flammability resistance and can be used in applications such as environmental protection and flame retardancy.

Author Contributions: M.J.M. and S.I.M. co-designed and guided the review as well as co-writing Sections 1–3, 6 and 7 of the article; J.S.S. and E.R.S. co-wrote Sections 4, 5 and 8; T.C.M. and M.J.M. co-wrote Sections 9 and 10; while M.J.M. and S.I.M. compiled the article together. All authors have read and agreed to the published version of the manuscript.

Funding: This research was funded by the National Research Foundation (NRF) of South Africa, grant number (s) 127278 and 114270.

Acknowledgments: The National Research Foundation (NRF) of South Africa is acknowledged for financial support.

Conflicts of Interest: The authors declare no conflict of interest.

References

1. Mittal, V. Polymer layered silicate nanocomposites: A review. *Materials* **2009**, *2*, 992–1057. [CrossRef]
2. Huang, G.; Zhuo, A.; Wang, L.; Wang, X. Preparation and flammability properties of intumescent flame retardant-functionalized layered double hydroxides/polymethyl methacrylate nanocomposites. *Mater. Chem. Phys.* **2011**, *130*, 714–720. [CrossRef]
3. Pereira, C.M.C.; Herrero, M.; Labajos, F.M.; Marques, A.T.; Rives, V. Preparation and properties of new flame retardant unsaturated polyester nanocomposites based on layered double hydroxides. *Polym. Degrad. Stab.* **2009**, *94*, 939–946. [CrossRef]
4. Du, L.; Qu, B.; Zhang, M. Thermal properties and combustion characterization of nylon 6/MgAl-LDH nanocomposites via organic modification and melt intercalation. *Polym. Degrad. Stab.* **2007**, *92*, 497–502. [CrossRef]
5. Benaddi, H.; Benachour, D.; GrOHens, Y. Preparation and characterization of polystyrene-MgAl layered double hydroxide nanocomposite using bulk polymerization. *J. Polym. Eng.* **2015**, *36*, 681–693. [CrossRef]
6. Dinari, M.; Rajabi, A.R. Structural, thermal and mechanical properties of polymer nanocomposites based on organosoluble polyimide with naphthyl pendent group and layered double hydroxide. *High Perform. Polym.* **2016**, *29*, 951–959. [CrossRef]
7. Elbasuney, S. Surface engineering of layered double hydroxide (LDH) nanoparticles for polymer flame retardancy. *Powder Technol.* **2015**, *277*, 63–73. [CrossRef]
8. Kaul, P.K.; Samson, A.J.; Selvan, G.T.; Enoch, I.V.M.V.; Selvakumar, P.M. Synergistic effect of LDH in the presence of organophosphate on thermal and flammable properties of an epoxy nanocomposite. *Appl. Clay Sci.* **2017**, *135*, 234–243. [CrossRef]
9. Manzi-Nshuti, C.; Songtipya, P.; Manias, E.; Jimenez-Gasco, M.M.; Hossenlopp, J.M.; Wilkie, C.A. Polymer nanocomposites using aluminium and magnesium aluminium oleate layered double hydroxides: Effects of LDH divalent metals on dispersion, thermal, mechanical and fire performance in various polymers. *Polymer* **2009**, *50*, 3564–3574. [CrossRef]
10. Scarpellini, D.; Falconi, C.; Gaudio, P.; Mattoccia, A.; Medaglia, P.G.; Orsini, A.; Pizzoferrato, R.; Richetta, M. Morphology of Zn/Al layered double hydroxidenanosheets grown onto aluminum thin films. *Microelectron. Eng.* **2014**, *126*, 129–133. [CrossRef]
11. Cavani, F.; Trifiro, F.; Vaccari, A. Hydrotalcite-type anionic clays: Preparation, properties and applications. *Catal. Today* **1991**, *11*, 173–301. [CrossRef]
12. Bröcker, F.J.; Kainer, L. German Patent, 2024282, 1970; UK Patent 1342020, 1971.
13. Evans, D.G.; Slade, R.C.T. Structural aspects of layered double hydroxides. In *Layered Double Hydroxides*; Springer: Berlin/Heidelberg, Germany, 2006; pp. 1–87.

14. Zamarreno, I.; Plana, F.; Vazquez, A.; Clague, D.A. Motukoreaite: A common alteration product in submarine basalts. *Am. Mineral.* **1989**, *74*, 1054–1058.
15. Drits, V.A.; Lisitsyna, N.A.; Cherkashin, V.I. New mineral varieties from the hydrotalcite-manas seite group as products of the low-temperature transformations of basalts and volcanogenic sedimentary-rocks of the ocean bottom. *Dokl. Akad. Nauk SSSR* **1985**, *284*, 443–447.
16. Halcom, Y.F.M. Layered Double Hydroxides: Morphology, Interlayer Anion and the Origins of Life. Ph.D. Thesis, University of North Texas, Denton, TX, USA, 2003.
17. Rives, V. Characterisation of layered double hydroxides and their decomposition products. *Mater. Chem. Phys.* **2002**, *75*, 19–25. [CrossRef]
18. Bouzaid, J.M.; Frost, R.L.; Martens, W.N. Thermal decomposition of the composite hydrotalcites of iowaite and woodallite. *J. Therm. Anal. Calorim.* **2007**, *89*, 511–519. [CrossRef]
19. Palmer, S.J.; Spratt, H.J.; Frost, R.L. Thermal decomposition of hydrotalcites with variable cationic ratios. *J. Therm. Anal. Calorim.* **2009**, *95*, 123–129. [CrossRef]
20. Tong, D.S.; Zhou, C.H.C.; Li, M.Y.; Yu, W.H.; Beltramini, J.; Lin, C.X.; Xu, Z.P.G. Structure and catalytic properties of Sn-containing layered double hydroxides synthesized in the presence of dodecylsulfate and dodecylamine. *Appl. Clay Sci.* **2010**, *48*, 569–574. [CrossRef]
21. Kloprogge, J.T.; Wharton, D.; Hickey, L.; Frost, R.L. Infrared and Raman study of interlayer anions CO_3^{2-}, NO^{3-}, SO_4^{2-} and ClO^{4-} in Mg/Al-hydrotalcite. *Am. Mineral.* **2002**, *87*, 623–629. [CrossRef]
22. Tao, Q.; Reddy, B.J.; He, H.; Frost, R.L.; Yuan, P.; Zhu, J. Synthesis and infrared spectroscopic characterization of selected layered double hydroxides containing divalent Ni and Co. *Mater. Chem. Phys.* **2008**, *112*, 869–875. [CrossRef]
23. Khan, A.I.; O'Hare, D. Intercalation chemistry of layered double hydroxides: Recent developments and applications. *J. Mater. Chem.* **2002**, *12*, 3191–3198. [CrossRef]
24. De Roy, A.; Forano, C.; El Malki, K.; Besse, J.-P. Anionic clays: Trends in pillaring chemistry. In *Expanded Clays and Other Microporous Solids*; Springer: Boston, MA, USA, 1992; pp. 108–169.
25. Brindley, G.W.; Kikkawa, S. Thermal behavior of hydrotalcite and of anion-exchanged forms of hydrotalcite. *Clay Clay Miner.* **1980**, *28*, 87–91. [CrossRef]
26. Komarneni, S.; Kozai, N.; Roy, R. Novel function for anionic clays: Selective transition metal cation uptake by diadochy. *J. Mater. Chem.* **1998**, *8*, 1329–1331. [CrossRef]
27. Bujoli-Doeuff, M.; Force, L.; Gadet, V.; Verdaguer, M.; El Malki, K.; De Roy, A.; Besse, J.P.; Renard, J.P. A new two-dimensional approach to molecular-based magnets: Nickel (II)-chromium (III) double hydroxide systems. *Mater. Res. Bull.* **1991**, *26*, 577–587. [CrossRef]
28. Adachi-Pagano, M.; Forano, C.; Besse, J.-P. Delamination of layered double hydroxides by use of surfactants. *Chem. Commun.* **2000**, 91–92. [CrossRef]
29. Choy, J.H.; Kewk, S.Y.; Jeong, Y.J.; Park, J.S. Inorganic layered double hydroxides as nonviral vectors. *Angew. Chem. Int. Ed.* **2000**, *39*, 4041. [CrossRef]
30. Ambrogi, V.; Fardella, G.; Grandolini, G.; Perioli, L. Intercalation compounds of hydrotalcite-like anionic clays with antiinflammatory agents—I. Intercalation and in vitro release of ibuprofen. *Int. J. Pharm.* **2001**, *220*, 23–32. [CrossRef]
31. Tyner, K.M.; Roberson, M.S.; Berghorn, K.A.; Li, L.; Gilmour, R.F., Jr.; Batt, C.A.; Giannelis, E.P. Intercalation, delivery, and expression of the gene encoding green fluorescence protein utilizing nanobiOHybrids. *J. Control. Release* **2004**, *100*, 399–409. [CrossRef]
32. Sharma, S.K.; Kushwaha, P.K.; Srivastava, V.K.; Bhatt, S.D.; Jasra, R.V. Effect of hydrothermal conditions on structural and textural properties of synthetic hydrotalcites of varying Mg/Al ratio. *Ind. Eng. Chem. Res.* **2007**, *46*, 4856–4865. [CrossRef]
33. Renaudin, G.; Francois, M.; Evrard, O. Order and disorder in the lamellar hydrated tetracalcium monocarboaluminate compound. *Cem. Concr. Res.* **1999**, *29*, 63–69. [CrossRef]
34. Costantino, U.; Marmottini, F.; Nocchetti, M.; Vivani, R. New Synthetic Routes to Hydrotalcite-Like Compounds—Characterisation and Properties of the obtained Materials. *Eur. J. Inorg. Chem.* **1998**, *1998*, 1439–1446. [CrossRef]
35. Ogawa, M.; Kaiho, H. Homogeneous precipitation of uniform hydrotalcite particles. *Langmuir* **2002**, *18*, 4240–4242. [CrossRef]

36. Livage, J.; Henry, M.; Sanchez, C. Sol-gel chemistry of transition metal oxides. *Prog. Solid State Chem.* **1988**, *18*, 259–341. [CrossRef]
37. Prinetto, F.; Ghiotti, G.; Graffin, P.; Tichit, D. Synthesis and characterization of sol–gel Mg/Al and Ni/Al layered double hydroxides and comparison with co-precipitated samples. *Micropor. Mesopor. Mater.* **2000**, *39*, 229–247. [CrossRef]
38. Aramendía, M.A.; Borau, V.; Jiménez, C.; Marinas, J.M.; Ruiz, J.R.; Urbano, F.J. Comparative study of Mg/M (III)(M = Al, Ga, In) layered double hydroxides obtained by coprecipitation and the sol–gel method. *J. Solid State Chem.* **2002**, *168*, 156–161. [CrossRef]
39. Jitianu, M.; Bǎlǎsoiu, M.; Zaharescu, M.; Jitianu, A.; Ivanov, A. Comparative study of sol-gel and coprecipitated Ni-Al hydrotalcites. *J. Sol-Gel Sci. Technol.* **2000**, *19*, 453–457. [CrossRef]
40. Lopez, T.; Bosch, P.; Ramos, E.; Gomez, R.; Novaro, O.; Acosta, D.; Figueras, F. Synthesis and Characterization of Sol–Gel Hydrotalcites. Structure and Texture. *Langmuir* **1996**, *12*, 189–192. [CrossRef]
41. Duan, X.; Evans, D.G. *Layered Double Hydroxides*; Springer Science & Business Media: Berlin/Heidelberg, Germany, 2006; Volume 119.
42. Carlino, S.; Hudson, M.J. A thermal decomposition study on the intercalation of tris-(oxalato) ferrate (III) trihydrate into a layered (Mg/Al) double hydroxide. *Solid State Ion.* **1998**, *110*, 153–161. [CrossRef]
43. Ukrainczyk, L.; Chibwe, M.; Pinnavaia, T.J.; Boyd, S.A. ESR study of cobalt (II) tetrakis (N-methyl-4-pyridiniumyl) porphyrin and cobalt (II) tetrasulfophthalocyanine intercalated in layered aluminosilicates and a layered double hydroxide. *J. Phys. Chem.* **1994**, *98*, 2668–2676. [CrossRef]
44. Kooli, F.; Depege, C.; Ennaqadi, A.; De Roy, A.; Besse, J.P. Rehydration of Zn-Al layered double hydroxides. *Clay Clay Miner.* **1997**, *45*, 92–98. [CrossRef]
45. Tagaya, H.; Sato, S.; Morioka, H.; Kadokawa, J.; Karasu, M.; Chiba, K. Preferential intercalation of isomers of naphthalenecarboxylate ions into the interlayer of layered double hydroxides. *Chem. Mater.* **1993**, *5*, 1431–1433. [CrossRef]
46. Chibwe, M.; Pinnavaia, T.J. Stabilization of a cobalt (II) phthalocyanine oxidation catalyst by intercalation in a layered double hydroxide host. *J. Chem. Soc. Chem. Commun.* **1993**, 278–280. [CrossRef]
47. Conterosito, E.; Palin, L.; Antonioli, D.; Riccardi, M.P.; Boccaleri, E.; Aceto, M.; Milanesio, M.; Gianotti, V. On the Rehydration of organic Layered Double Hydroxides to form Low-ordered Carbon/LDH Nanocomposites. *Inorganics* **2018**, *6*, 79. [CrossRef]
48. Zhu, K.; Gao, Y.; Tan, X.; Chen, C. Polyaniline-modified Mg/Al layered double hydroxide composites and their application in efficient removal of Cr (VI). *ACS Sustain. Chem.* **2016**, *4*, 4361–4369. [CrossRef]
49. Youssef, A.M.; Moustafa, H.A.; Barhoum, A.; Abdel Hakim, A.E.-F.A.; Dufresne, A. Evaluation of the morphological, electrical and antibacterial properties of polyaniline nanocomposite based on Zn/Al-layered double hydroxides. *ChemistrySelect* **2017**, *2*, 8553–8566. [CrossRef]
50. Zheng, Y.; Chen, Y. Preparation of polypropylene/Mg- Allayered double hydroxides nanocomposites through wet pan-milling: Formation of a second-staging structure in LDHs intercalates. *RSC Adv.* **2017**, *7*, 1520–1530. [CrossRef]
51. Gerds, N.; Katiyar, V.; Koch, C.B.; Hansen, H.C.B.; Plackett, D.; Larsen, E.H.; Risbo, J. Degradation of L-polylactide during melt processing with layered double hydroxides. *Polym. Degrad. Stabil.* **2012**, *97*, 2002–2009. [CrossRef]
52. Karami, Z.; Ganjali, M.R.; Dehaghani, M.Z.; Aghazadeh, M.; Jouyandeh, M.; Esmaeili, A.; Habibzadeh, S.; MOHaddespour, A.; Inamuddin, F.K.; Haponiuk, J.Z.; et al. Kinetics of cross-linking reaction of epoxy resin with hydroxyapatite-functionalized layered double hydroxides. *Polymers* **2020**, *12*, 1157. [CrossRef]
53. Zhao, W.; Du, Y.C.; Sun, Y.B.; Wang, J.C. Study on preparation of layered double hydroxides (LDHs) and properties of EPDM/LDHs composites. *Plast. Rubber Compos.* **2014**, *43*, 192–201. [CrossRef]
54. Chen, W.; Qu, B. LLDPE/ZnAl LDH-exfoliated nanocomposites: Effects of nanolayers on thermal and mechanical properties. *J. Mater. Chem.* **2004**, *14*, 1705–1710. [CrossRef]
55. Magagula, B.; Nhlapho, N.; Focke, W.W. Mn_2Al-LDH- and Co_2Al-LDH-stearate as photodegradants for LDPE film. *Polym. Degrad. Stabil.* **2009**, *94*, 947–954. [CrossRef]
56. Ciou, C.-Y.; Li, S.-Y.; Wu, T.-M. Morphology and degradation behavior of poly(3-hydroxybutyrate-co-3-hydroxyvalerate)/layered double hydroxides composites. *Eur. Polym. J.* **2014**, *59*, 136–143. [CrossRef]

57. Huang, S.-C.; Deng, C.; Wang, S.-X.; Wei, W.-C.; Chen, H.; Wang, Y.-Z. Electrostatic action induced interfacial accumulation of layered double hydroxides towards highly efficient flame retardance and mechanical enhancement of thermoplastic polyurethane/ammonium polyphosphate. *Polym. Degrad. Stabil.* **2019**, *165*, 126–136. [CrossRef]
58. Costantino, U.; Bugatti, V.; Gorrasi, G.; Montanari, F.; Nocchetti, M.; Tammaro, L.; Vittoria, V. New Polymeric Composites Based on Poly (ϵ-caprolactone) and Layered Double Hydroxides Containing Antimicrobial Species. *ACS Appl. Mater. Interfaces* **2009**, *1*, 668–677. [CrossRef]
59. Bugatti, V.; Viscusi, G.; Di Bartolomeo, A.; Iemmo, L.; Zampino, D.C.; Vittoria, V.; Gorrasi, G. Ionic liquid as dispersing agent of LDH-carbon nanotubes into a biodegradable vinyl alcOHol polymer. *Polymers* **2020**, *12*, 495. [CrossRef]
60. Marek, A.A.; Verney, V.; Totaro, G.; Sisti, L.; Celli, A.; Cionci, N.B.; Gioia, D.D.; Massacrier, L.; Leroux, F. Organo-modified LDH fillers endowing multi-functionality to bio-based poly (butylene succinate): An extended study from the laboratory to possible market. *Appl. Clay Sci.* **2020**, *188*, 105502. [CrossRef]
61. Youssef, A.M.; Bujdosó, T.; Hornok, V.; Papp, S.; Dékány, I. Structural and thermal properties of polystyrene nanocomposites containing hydrophilic and hydrophobic layered double hydroxides. *Appl. Clay Sci.* **2013**, *77*, 46–51. [CrossRef]
62. Leng, J.; Kang, N.; De-Yi, W.; Falkenhagen, J.; Thünemann, A.F.; Schönhals, A. Structure-property relationship of nanocomposites based on polylactide and layered double hydroxides-comparison of MgAl and NiAl LDH as nanofiller. *Macromol. Chem. Phys.* **2017**, *218*, 1700232. [CrossRef]
63. Quispe-Dominguez, R.; Naseem, S.; Leuteritz, A.; Kuehnert, I. Synthesis and characterization of MgAl-DBS LDH/PLA composite by sonication-assisted masterbatch (SAM) melt mixing method. *RSC Adv.* **2019**, *9*, 658–667. [CrossRef]
64. Nagendra, B.; MOHan, K.; Gowd, E.B. Polypropylene/layered double hydroxide (LDH) nanocomposites: Influence of LDH particle size on the crystallization behaviour of polypropylene. *ACS Appl. Mater. Interfaces* **2015**, *7*, 12399–12410. [CrossRef]
65. Guo, S.; Zhang, C.; Peng, H.; Wang, W.; Liu, T. Structural characterization, thermal and mechanical properties of polyurethane/CoAl layered double hydroxide nanocomposites prepared via in situ polymerization. *Compos. Sci. Technol.* **2011**, *71*, 791–796. [CrossRef]
66. Singh, M.; Somvanshi, D.; Singh, R.K.; Mahanta, A.K.; Maiti, P.; Misra, N.; Paik, P. Functionalized polyvinyl chloride/layered double hydroxide nanocomposites and its thermal and mechanical properties. *J. Appl. Polym. Sci.* **2020**, *137*, 48894. [CrossRef]
67. Barik, S.; Behera, L.; Badamali, S.K. Assessment of thermal and antimicrobial properties of PAN/Zn-Al layered double hydroxide nanocomposites. *Compos. Interfaces* **2017**, *24*, 579–591. [CrossRef]
68. Barik, S.; Badamali, S.K.; Behera, L.; Jena, P.K. Mg–Al LDH reinforced PMMA nanocomposites: A potential material for packaging industry. *Compos. Interfaces* **2018**, *25*, 369–380. [CrossRef]
69. Xie, J.; Zhang, K.; Zhao, Q.; Wang, Q.; Xu, J. Large-scale fabrication of linear low-density polyethylene/layered double hydroxides composite films with enhanced heat retention, thermal, mechanical, optical and water vapor barrier properties. *J. Solid State Chem.* **2016**, *243*, 62–69. [CrossRef]
70. Dong, S.; Jia, Y.; Xu, X.; Luo, J.; Han, J.; Sun, X. Crystallization and properties of poly(ethylene terephthalate)/layered double hydroxide nanocomposites. *J. Colloid Interface Sci.* **2018**, *539*, 54–64. [CrossRef] [PubMed]
71. Kumar, M.; Chaudhary, V.; Suresh, K.; Pugazhenthi, G. Synthesis and characterization of exfoliated PMMA/Co–Al LDH nanocomposites via solvent blending technique. *RSC Adv.* **2015**, *5*, 39810–39820. [CrossRef]
72. Qiu, L.; Chen, W.; Qu, B. Morphology and thermal stabilization mechanism of LLDPE/MMT and LLDPE/LDH nanocomposites. *Polymer* **2006**, *47*, 922–930. [CrossRef]
73. Suresh, K.; Kumar, M.; Pugazhenthi, G.; Uppaluri, R. Enhanced mechanical and thermal properties of polystyrene nanocomposites prepared using organo-functionalized NiAl layered double hydroxide via melt intercalation technique. *J. Sci. Adv. Mater. Dev.* **2017**, *2*, 245–254.
74. Guo, B.Z.; Zhao, Y.; Huang, Q.T.; Jiao, Q.Z. A new method to prepare exfoliated UV-cured polymer/LDH nanocomposites via nanoplatelet-like LDHs modified with N-Lauroyl-glutamate. *Compos. Sci. Technol.* **2013**, *81*, 37–41. [CrossRef]

75. Hajibeygi, M.; Shabanian, M.; Khonakdar, H.A. Zn-Al LDH reinforced nanocomposites based on polyamide containing imide group: From synthesis to properties. *Appl. Clay Sci.* **2015**, *114*, 256–264. [CrossRef]
76. Hajibeygi, M.; Shafiei-Navid, S.; Shabanian, M.; Vahabi, H. Novel poly(amide-azomethine) nanocomposites reinforced with polyacrylic acid-co-2-acrylamido-2-methylpropanesulfonic acid modified LDH: Synthesis and properties. *Appl. Clay Sci.* **2018**, *157*, 165–176. [CrossRef]
77. Huang, G.; Chen, S.; Song, P.; Lu, P.; Wu, C.; Liang, H. Combination effects of graphene and layered double hydroxides on intumescent flame-retardant poly (methyl methacrylate) nanocomposites. *Appl. Clay Sci.* **2014**, *88*, 78–85. [CrossRef]
78. Wang, Q.; Zhang, X.; Wang, C.J.; Zhu, J.; Guo, Z.; O'Hare, D. Polypropylene/layered double hydroxide nanocomposites. *J. Mater. Chem.* **2012**, *22*, 19113–19121. [CrossRef]
79. Kredatusová, J.; Beneš, H.; Livi, S.; Pop-Georgievski, O.; Ecorchard, P.; Abbrent, S.; Pavlova, E.; Bogdał, D. Influence of ionic liquid-modified LDH on microwave-assisted polymerization of ε-Caprolactone. *Polymer* **2016**, *100*, 86–94. [CrossRef]
80. Jaerger, S.; Zawadzki, S.F.; Leuteritz, A.; Wypych, F. New alternative to produce colored polymer nanocomposites: Organophilic Ni/Al and Co/Al layered double hydroxide as fillers into low-density polyethylene. *J. Braz. Chem. Soc.* **2017**, *28*, 2391–2401. [CrossRef]
81. Zhang, S.; Liu, X.; Gu, X.; Jiang, P.; Sun, J. Flammability and thermal behavior of polypropylene composites containing dihydrogen phosphate anion-intercalated layered double hydroxides. *Polym. Compos.* **2015**, *36*, 2230–2237. [CrossRef]
82. Qian, Y.; Jiang, K.; Li, L. Improving the flame retardancy of ethylene vinyl acetate composites by incorporating layered double hydroxides based on bayer red mud. *e-Polymers* **2019**, *19*, 129–140. [CrossRef]
83. Zoromba, M.S.; Nour, M.A.; Eltamimy, H.E.; Abd El-Maksoud, S.A. Effect of modified layered double hydroxide on the flammability and mechanical properties of polypropylene. *Sci. Eng. Compos. Mater.* **2018**, *25*, 101–108. [CrossRef]
84. Lee, J.-H.; Zhang, W.; Ryu, H.-J.; Choi, G.; Choi, J.Y.; Choy, J.-H. Enhanced thermal stability and mechanical property of EVA nanocomposites upon addition of organo-intercalated LDH nanoparticles. *Polymer* **2019**, *177*, 274–281. [CrossRef]
85. Zhang, Z.; Qin, J.; Zhang, W.; Pan, Y.-T.; Wang, D.-Y.; Yang, R. Synthesis of a novel dual layered double hydroxide hybrid nanomaterial and its application in epoxy nanocomposites. *Chem. Eng. J.* **2020**, *381*, 122777. [CrossRef]
86. Qian, Y.; Qiao, P.; Li, L.; Han, H.; Zhang, H.; Chang, G. Hydrothermal synthesis of lanthanum-doped MgAl-layered double hydroxide/graphene oxide hybrid and its application as flame retardant for thermoplastic polyurethane. *Adv. Polym.* **2020**, *2020*, 1018093. [CrossRef]
87. Li, L.; Qian, Y.; Qiao, P.; Han, H.; Zhang, H. Preparation of LDHs based on bittern and its flame-retardant properties in EVA/LDHs properties. *Adv. Polym.* **2019**, *2019*, 4682164.
88. Qiu, L.; Gao, Y.; Zhang, C.; Yan, Q.; O'Hare, D.; Wang, Q. Synthesis of highly efficient flame-retardant polypropylene with surfactant intercalated layered double hydroxides. *Dalton Trans.* **2017**, *47*, 2965–2975. [CrossRef]
89. Nagendra, B.; Rosely, C.V.S.; Leuteritz, A.; Reuter, U. Polypropylene/layered double hydroxide nanocomposites: Influence of LDH intralayer metal constituents on the properties of polypropylene. *ACS Omega* **2017**, *2*, 20–31. [CrossRef]
90. Gao, Y.; Zhang, Y.; Williams, G.R.; O'Hare, D.; Wang, Q. Layered double hydroxide-oxidized carbon nanotube hybrids as highly efficient flame retardant nanofillers for polypropylene. *Sci. Rep.* **2016**, *6*, 35502. [CrossRef]
91. Gao, Y.; Wang, Q.; Lin, W. Ammonium polyphosphate intercalated layered double hydroxide and zinc borate as highly efficient flame Retardant nanofillers for polypropylene. *Polymers* **2018**, *10*, 1114. [CrossRef]
92. Li, L.; Jiang, K.; Qian, Y.; Han, H.; Qiao, P.; Zhang, H. Effect of organically intercalation modified layered double hydroxides-graphene oxide hybrids on flame retardancy of thermoplastic polyurethane nanocomposites. *J. Therm. Anal. Calorim.* **2020**, 1–11. [CrossRef]
93. Wang, B.-N.; Chen, N.-Y.; Yang, B.-J. Modification and compounding of CaMgAl-Layered Double Hydroxides and their application in the flame retardance of acrylonitrile-butadiene-styrene resin. *Polymers* **2019**, *11*, 1623. [CrossRef]
94. Dou, Y.; Zhou, A.; Pan, T.; Han, J.; Wei, M.; Evans, D.G.; Duan, X. Humidity-triggered self-healing films with excellent oxygen barrier performance. *Chem. Commun.* **2014**, *50*, 7136–7138. [CrossRef]

95. Pan, T.; Xu, S.; Dou, Y.; Liu, X.; Li, Z.; Han, J.; Yan, H.; Wei, M. Remarkable oxygen barrier films based on a layered double hydroxide/chitosan hierarchical structure. *J. Mater. Chem. A* **2015**, *3*, 12350–12356. [CrossRef]
96. Han, J.; Dou, Y.; Wei, M.; Evans, D.G.; Duan, X. Erasable nanoporous antireflection coatings based on the reconstruction effect of layered double hydroxides. *Angew. Chem. Int. Ed.* **2010**, *49*, 2171–2174. [CrossRef] [PubMed]
97. Gu, Y.; Lu, Z.; Chang, Z.; Liu, J.; Lei, X.; Li, Y.; Sun, X. NiTi layered double hydroxide thin films for advanced pseudocapacitor electrodes. *J. Mater. Chem. A* **2013**, *1*, 10655–10661. [CrossRef]
98. Xu, Z.P.; Braterman, P.S. High affinity of dodecylbenzene sulfonate for layered double hydroxide and resulting morphological changes. *J. Mater. Chem.* **2003**, *13*, 268–273. [CrossRef]
99. Merchán, M.; Ouk, T.S.; Kubát, P.; Lang, K.; Coelho, C.; Verney, V.; Commereuc, S.; Leroux, F.; Sol, V.; Taviot-Guého, C. Photostability and photobactericidal properties of porphyrin-layered double hydroxide–polyurethane composite films. *J. Mater. Chem. B* **2013**, *1*, 2139–2146. [CrossRef] [PubMed]
100. Yoo, J.T.; Lee, S.B.; Lee, C.K.; Hwang, S.W.; Kim, C.R.; Fujigaya, T.; Nakashima, N.; Shim, J.K. Graphene oxide and laponite composite films with high oxygen-barrier properties. *Nanoscale* **2014**, *6*, 10824–10830. [CrossRef]
101. Huang, H.-D.; Liu, C.-Y.; Li, D.; Chen, Y.-H.; Zhong, G.-J.; Li, Z.-M. Ultra-low gas permeability and efficient reinforcement of cellulose nanocomposite films by well-aligned graphene oxide nanosheets. *J. Mater. Chem. A* **2014**, *2*, 15853–15863. [CrossRef]
102. Wu, C.-N.; Yang, Q.; Takeuchi, M.; Saito, T.; Isogai, A. Highly tough and transparent layered composites of nanocellulose and synthetic silicate. *Nanoscale* **2014**, *6*, 392–399. [CrossRef]
103. Möller, M.W.; Lunkenbein, T.; Kalo, H.; Schieder, M.; Kunz, D.A.; Breu, J. Barrier Properties of Synthetic Clay with a Kilo-Aspect Ratio. *Adv. Mater.* **2010**, *22*, 5245–5249. [CrossRef]
104. Dou, Y.; Xu, S.; Liu, X.; Han, J.; Yan, H.; Wei, M.; Evans, D.G.; Duan, X. Transparent, flexible films based on layered double hydroxide/cellulose acetate with excellent oxygen barrier property. *Adv. Funct. Mater.* **2014**, *24*, 514–521. [CrossRef]
105. Huang, Y.-W.; Cheng, S. Carboxylate-intercalated layered double hydroxides for H2 sorption. *J. Mater. Chem. A* **2014**, *2*, 13452–13463. [CrossRef]
106. Wang, M.; Li, H.; Du, C.; Liang, Y.; Liu, M. Preparation and Barrier Properties of Nanocellulose/Layered Double Hydroxide Composite Film. *BioResources* **2018**, *13*, 1055–1064. [CrossRef]
107. Wang, L.; Dou, Y.; Wang, J.; Han, J.; Liu, L.; Wei, M. Layer-by-layer assembly of layered double hydroxide/rubber multilayer films with excellent gas barrier property. *Compos. Part A Appl. Sci. Manuf.* **2017**, *102*, 314–321. [CrossRef]
108. Yang, W.; Xia, Y.; Liu, X.; Yang, J.; Liu, Y. Layered double hydroxides/reduced graphene oxide nanocomposites with enhanced barrier properties. *Polym. Compos.* **2018**, *39*, 3841–3848. [CrossRef]
109. Botan, R.; Pinheiro, I.F.; Ferreira, F.V.; Lona, L.M.F. Correlation between water absorption and mechanical properties of polyamide 6 filled layered double hydroxides (LDH). *Mater. Res. Express* **2018**, *5*, 65004. [CrossRef]
110. Kotal, M.; Srivastava, S.K.; Bhowmick, A.K.; Chakraborty, S.K. Morphology and properties of stearate-intercalated layered double hydroxide nanoplatelet-reinforced thermoplastic polyurethane. *Polym. Int.* **2011**, *60*, 772–780. [CrossRef]
111. Feng, J.; Liao, Z.; Zhu, J.; Su, S. Comparison of morphology and mechanical properties of peroxide-cured acrylonitrile butadiene rubber/LDH composites prepared from different organically modified LDHs. *J. Appl. Polym. Sci.* **2013**, *127*, 3310–3317. [CrossRef]
112. Suresh, K.; Pugazhenthi, G.; Uppaluri, R. Properties of polystyrene (PS)/Co-Al LDH nanocomposites prepared by melt intercalation. *Mater. Today* **2019**, *9*, 333–350. [CrossRef]
113. Dong, Y.; Zhu, Y.; Dai, X.; Zhao, D.; Zhou, X.; Qi, Y.; Koo, J.H. Ammonium alcOHol polyvinyl phosphate intercalated LDHs/epoxy nanocomposites. *J. Therm. Anal. Calorim.* **2015**, *122*, 135–144. [CrossRef]
114. Mao, L.; Liu, J.-Y.; Zheng, S.-J.; Wu, H.-Q.; Liu, Y.-J.; Li, Z.-H.; Bai, Y.-K. Mussel-inspired nano-silver loaded layered double hydroxides embedded into a biodegradable polymer matrix for enhanced mechanical and gas barrier properties. *RSC Adv.* **2019**, *9*, 5834–5843. [CrossRef]
115. Moyo, L.; Focke, W.W.; Heidenreich, D.; Labuschagne, F.J.W.J.; Radusch, H.J. Properties of layered double hydroxide micro-and nanocomposites. *Mater. Res. Bull.* **2013**, *48*, 1218–1227. [CrossRef]

116. Qin, Y.; Wang, L.; Wang, X. A high performance sensor based on PANI/ZnTi-LDHs nanocomposite for trace NH3 detection. *Org. Electron.* **2019**, *66*, 102–109. [CrossRef]
117. Otoukesh, M.; Es'haghi, Z.; Feizy, J.; Nerin, C. Graphene oxide/Layered Double Hydroxides@ Sulfonated Polyaniline: A sorbent for ultrasonic assisted dispersive solid phase extraction of phthalates in distilled herbal beverages. *J. Chromatogr. A* **2020**, *1625*, 461307. [CrossRef]
118. Shamitha, C.; Mahendran, A.; Anandhan, S. Effect of polarization switching on piezoelectric and dielectric performance of electrospun nanofabrics of poly (vinylidene fluoride)/Ca–Al LDH nanocomposite. *J. Appl. Polym. Sci.* **2020**, *137*, 48697. [CrossRef]
119. Wang, H.; Wu, J.; Zheng, L.; Cheng, X. Preparation and properties of ZnAl layered double hydroxide/Polycaprolactone nanocomposites for use in drug delivery. *Polym. Plast. Technol.* **2019**, *58*, 1027–1035. [CrossRef]
120. Chakraborti, M.; Jackson, J.K.; Plackett, D.; Gilchrist, S.E.; Burt, H.M. The application of layered double hydroxide clay (LDH)-poly (lactide-co-glycolic acid)(PLGA) film composites for the controlled release of antibiotics. *J. Mater. Sci. Mater. Med.* **2012**, *23*, 1705–1713. [CrossRef] [PubMed]
121. Kang, N.J.; Wang, D.Y.; Kutlu, B.; Zhao, P.C.; Leuteritz, A.; Wagenknecht, U.; Heinrich, G. A new approach to reducing the flammability of layered double hydroxide (LDH)-based polymer composites: Preparation and characterization of dye structure-intercalated LDH and its effect on the flammability of polypropylene-grafted maleic anhydride/d-LDH composites. *ACS Appl. Mater. Interfaces* **2013**, *5*, 8991–8997.
122. Troutier-Thuilliez, A.L.; Taviot-Guého, C.; Cellier, J.; Hintze-Bruening, H.; Leroux, F. Layered particle-based polymer composites for coatings: Part, I. Evaluation of layered double hydroxides. *Prog. Org. Coat.* **2009**, *64*, 182–192. [CrossRef]
123. Yang, B.; Cai, J.; Wei, S.; Nie, N.; Liu, J. Preparation of Chitosan/NiFe-Layered Double Hydroxides Composites and Its Fenton-Like Catalytic oxidation of Phenolic Compounds. *J. Polym. Environ.* **2020**, *28*, 343–353. [CrossRef]
124. Zhang, S.; Yan, Y.; Wang, W.; Gu, X.; Li, H.; Li, J.; Sun, J. Intercalation of phosphotungstic acid into layered double hydroxides by reconstruction method and its application in intumescent flame retardant poly (lactic acid) composites. *Polym. Degrad. Stabil.* **2018**, *147*, 142–150. [CrossRef]
125. Jia, X.; Gao, S.; Liu, T.; Li, D.; Tang, P.; Feng, Y. Fabrication and bifunctional electrocatalytic performance of ternary CoNiMn layered double hydroxides/polypyrrole/reduced graphene oxide composite for oxygen reduction and evolution reactions. *Electrochim. Acta* **2017**, *245*, 59–68. [CrossRef]
126. Alnaqbi, M.A.; Samson, J.A.; Greish, Y.E. Electrospun Polystyrene/LDH Fibrous Membranes for the Removal of Cd2+ Ions. *J. Nanomater.* **2020**, *2020*, 1–12. [CrossRef]
127. Singh, M.; Singh, R.K.; Singh, S.K.; Mahto, S.K.; Misra, N. In vitro biocompatibility analysis of functionalized poly (vinyl chloride)/layered double hydroxide nanocomposites. *RSC Adv.* **2018**, *8*, 40611–40620. [CrossRef]

© 2020 by the authors. Licensee MDPI, Basel, Switzerland. This article is an open access article distributed under the terms and conditions of the Creative Commons Attribution (CC BY) license (http://creativecommons.org/licenses/by/4.0/).